Analysis of
Distributional Data

Analysis of Distributional Data

Edited by
Paula Brito
Sónia Dias

CRC Press
Taylor & Francis Group
Boca Raton London New York

CRC Press is an imprint of the
Taylor & Francis Group, an **informa** business

A CHAPMAN & HALL BOOK

First edition published 2022
by CRC Press
6000 Broken Sound Parkway NW, Suite 300, Boca Raton, FL 33487-2742

and by CRC Press
4 Park Square, Milton Park, Abingdon, Oxon, OX14 4RN

ISBN: 978-1-498-72545-3 (hbk)
ISBN: 978-1-032-25571-2 (pbk)
ISBN: 978-1-315-37054-5 (ebk)

DOI: 10.1201/9781315370545

Typeset in CMR10
by KnowledgeWorks Global Ltd.

Publisher's note: This book has been prepared from camera-ready copy provided by the authors.

Contents

Preface

In a time when increasingly larger and complex data collections are being produced, it is clear that new and adaptive forms of data representation and analysis have to be conceived and implemented. Social networks, but also governments and other institutions are producing large amounts of open data. Smart devices allow for the collection of data on households and individuals; people talk of "datafication" of business processes and relations. Smart Statistics is presented as the future system of official statistics, where data capturing, processing and analysis will be embedded in the system itself. "Data scientist" is referred to as the most interesting job of the 21st century, and the demand for these professionals keeps increasing.

However, the large amounts of available raw data, given their size and high entropy, may often not be the researcher direct concern and only an analysis at a higher level, after aggregation, may put in evidence patterns of interest – consider, e.g., high-frequency data streams captured by sensors, internet flows or large networks. In order to obtain descriptions at the appropriate level that reveal the properties of the process or system under analysis, some data aggregation prior to the analysis is often necessary. Such is also the case of very large surveys, where often interest lays on specific groups of the inquired population (e.g. regions, social groups) – there again data have to be aggregated before analysis may proceed. Analysis of official statistics data also falls into this framework, as confidentiality issues prevent the dissemination and analysis of the microdata.

The standard model for data representation of classical statistics and multivariate data analysis, where the basic units under analysis are single individuals, each one taking one single value for each numerical and/or categorical variable, is somehow restricted to take into account variability inherent to such data. Rather than resorting to a common approach consisting in reducing the data to central statistics (an average or a mode), which would lead to a loss of important information, other forms of realizations, taking into account the data intrinsic variability, may be considered.

In this book, we focus on the special case where individual units are described by distributions. Distributional data may result from the aggregation of large amounts of open/collected/generated data, or it may be directly available in a structured or unstructured form, describing the variability of some features. The book provides models and methods for the representation, analysis, interpretation, and organization of distributional data, taking into account its specific nature. Distributional Data Analysis, offering the possibility of aggregating data at the user's chosen degree of granularity while keeping the information on the intrinsic variability, and then analyze the resulting data arrays, will certainly play an important role in the context of Data Science. Conceived as an edited book, gathering contributions from multiple authors, the book presents alternative representations and analysis' methods for distributional data of different types. We have organized the proposed approaches in the following five parts:

- Data Representation and Exploratory Analysis – introducing the main concepts and measures for distributional-valued variables, uni- and bivariate descriptive statistics, and the quantile approach for the analysis of distributional data;

- Clustering and Classification – where alternative approaches for the clustering or classification of distributional data of different types are presented;

- Dimension Reduction – proposing methods for the representation in low-dimensional spaces;

- Regression and Forecasting – presenting regression models adapted to numerical distributional variables, as well as forecasting approaches for distributional time series.

The different chapters present applications to show how the proposed methods work in practice, and how results are to be interpreted. Furthermore, they often provide information about available software.

The chapters in this book present recent developments in the analysis of distributional data, constituting a fine collection of papers that characterize current research in this up-to-date, developing area. Combining new methodological advances with a wide variety of applications, this volume is certainly of great value for researchers and practitioners of data analytics alike.

The editors would like to thank all who have contributed to the design and production of this book, to all authors for their contributions and cooperation, as well as to Taylor and Francis, in particular, Dr. Robert Calver and Vaishali Singh, for their help concerning all aspects of publication.

Paula Brito
Sónia Dias

About the Editors

Paula Brito is a Professor at the Faculty of Economics of the University of Porto, and a member of the Artificial Intelligence and Decision Support Research Group (LIAAD) of INESC TEC, Portugal. She holds a doctorate degree in Applied Mathematics from the University Paris Dauphine and a Habilitation in Applied Mathematics from the University of Porto. Her current research focuses on the analysis of multidimensional complex data, known as symbolic data, for which she develops statistical approaches and multivariate analysis methodologies. In this context, she has been involved in two European research projects. Paula Brito has been president of the International Association for Statistical Computing (IASC-ISI) in 2013–2015 and of the Portuguese Association for Classification and Data Analysis for the term 2021–2023. She has been invited speaker at several international conferences and is a regular member of international program committees. Paula Brito has been chair of COMPSTAT 2008 and is chair of the IFCS 2022 conference.

Sónia Dias is a Professor in the area of Mathematics at the School of Technology and Management of the Polytechnic Institute of Viana do Castelo, and a member of the Laboratory in Artificial Intelligence and Decision Support (LIAAD) of INESC TEC, Portugal. She holds a PhD in Applied Mathematics from the University of Porto (2014). Her main scientific areas of research are Data Analysis, Symbolic Data Analysis (analysis of multidimensional complex data) and Statistical/Mathematical Applications. Under this context, she has participated in several conferences and published articles in international journals and proceedings. She was a member of the organizing committee of the international Symbolic Data Analysis Workshop - SDA2018 and is a member of the organizing committee of the IFCS 2022 conference.

List of Figures

List of Tables

Contributors

Javier Arroyo
Instituto de Tecnología del
   Conocimiento, Universidad
   Complutense de Madrid
Madrid, Spain

Vladimir Batagelj
Institute of Mathematics, Physics
   and Mechanics
Ljubljana, Slovenia

Andrej Marušič Institute, University
   of Primorska
Koper, Slovenia

National Research University Higher
   School of Economics
Moscow, Russia

Paula Brito
Faculty of Economics, University of
   Porto
Porto, Portugal

LIAAD, INESC TEC
Porto, Portugal

Marie Chavent
Inria Bordeaux Sud-Ouest, ASTRAL
   team-IMB, UMR CNRS 5251,
   University of Bordeaux
Bordeaux, France

Meiling Chen
School of Statistics, Capital
   University of Economics and
   Business
Beijing, China

Sónia Dias
School of Technology and
   Management, Polytechnic
   Institute of Viana do Castelo
Viana do Castelo, Portugal

LIAAD, INESC TEC
Porto, Portugal

Edwin Diday
Ceremade, University Paris
   Dauphine-PSL
Paris, France

Richard Emilion
Denis Poisson Institute, University of
   Orléans
Orléans, France

Peter Filzmoser
Institute of Statistics and
   Mathematical Methods in
   Economy, TU Wien
Vienna, Austria

Patrick J.F. Groenen
Econometric Institute, Erasmus
   University Rotterdam
Rotterdam, The Netherlands

Karel Hron
Department of Mathematical
   Analysis and Applications of
   Mathematics, Palacký University
Olomouc, Czech Republic

Manabu Ichino
Tokyo Denki University
Tokyo, Japan

Antonio Irpino
Department of Mathematics and
    Physics, University of Campania
    "L. Vanvitelli"
Caserta, Italy

Nataša Kejžar
Institute for Biostatistics and
    Medical Informatics, Faculty of
    Medicine, University of Ljubljana
Ljubljana, Slovenia

Simona Korenjak-Černe
School of Economics and Business,
    University of Ljubljana
Ljubljana, Slovenia

Institute of Mathematics, Physics
    and Mechanics
Ljubljana, Slovenia

Sun Makosso-Kallyth
SM Analytic Canada
Richmond Hill, Canada

Alessandra Menafoglio
Department of Mathematics,
    Politecnico di Milano, MOX
Milano, Italy

Ivana Pavlů
Department of Mathematical
    Analysis and Applications of
    Mathematics, Palacký University
Olomouc, Czech Republic

Yoshikazu Terada
Division of Mathematical Science,
    Graduate School of Engineering
    Science, Osaka University
Osaka, Japan

Rosanna Verde
Department of Mathematics and
    Physics, University of Campania
    "L. Vanvitelli"
Caserta, Italy

Huiwen Wang
School of Economics and
    Management, Beihang University
Beijing, China

Part I

Data Representation and Exploratory Analysis

1

Fundamental Concepts about Distributional Data

Sónia Dias

School of Technology and Management, Polytechnic Institute of Viana do Castelo & LIAAD-INESC TEC, Portugal

Paula Brito

Faculty of Economics, University of Porto & LIAAD-INESC TEC, Porto, Portugal

CONTENTS

In the classical data framework, one numerical value or one category is associated to each individual, such data is known as microdata. However, the interest of many studies is based on groups of records gathered according to a set of characteristics of the individuals, leading to macrodata. *Symbolic Data Analysis (SDA)* emerges with the aim to allow working with more complex data tables where the cells include more accurate and complete information. These cells may contain finite sets of values/categories, intervals, or distributions. The classification of the symbolic variables is defined according to the kind of observations. In distributional data we work with histogram-valued variables, where to each entity under analysis corresponds an empirical distribution that can be represented by a histogram or a quantile function. Accordingly, it is necessary to study the operations and distances that may be used when we work with these kinds of elements, to be possible to extend the classical statistics concepts and methods to histogram-valued variables.

DOI: 10.1201/9781315370545-1

1.1   Introduction

The extensive and complex data that emerged in the last decades made it necessary to extend and generalize classical concepts such as the data array and variables. Data tables where the cells contain a single quantitative or categorical value were no longer sufficient. More complex data tables were needed, with variables that express the variability of the records associated with each observed unit.

When we want to study a characteristic that floats over a period of time or is associated with a specific group/class of individuals, the "value" that best describes this characteristic is not a real value or category but a set/distribution/range of values. In these cases, we are in the presence of data with variability. Examples of this type of data are the temperature or stock values that fluctuate during a day, week, or month; the prices of a product in some regions; the age of the patients in various healthcare centers; and the weight or height of the players of a football team. In some situations the variability of the data might emerge due to the aggregation of single observations. A possible classical solution to analyze these data is to reduce the collection of records associated with each individual or class of individuals to one value, typically the mean, mode, or maximum/minimum; however, with this option the variability across the records is lost.

Considering the set of values or the distribution associated with each unit allows accounting for the variability of the records and hence perform more accurate studies. This perspective is in agreement with the opinion of Schweizer that around 30 years ago advocated that "distributions are the numbers of the future". Following in his footsteps, Diday [11] generalized the classical concept of variables in Multivariate Data Analysis and introduced *SDA*. In recent years, the development of statistical concepts and methods that use distributions instead of real numbers or categories is expanding at a steady rate [5–7, 16]. As distributions are more complex elements than the single values, it became necessary to study the best ways to represent them. Initially, histograms were considered the best option, but more recent studies opted to use the quantile functions, the inverse of the cumulative distribution functions. Moreover, due to the different nature of these variables, the need arose to select measures to evaluate the (dis)similarity between distributions. According to the works of Irpino and Verde [14] and Arroyo and Maté [3], the Mallows distance is considered a "good choice".

This chapter introduces the fundamental concepts of *SDA* that support the research and the development of the concepts and methods for distributional data.

1.2 The framework of distributional data

1.2.1 Definition and classification of symbolic variables

Symbolic Data Analysis studies complex data tables where each cell expresses the variability of each observed unit. The *symbolic data tables* may be given as such, but frequently, the variability of the data emerges due to the aggregation of observations from classical data tables where each individual, termed first-level unit, is described by classical variables – micro-data. According to Arroyo and Maté [3] the aggregation of the data may be:

- temporal – if the time is the aggregation criterion and the records are grouped over one unit of time, for example one day. In this case, the entities under analysis are the original first-level units, now characterized by sets of values originating from the records collected over a unit of time.

- contemporary – if the records are collected at the same temporal instant or the temporal instant is not relevant. In situations where the aggregation is contemporary, the entities – higher-level units – are classes of individuals (sets of first-level units) grouped according to specific characteristics. The variables describing both the higher-level and the respective first-level units are the same; however, the "values" that the variables take for each higher-level unit are now sets of values or distributions obtained from the respective first-level units.

Henceforth in this framework, when we use the terms unit/ individual/ observation, we will be referring to a first-level unit or to a higher-level unit according to the kind of prior aggregation of the microdata used to build the symbolic data table.

In each cell of a symbolic data table is a *"symbolic value"*; therefore this kind of data requires a more general concept of variable as it was formally presented in Chapter 3 of the book from Bock and Diday [6].

Definition 1 A symbolic variable Y, with underlying domain $\mathcal{Y}$, is defined by a mapping

$$Y: \quad S \quad \rightarrow \quad \mathbb{B}$$
$$s_i \quad \mapsto \quad Y(s_i) = \xi_i$$

defined on a set S of statistical entities.

We have $S \equiv \Omega = \{1, 2, \ldots, n\}$ when the individuals are first-level units or $S = \{C_1, C_2, \ldots C_n\}$ with $C_i \subseteq \Omega$ when the individuals are higher-level units (classes/concepts or categories).

Each unit s_i in S takes its "values" in a set $\mathbb{B}$. According to the type of realization of the symbolic variables, the set $\mathbb{B}$ will be $\mathbb{B} = \mathcal{Y}$ (classical variables); $\mathbb{B} = \{\mathcal{D} : \mathcal{D} \subseteq \mathcal{Y} \wedge \mathcal{D} \neq \varnothing\}$ (multi-valued variables); $\mathbb{B}$ a set of intervals of values in $\mathcal{Y} \subseteq \mathbb{R}$ or $\mathbb{B}$ a family of distributions on $\mathcal{Y}$. So, ξ_i, $i \in \{1, 2, \ldots, n\}$ may be a real value/category; a finite set of values/categories; an interval or a distribution.

Notation 1 Analogously to the classical statistics, when we have j symbolic variables, with $j \in \{1, 2, \ldots, p\}$, we denote each variable by Y_j and the "value" that the variable Y_j takes for an unit s_i by $Y_j(s_i)$ or to simplify the notation by Y_{ij}. When we have only one variable Y, the value that the variable takes on the unit i, may be denoted by $Y(s_i)$ or Y_i.

The symbolic data table that organizes the records of the p symbolic variables for n units is a matrix with n rows and p columns, as illustrated in Table 1.1.

TABLE 1.1

Symbolic data table.

	Y_1	Y_2	$\ldots$	Y_j	$\ldots$	Y_p
s_1	ξ_{11}	ξ_{12}	$\ldots$	ξ_{1j}	$\ldots$	ξ_{1p}
s_2	ξ_{21}	ξ_{22}	$\ldots$	ξ_{2j}	$\ldots$	ξ_{2p}
$\vdots$						
s_i	ξ_{i1}	ξ_{i2}	$\ldots$	ξ_{ij}	$\ldots$	ξ_{ip}
$\vdots$						
s_n	ξ_{n1}	ξ_{n2}	$\ldots$	ξ_{nj}	$\ldots$	ξ_{np}

To each row of the table corresponds an unit (individual or class of individuals) that contains its *symbolic description*, and each column corresponds to a *symbolic variable*.

Definition 2 Consider the p symbolic variables $Y_1, Y_2, \ldots, Y_p$, and let $\mathbb{B}_j$, with $j \in \{1, 2, \ldots, p\}$ be the sets where each variable Y_j takes its "symbolic value" for each unit s_i. A symbolic description of an unit $s_i \in S$ is given by the description vector $d_i = (\xi_{i1}, \ldots, \xi_{ip}) \in \mathbb{B}_1 \times \ldots \times \mathbb{B}_p$, with $i \in \{1, 2, \ldots, n\}$.

Hereafter, we present two situations where symbolic data tables may be used to express adequately all information associated with each individual.

Example 1 Consider the classical data table, in Table 1.2, that contains information about the diary level of hematocrit and hemoglobin of a set of patients attending a healthcare center during one month.

Aggregating (temporal aggregation) the values associated with each patient we build a symbolic data table, Table 1.3 [5]. In this table, to each unit, the patient, corresponds a distribution of values that describes the variability of the recorded values of hematocrit and hemoglobin.

TABLE 1.2
Classical data table (microdata) with the records of the level of
hematocrit and hemoglobin of each patient per day.

Patients	Level of hematocrit (Y)				Level of hemoglobin (X)			
	Day 1	Day 2	...	Day 30	Day 1	Day 2	...	Day 30
1	35.68	39.61	...	34.54	12.40	12.19	...	11.54
2	40.83	36.69	...	39.45	12.67	13.04	...	12.07
3	46.45	47.97	...	48.68	12.38	13.63	...	16.16
4	42.62	38.34	...	39.89	14.26	13.58	...	12.89
5	48.65	46.32	...	39.19	14.61	13.80	...	16.24
6	46.58	39.70	...	39.12	13.98	14.54	...	13.81
7	47.64	46.09	...	48.25	14.81	15.55	...	14.68
8	43.68	39.84	...	38.40	13.27	13.68	...	13.67
9	38.88	29.06	...	41.64	10.97	11.98	...	13.56
10	47.54	50.60	...	49.82	15.95	15.64	...	16.01

TABLE 1.3
Symbolic data table (macrodata) when the levels of hematocrit and
hemoglobin are symbolic variables.

Patients	Level of hematocrit (Y)	Level of hemoglobin (X)
1	$\{[33.29; 37.52[, 0.6; [37.52; 39.61], 0.4\}$	$\{[11.54; 12.19[, 0.4; [12.19; 12.8], 0.6\}$
2	$\{[36.69; 39.11[, 0.3; [39.11; 45.12], 0.7\}$	$\{[12.07; 13.32[, 0.5; [13.32; 14.17], 0.5\}$
3	$\{[36.70; 42.64[, 0.5; [42.64; 48.68], 0.5\}$	$\{[12.38; 14.2[, 0.3; [14.2; 16.16], 0.7\}$
4	$\{[36.38; 40.87[, 0.4; [40.87; 47.41], 0.6\}$	$\{[12.38; 14.26[, 0.5; [14.26; 15.29], 0.5\}$
5	$\{[39.19; 50.86], 1\}$	$\{[13.58; 14.28[, 0.3; [14.28; 16.24], 0.7\}$
6	$\{[39.7; 44.32[, 0.4; [44.32; 47.24], 0.6\}$	$\{[13.81; 14.5[, 0.4; [14.5; 15.2], 0.6\}$
7	$\{[41.56; 46.65[, 0.6; [8.8146.65; 4], 0.4\}$	$\{[14.34; 14.81[, 0.5; [14.81; 15.55], 0.5\}$
8	$\{[38.4; 42.93[, 0.7; [42.93; 45.22], 0.3\}$	$\{[13.27; 14.0[, 0.6; [14.0; 14.6], 0.4\}$
9	$\{[28.83; 35.55[, 0.5; [35.55; 41.98], 0.5\}$	$\{[9.92; 11.98[, 0.4; [11.98; 13.8], 0.6\}$
10	$\{[44.48; 52.53], 1\}$	$\{[15.37; 15.78[, 0.3; [15.78; 16.75], 0.7\}$

Example 2 Consider the symbolic data tables, Tables 1.4 and 1.5, containing
information about patients (adults) attending three healthcare centers during
a fixed period of time. Notice that in this example the entities under analysis are the healthcare centers (higher-level units), for each of which we have
aggregated information (contemporary aggregation), and not the individual
patients of each center (first-level units).

TABLE 1.4
Symbolic data table with information of three healthcare centers (part 1).

Healthcare center	Gender Y_1	Age Y_2	Education degree Y_3
A	$\{F, \frac{1}{2}; M, \frac{1}{2}\}$	$[25, 83]$	$\{9^{th} grade, 1/2; Higher\ education, 1/2\}$
B	$\{F, \frac{2}{3}; M, \frac{1}{3}\}$	$[18, 90]$	$\{6^{th} grade, 1/4; 9^{th} grade, 1/4;$ $12^{th} grade, 1/4; Higher\ education, 1/4\}$
C	$\{F, \frac{2}{5}; M, \frac{3}{5}\}$	$[20, 74]$	$\{4^{th} grade, 1/3; 9^{th} grade, 1/3; 12^{th} grade, 1/3\}$

According to the information in Tables 1.4 and 1.5, we know that in the
healthcare center A, the age of patients ranged from 25 to 83 years old, in

TABLE 1.5
Symbolic data table with information of three healthcare centers (part 2).

Healthcare center	Nº of emergency appointments $-Y_4$	Waiting time for an appointment (min) $-Y_5$	Emergency 24h $-Y_6$
A	$\{1, 2, 3\}$	$\{[15, 30[, 0.1; [30, 45[, 0.6; [45, 90], 0.3\}$	No
B	$\{0, 1, 4, 5, 10\}$	$\{[0, 15[, 0.8; [15, 45[, 0.2\}$	Yes
C	$\{0, 1, 3, 7\}$	$\{[0, 15[, 0.6; [30, 60], 0.4\}$	Yes

healthcare center B, it ranged from 18 to 90 years old, and in healthcare center C, the age of patients ranged from 20 to 74 years old. Considering another variable that records the waiting time for a medical appointment we may conclude that healthcare center A is the most problematic. 10% of the patients wait between 15 and 30 minutes for a medical appointment; 60% wait between 30 and 45 minutes; and 30% wait between 45 and 90 minutes for the medical appointment.

The symbolic description of healthcare center A is given by the description vector

$$d_A = \Big\{ \{F, 1/2; M, 1/2\}, [25, 83], \{9^{th} \text{ grade}, 1/2; \text{Higher education}, 1/2\},$$
$$\{1, 2, 3\}, \{[15, 30[, 0.1; [30, 45[, 0.6; [45, 90], 0.3\}, No \Big\}.$$

Similarly to the case of classical variables, symbolic variables may also be classified as quantitative or qualitative. Within these two categories, the classification of the variables is done according to the kind of elements in $\mathbb{B}$ (see Definition 1). According to Bock and Diday [6] and Billard and Diday [5], the symbolic variables may be classified as follows:

- **Single-valued variables** – when $\mathbb{B} = \mathcal{Y}$ we are in the presence of classical variables.

    - If $\mathcal{Y} \subseteq \mathbb{R}$, we have a **single-valued quantitative variable**.
    - If $\mathcal{Y}$ is a finite set of categories, the variable Y is classified as **single-valued categorical variable**.

- **Multi-valued variables** – when $\mathbb{B} = \mathcal{P}(\mathcal{Y}) = \{\mathcal{D} : \mathcal{D} \subseteq \mathcal{Y} \wedge \mathcal{D} \neq \emptyset\}$ is the set of non-empty subsets of $\mathcal{Y}$.

    - If $\mathcal{Y} \subseteq \mathbb{R}$ and the "values" of $Y(s_i)$ are finite sets of real numbers, we have a **multi-valued quantitative variable**.
    - If the "values" of $Y(s_i)$ are finite sets of categories in $\mathcal{Y}$, the symbolic variable Y is classified as **multi-valued categorical variable**.

- **Interval-valued variable** – when $\mathbb{B}$ is a set of intervals of values in $\mathcal{Y} \subseteq \mathbb{R}$. In this case, the "values" of $Y(s_i)$ are intervals of real numbers.

- **Modal-valued variable** – when $\mathbb{B}$ is a set of distributions on $\mathcal{Y}$. A particular outcome of modal-valued variables takes the form

$$Y(s_i) = \{\eta_{i\ell}, p_{i\ell}; \ell \in \{1, \ldots, m_i\}\}$$

where for each unit s_i, $p_{i\ell}$ is a non-negative measure (weight, probability, relative frequency) associated with $\eta_{i\ell}$ and m_i is the number of $\eta_{i\ell}$ taken by $Y(s_i)$; $\eta_{i\ell}$ may be finite or infinite in number and categorical or quantitative in value.

- If $\mathcal{Y} \subseteq \mathbb{R}$ or $\mathcal{Y}$ is a finite set of categories and $\{\eta_{i1}; \ldots; \eta_{im_i}\} \subseteq \mathcal{Y}$, Y is a **modal-valued quantitative variable** or **modal-valued categorical variable**, respectively.

- If, for each unit s_i, the "values" $\eta_{i\ell}$ with $\ell \in \{1, \ldots, m_i\}$ are ordered and disjoint intervals of values in $\mathcal{Y} \subseteq \mathbb{R}$ and $\sum_{\ell=1}^{m_i} p_{i\ell} = 1$, the symbolic variable Y is a **histogram-valued variable**. In this case, the values attained by the variable for each unit are empirical distributions or, more specifically, histograms, where the values in each subinterval are assumed to be uniformly distributed.

Example 3 The symbolic variables in Tables 1.4 and 1.5 may be classified as follows:

- Gender and Education degree are modal-valued categorical variables;

- Age is an interval-valued variable;

- Number of emergency medical appointments is a multi-valued quantitative variable;

- Waiting time for a medical appointment is a histogram-valued variable;

- Emergency medical appointment is a single-valued categorical variable.

Since the eighties of the last century, *SDA* has achieved considerable development of new statistical techniques to analyze multi-valued data (see, for instance [5, 6, 9]). Recently, there has been a growing interest in the specific study of histogram-valued variables.

1.2.2 Histogram-valued variables

Quantitative distributional variables are a special kind of modal-valued variables, the histogram-valued variables.

Definition 3 [9] Consider a symbolic variable $Y : S \to \mathbb{B}$. The set of units S may be $S \equiv \Omega = \{1, 2, \ldots, n\}$ when the individuals are first-level units or $S \equiv E = \{C_1, C_2, \ldots, C_n\}$ with $C_i \subseteq \Omega$ when the individuals are higher-level units. Consider also the quantitative (single-value) variable $\dot{Y}$ defined on Ω. If the aggregation of the observations is temporal, to each unit $s_i \in S(\equiv \Omega)$ corresponds the empirical distribution of the values that $\dot{Y}$ takes within a certain period of time. If the aggregation is contemporary, to each unit s_i corresponds the empirical distribution of values of $\dot{Y}$ in C_i.

In these cases, the outcome associated with an unit s_i may take the form:

$$Y(s_i) = \{I_{i1}, p_{i1}; I_{i2}, p_{i2}; \ldots; I_{im_i}, p_{im_i}\}$$

where $I_{i\ell}$, represents the subinterval ℓ for the unit i; $p_{i\ell}$ is the weight associated with the subinterval $I_{i\ell}$ and $\sum_{\ell=1}^{m_i} p_{i\ell} = 1$ with m_i the number of subintervals for the ith unit. Within each subinterval $I_{i\ell}$ that define the empirical distribution, the Uniform distribution is assumed [4].

When $m_i = 1$ for each unit s_i, to $Y(s_i)$ corresponds the interval $[l_i, u_i]$, with frequency $p_i = 1$. In this case, the histogram-valued variable is then reduced to the particular case of an interval-valued variable. Consequently, the values in each interval are assumed to be uniformly distributed. However, interval-valued variables are not originally defined as a particular case of histogram-valued variables [5,6] and for this reason the distribution within the intervals is not necessarily Uniform.

Notation 2 When we have $j \in \{1, 2, \ldots, p\}$ histogram-valued variables, the outcome associated with variable Y_j for a unit s_i is represented by

$$Y_j(s_i) = Y_{ij} = \{I_{ij1}, p_{i1}; I_{ij2}, p_{i2}; \ldots; I_{ij\ell}, p_{i\ell}; \ldots; I_{ijm_i}, p_{im_i}\}.$$

The empirical distribution $Y(s_i)$ may be represented in different ways.

1. *Histogram* – Histograms are an usual representation of empirical distributions. The outcome associated with each unit s_i of a histogram-valued variable may take the form:

$$H_i = \{[l_{i_1}, u_{i_1}[, p_{i1}; \ldots; [l_{im_i}, u_{im_i}], p_{im_i}\} \qquad (1.1)$$

where $l_{i\ell}$ is the lower bound and $u_{i\ell}$ the upper bound of the subinterval $I_{i\ell} = [l_{i\ell}, u_{i\ell}[$.

The subinterval $I_{i\ell}$ may also be represented by the centers and half ranges of the subinterval. So, we have:

$$H_i = \left\{ [c_{i1} - r_{i1}, c_{i1} + r_{i1}[, p_{i1}; \ldots; [c_{im_i} - r_{im_i}, c_{im_i} + r_{im_i}], p_{im_i} \right\} \qquad (1.2)$$

where $c_{i\ell} = \frac{u_{i\ell} + u_{i\ell}}{2}$ is the center and $r_{i\ell} = \frac{u_{i\ell} - l_{i\ell}}{2}$ is the half range of the subinterval $I_{i\ell} = [c_{i\ell} - r_{i\ell}, c_{i\ell} + r_{i\ell}[$.

Note that, for H_i to be a histogram it is necessary that all $I_{i\ell}$ will be intervals i.e. $l_{i\ell}, \leq u_{i\ell};\ u_{i\ell} \leq l_{i(\ell+1)}$ and the condition $\sum_{\ell=1}^{m_i} p_{i\ell} = 1$ will be verified.

2. *Cumulative distribution function* – As usual, the empirical distribution $Y(s_i)$ may be represented by a cumulative distribution function, $F_i(x)$. As the Uniform distribution within subintervals is assumed, the cumulative distribution function is given by

$$
F_i(x) = \begin{cases}
0 & if \quad x \leq l_{i1} \\[2mm]
\frac{x-l_{i1}}{u_{i1}-l_{i1}} p_{i1} & if \quad l_{i1} \leq x < u_{i1} \\[2mm]
p_{i1} + \frac{x-l_{i2}}{u_{i2}-l_{i2}} p_{i2} & if \quad l_{i2} \leq x < u_{i2} \\[2mm]
\vdots & \\[2mm]
1 & if \quad x \geq u_{im_i}
\end{cases} \tag{1.3}
$$

3. *Quantile function [14]* – In earlier studies with histogram-valued variables [10,15], the "symbolic observation values" associated with each unit are represented by the inverse of the cumulative empirical distribution function, $\Psi_i(t)$ with $t \in [0,1]$, called quantile function. As the Uniform distribution is assumed for the values within the subintervals, the quantile function is defined as follows:

$$
\Psi_i(t) = \begin{cases}
l_{i1} + \frac{t}{w_{i1}} \underbrace{\left(u_{i1} - l_{i1} \right)}_{a_{i1}} & if \quad 0 \leq t < w_{i1} \\[4mm]
l_{i2} + \frac{t-w_{i1}}{w_{i2}-w_{i1}} \underbrace{\left(u_{i2} - l_{i2} \right)}_{a_{i2}} & if \quad w_{i1} \leq t < w_{i2} \\[4mm]
\vdots & \\[4mm]
l_{im_i} + \frac{t-w_{i(m_i-1)}}{1-w_{i(m_i-1)}} \underbrace{\left(u_{im_i} - l_{im_i} \right)}_{a_{im_i}} & if \quad w_{i(m_i-1)} \leq t \leq 1
\end{cases}
$$

$$\tag{1.4}$$

or,

$$\Psi_i(t) = \begin{cases} c_{i1} + \left(\frac{2t}{w_{i1}} - 1\right) r_{i1} & \text{if} \quad 0 \le t < w_{i1} \\ c_{i2} + \left(\frac{2(t-w_{i1})}{w_{i2}-w_{i1}} - 1\right) r_{i2} & \text{if} \quad w_{i1} \le t < w_{i2} \\ \vdots \\ c_{im_i} + \left(\frac{2(t-w_{i(m_i-1)})}{1-w_{i(m_i-1)}} - 1\right) r_{im_i} & \text{if} \quad w_{i(m_i-1)} \le t \le 1 \end{cases}$$

(1.5)

where $w_{i\ell} = \begin{cases} 0 & \text{if} \quad \ell = 0 \\ \sum\limits_{h=1}^{\ell} p_{ih} & \text{if} \quad \ell = 1, \dots, m_i \end{cases}$ and m_i is the number of subintervals in Y_i.

Note that in a histogram the lower bound of each subinterval is always less than or equal to the upper bound, $l_{i\ell} \le u_{i\ell}$ and the upper bound of the following subinterval is always greater or equal to the previous, $u_{i\ell} \le l_{i(\ell+1)}$. Consequently, the quantile function that represents the empirical distribution is always a non-decreasing function defined in $[0, 1]$.

If any of the weights $p_{i\ell}$ with $\ell > 1$ is null, the function $F_i(x)$ will not have inverse with domain $[0, 1]$. Consequently, the function $\Psi_i(t)$ is not continuous and has $m_i - 1$ pieces. In this case, it is not possible to calculate the value of $\Psi_i(w_{i\ell-1})$ but only $\lim\limits_{t \to w_{i\ell-1}^-} \Psi_i(t)$ and $\lim\limits_{t \to w_{i\ell-1}^+} \Psi_i(t)$.

When we work with histogram-valued variables, it is important to note that:

- the subintervals in histograms must be ordered and disjoint. If these last conditions do not occur, it is possible to rewrite them in the required form using the *Williamson algorithm* [20].

- in histogram-valued variables, for different observations, the number of subintervals in the histograms or equivalently the number of pieces in quantile functions may be different. However, in some situations, it is necessary to rewrite a set of histograms (quantile functions) with the same number of subintervals (pieces), and the weight associated with each subinterval (the domain of each piece) has to be the same in all histograms (functions) but not equal in all subintervals of each histogram (all pieces of each quantile function) i.e., the histograms are not necessarily equiprobable histograms. To represent the distributions as quantile functions or histograms in these conditions, it may be necessary to apply the *Irpino and Verde procedure* [14].

Algorithm to build ordered and disjoint histograms [20].

Consider the histogram

$$H_Z = \{[l_{Z_1}, u_{Z_1}[, p_1; [l_{Z_2}, u_{Z_2}[, p_2; \dots; [l_{Zm}, u_{Zm}], p_m\}$$

where the subintervals are not ordered nor disjoint. If the histogram is not disjoint, we can rewrite it as a histogram with ordered and disjoint subintervals. For this, the process is the following:

1. *Order the subintervals:* The specified order is given by the relation $\leq_H$ and builds the histogram H_{Z_O}. The relation $\leq_H$ is defined by

$$I_{Z_\ell}, p_\ell \leq_H I_{Z_\gamma}, p_\gamma$$

if

$$l_{Z_\ell} < l_{Z_\gamma} \text{ or } l_{Z_\ell} = l_{Z_\gamma} \text{ and } u_{Z_\ell} < u_{Z_\gamma}.$$

2. *Construction of the disjoint histogram H_{Z_D}* : The next step is to build a disjoint histogram H_{Z_D} where

$$H_{Z_D} = \left\{ \left[l_{Z_{D_1}}, u_{Z_{D_1}} \right[, p_1^*; \left[l_{Z_{D_2}}, u_{Z_{D_2}} \right[, p_2^*; \ldots; \left[l_{Z_{D_m}}, u_{Z_{D_m}} \right], p_m^* \right\}$$

and

$$\left[l_{Z_{D_\ell}}, u_{Z_{D_\ell}} \right[\cap \left[l_{Z_{D_\gamma}}, u_{Z_{D_\gamma}} \right[= \emptyset \text{ for all } \ell \neq \gamma.$$

The histogram H_{Z_D} will have the property $u_{Z_{D_\ell}} = l_{Z_{D_{\ell+1}}}$, for $\ell \in \{1, \ldots, m-1\}$. However it will not be an equiprobable histogram (i.e. there may exist an ℓ and γ, $\ell \neq \gamma$, such that $p_\ell^* \neq p_\gamma^*$.) The procedure used to form the histogram H_{Z_D} from the ordered histogram H_{Z_O} is as follows:

$$
\begin{aligned}
&\textbf{for } (\ell := 0; \ell < m; \ell++) \{ \\
&\quad l_{Z_{D_\ell}} := l_{Z_{O_\ell}}; u_{Z_{D_\ell}} := u_{Z_{O_{\ell+1}}}; p_\ell^* = 0; \\
&\quad \textbf{for } (\gamma = \ell; \gamma \geq 0; \gamma--) \{ \\
&\quad\quad \textbf{if } u_{Z_{O_\gamma}} > l_{Z_{D_\ell}} \text{ and } l_{Z_{O_\gamma}} > u_{Z_{D_\ell}} \{ \\
&\quad\quad\quad \textbf{if } u_{Z_{D_\ell}} > u_{Z_{O_\gamma}} \text{ and } l_{Z_{D_\ell}} \geq l_{Z_{O_\gamma}} \\
&\quad\quad\quad\quad p_\ell^* += p_\gamma \frac{u_{Z_{O_\gamma}} - l_{Z_{D_\ell}}}{u_{Z_{O_\gamma}} - l_{Z_{O_\gamma}}} \\
&\quad\quad\quad \textbf{else if } u_{Z_{D_\ell}} \leq l_{Z_{O_j}} \text{ and } l_{Z_{D_\ell}} > u_{Z_{O_\gamma}} \\
&\quad\quad\quad\quad p_\ell^* += p_\gamma \frac{u_{Z_{D_\ell}} - l_{Z_{D_\ell}}}{u_{Z_{O_\gamma}} - l_{Z_{O_\gamma}}} \\
&\quad\quad\quad \} \\
&\quad\quad \} \\
&\quad \}
\end{aligned}
$$

Procedure to rewrite a set of histograms with equal number of subintervals [14].

Consider the empirical distributions $Y(s_i)$ with $i \in \{1, 2, \ldots, n\}$, each of them represented by a histogram or a quantile function with m_i subintervals/pieces, as in expressions (1.1) or (1.4), respectively.

Let W be the set of the cumulative weights of the n distributions:

$$W = \{w_{10}, \ldots, w_{1m_1}; w_{20}, \ldots, w_{2m_2}; \ldots; w_{n0}, \ldots, w_{nm_n}\}$$

To rewrite the histograms/quantile functions, we need to sort W without repetitions. The sorted values may be represented by

$$Z = \{w_0, w_1, \ldots, w_\ell, \ldots, w_m\}$$

where $\ell \in \{0, \ldots, m\}$; $w_0 = 0$, $w_m = 1$ and $\max\{m_1, \ldots, m_n\} \leq m \leq \sum_{i=1}^{n} m_i - 1$.

Each distribution $Y(s_i)$ may be rewritten as a histogram where the subintervals, with Uniform distribution, have as lower bound $\Psi_i(w_{\ell-1})$ and as upper bound $\Psi_i(w_\ell)$; so, we may rewrite each $Y(s_i)$ as the quantile function

$$\Psi_i(t) = \begin{cases} \Psi_i(w_0) + \frac{t}{w_1}\left(\Psi_i(w_1) - \Psi_i(w_0)\right) & \text{if} \quad 0 \leq t < w_1 \\ \Psi_i(w_1) + \frac{t-w_1}{w_2-w_1}\left(\Psi_i(w_2) - \Psi_i(w_1)\right) & \text{if} \quad w_1 \leq t < w_2 \\ \vdots \\ \Psi_i(w_{m-1}) + \frac{t-w_{m-1}}{1-w_{m-1}}\left(\Psi_i(w_m) - \Psi_i(w_{m-1})\right) & \text{if} \quad w_{m-1} \leq t \leq 1 \end{cases}$$

$$(1.6)$$

or as the histogram:

$$H_i = \left\{ \left[\Psi_i(0), \Psi_i(w_1)\right[, p_1; \left[\Psi_i(w_1), \Psi_i(w_2)\right[, p_2; \ldots; \left[\Psi_i(w_{m-1}), \Psi_i(1)\right], p_m \right\}$$

$$(1.7)$$

with $p_\ell = w_\ell - w_{\ell-1}$ and $\ell \in \{1, \ldots, m\}$.

The next example illustrates the possible representations of the "symbolic observation values" associated with histogram-valued variables. Moreover, the procedure to rewrite a set of histograms with equal number of subintervals and the same set of cumulative weights will be exemplified. The Williamson algorithm will be illustrated only in the next section.

Example 4 Consider the histogram-valued variable Y_5, "Waiting time for a medical appointment" in Table 1.5. The histogram values associated with the three healthcare centers are represented in Figure 1.1.

Alternatively, these histograms may be represented by their quantile functions (see Figure 1.2):

$$\Psi_{Y_5(A)}(t) = \begin{cases} 15 + \frac{t}{0.1} \times 15 & \text{if} \quad 0 \leq t < 0.1 \\ 30 + \frac{t-0.1}{0.6} \times 15 & \text{if} \quad 0.1 \leq t < 0.7 \\ 45 + \frac{t-0.7}{0.3} \times 45 & \text{if} \quad 0.7 \leq t \leq 1 \end{cases}$$

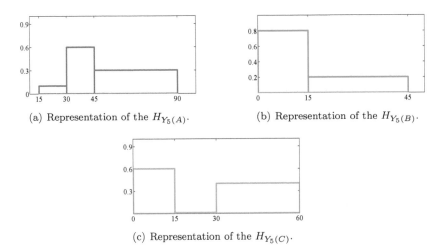

(a) Representation of the $H_{Y_5(A)}$.

(b) Representation of the $H_{Y_5(B)}$.

(c) Representation of the $H_{Y_5(C)}$.

FIGURE 1.1
Representation of the histograms associated with each healthcare center for the histogram-valued variable Y_5 in Table 1.5.

$$\Psi_{Y_5(B)}(t) = \begin{cases} \frac{t}{0.8} \times 15 & if \quad 0 \leq t < 0.8 \\ 15 + \frac{t-0.8}{0.2} \times 30 & if \quad 0.8 \leq t \leq 1 \end{cases}$$

$$\Psi_{Y_5(C)}(t) = \begin{cases} \frac{t}{0.6} \times 15 & if \quad 0 \leq t < 0.6 \\ 30 + \frac{t-0.6}{0.4} \times 30 & if \quad 0.6 < t \leq 1 \end{cases}$$

For the healthcare center C, the quantile function $\Psi_{Y_5(C)}(t)$ is an example of a non-continuous quantile function.

The quantile functions $\Psi_{Y_5(A)}(t), \Psi_{Y_5(B)}(t)$ and $\Psi_{Y_5(C)}(t)$ (or the respective histograms) may be rewritten with the same number of pieces and same set of cumulative weights, using the process proposed by Irpino and Verde [14] and described in Section 1.2.2.

Considering the three quantile functions, the set of the cumulative weights is $W = \{0, 0.1, 0.7, 1; 0, 0.8, 1; 0, 0.6, 1\}$. Selecting the weights without repetition, we have $Z = \{0, 0.1, 0.6, 0.7, 0.8, 1\}$. Using these weights, we can rewrite the quantile functions $\Psi_{Y_5(A)}(t), \Psi_{Y_5(B)}(t)$ and $\Psi_{Y_5(C)}(t)$ in the required conditions, as follows:

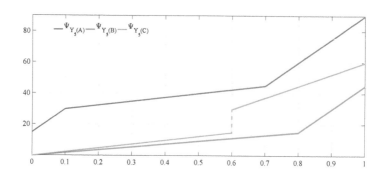

FIGURE 1.2
Representation of the quantile functions $\Psi_{Y_5(A)}$, $\Psi_{Y_5(B)}$, $\Psi_{Y_5(C)}$ in Table 1.5.

$$\Psi_{Y_5(A)}(t) = \begin{cases} 15.00 + \frac{t}{0.1} \times 15.00 & \text{if} \quad 0 \leq t < 0.1 \\ 30.00 + \frac{t-0.1}{0.5} \times 12.50 & \text{if} \quad 0.1 \leq t < 0.6 \\ 42.50 + \frac{t-0.6}{0.1} \times 2.50 & \text{if} \quad 0.6 \leq t < 0.7 \\ 45.00 + \frac{t-0.7}{0.1} \times 15.00 & \text{if} \quad 0.7 \leq t < 0.8 \\ 60.00 + \frac{t-0.8}{0.2} \times 30.00 & \text{if} \quad 0.8 \leq t \leq 1 \end{cases}$$

$$\Psi_{Y_5(B)}(t) = \begin{cases} \frac{t}{0.1} \times 1.875 & \text{if} \quad 0 \leq t < 0.1 \\ 1.875 + \frac{t-0.1}{0.5} \times 9.375 & \text{if} \quad 0.1 \leq t < 0.6 \\ 11.250 + \frac{t-0.6}{0.1} \times 1.875 & \text{if} \quad 0.6 \leq t < 0.7 \\ 13.125 + \frac{t-0.7}{0.1} \times 1.875 & \text{if} \quad 0.7 \leq t < 0.8 \\ 15.000 + \frac{t-0.8}{0.1} \times 30.000 & \text{if} \quad 0.8 \leq t \leq 1 \end{cases}$$

$$\Psi_{Y_5(C)}(t) = \begin{cases} \frac{t}{0.1} \times 2.50 & \text{if} \quad 0 \leq t < 0.1 \\ 2.50 + \frac{(t-0.1)}{0.5} \times 12.50 & \text{if} \quad 0.1 \leq t < 0.6 \\ 30.00 + \frac{t-0.6}{0.1} \times 7.50 & \text{if} \quad 0.6 < t < 0.7 \\ 37.50 + \frac{t-0.7}{0.1} \times 7.50 & \text{if} \quad 0.7 \leq t < 0.8 \\ 45.00 + \frac{t-0.8}{0.1} \times 15.00 & \text{if} \quad 0.8 \leq t \leq 1 \end{cases}$$

1.3 Operations with distributions

In classical statistics the data are real numbers and the standard arithmetic is used to operate with them. In this case, we need to operate with distributions that may be represented by histograms or quantile functions. In order to operate with these elements it is necessary to know the operations defined between them and study their properties.

1.3.1 Histogram Arithmetic

The histogram arithmetic was a generalization of the interval arithmetic proposed by Colombo and Jaarsma [8] and that afterwards was studied by Williamson [20].

Definition 4 Consider the histogram H_{Y_1} with m_1 subintervals,

$$H_{Y_1} = \left\{ [l_{11}, u_{11}[, p_{11}; [l_{12}, u_{12}[, p_{12}; \ldots, [l_{1m_1}, u_{1m_1}], p_{1m_1} \right\}^1$$

and the histogram H_{Y_2} with m_2 subintervals,

$$H_{Y_2} = \left\{ [l_{21}, u_{21}[, p_{21}; [l_{22}, u_{22}[, p_{22}; \ldots, [l_{2m_2}, u_{2m_2}], p_{2m_2} \right\}^1.$$

The operations $\Box$ with $\Box \in \{+, -, \times, \div\}$ between the histograms H_{Y_1} and H_{Y_2} (when $\Box = \div$ it is assumed that $0 \notin H_{Y_2}$), produce the histogram $H_{Y_1} \Box H_{Y_2}$ with $m = m_1 \times m_2$ subintervals and are defined by

$$l_{(\ell_1-1)m_2+\ell_2} = \min \left\{ l_{1\ell_1} \Box l_{2\ell_2}, u_{1\ell_1} \Box l_{2\ell_2}, l_{1\ell_1} \Box u_{2\ell_2}, u_{1\ell_1} \Box u_{2\ell_2} \right\}$$

$$u_{(\ell_1-1)m_2+\ell_2} = \max \left\{ l_{1\ell_1} \Box l_{2\ell_2}, u_{1\ell_1} \Box l_{2\ell_2}, l_{1\ell_1} \Box u_{2\ell_2}, u_{1\ell_1} \Box u_{2\ell_2} \right\}$$

$$p_{(\ell_1-1)m_2+\ell_2} = p_{1\ell_1} \times p_{2\ell_2}.$$

with $\ell_1 \in \{1, \ldots, m_1\}$ and $\ell_2 \in \{1, \ldots, m_2\}$.

When operating with histograms with m_1 and m_2 subintervals, we obtain a new histogram with $m_1 \times m_2$ subintervals, hence this new histogram may have a larger number of subintervals. The resulting histogram is not necessarily ordered and the subintervals can intersect each other, however it may be re-organized applying the Williamson algorithm (see Section 1.2.2).

We may particularize Definition 4, to define operations between histograms and real values. Consider a real number α, which may be represented by

---

[1]In this case, we are working with one histogram H_{Y_j} and not with a set of histograms $H_{Y_{ij}}$, because of this in histogram $H_{Y_j} = \left\{ [l_{j1}, u_{j1}[, p_{j1}; \ldots; [l_{j\ell}, u_{j\ell}[, p_{j\ell}; \ldots; [l_{jm_j}, u_{jm_j}], p_{jm_j} \right\}$, $l_{j\ell}$ and $u_{j\ell}$ represent respectively, the lower and upper bound of the ℓ-th subinterval of the histogram H_{Y_j}.

$\{[\alpha, \alpha], 1\}$. The addition between the histogram H and the real number α results in the following histogram:

$$H + \alpha = \left\{ [l_1 + \alpha, u_1 + \alpha[\,, p_1; \ldots; [l_{m_1} + \alpha, u_{m_1} + \alpha]\,, p_{m_1} \right\}.$$

In this case, the histogram is still ordered, disjointed and the range of the subintervals of the histograms H and $H + \alpha$ is the same, for all $\alpha \in \mathbb{R}$. If α is a positive number, the histogram $H + \alpha$ results from the translation of α units of the histogram H to the right; if α is a negative number, the histogram $H + \alpha$ results from the translation of α units of the histogram H to the left.

When we multiply one histogram H by a positive real number α, the resulting histogram is:

$$\alpha H = \left\{ [\alpha l_1, \alpha u_1[\,, p_1; [\alpha l_2, \alpha u_2[\,, p_2; \ldots; [\alpha l_{m_1}, \alpha u_{m_1}]\,, p_{m_1} \right\}$$

but if α is a negative real number, we obtain the histogram

$$\alpha H_Y = \left\{ [\alpha u_{m_1}, \alpha l_{m_1}[\,, p_{m_1}; \ldots; [\alpha u_2, \alpha l_2[\,, p_2; [\alpha u_1, \alpha l_1]\,, p_1 \right\}.$$

The result of multiplying a histogram by a real number α is a new ordered and disjoint histogram. The histogram αH with α negative and the histogram αH, with α positive, are symmetric in relation to the yy-axis. So, we may define symmetric histogram as follows:

Definition 5 Consider the histogram H. Using the histogram arithmetic, if we multiply H by the real number -1, we obtain its symmetric histogram $-H$. The histograms H and $-H$ are symmetric relatively to the yy-axis.

Example 5 Consider the histograms

$$H_X = \left\{ [1, 3[\,, 0.1; [3, 5[\,, 0.6; [5, 8]\,, 0.3 \right\}$$

and

$$H_Y = \left\{ [0, 1[\,, 0.8; [1, 4]\,, 0.2 \right\}.$$

These histograms will be used to exemplify some operations with histograms.

Addition

The addition of the histograms H_X and H_Y is not an ordered and disjoint histogram. However, applying the *Williamson algorithm* (see Section 1.2.2) it is possible to rewrite them. The histograms involved in the operation as well as the resulting histogram are represented in Figure 1.3.

Consider the addition operation presented in Definition 4 where we obtain an ordered but not disjoint histogram:

$$H_X + H_Y = \left\{ [1, 4[\,, 0.1 \times 0.8; [2, 7[\,, 0.1 \times 0.2; [3, 6[\,, 0.6 \times 0.8; [4, 9[\,, 0.6 \times 0.2; \\ [5, 9[\,, 0.3 \times 0.8; [6, 12]\,, 0.3 \times 0.2 \right\}.$$

The disjoint histogram is the following:

$$H_X + H_Y = \{\, [1,2[\,, (0.1 \times 0.3) \times \tfrac{1}{3}; [2,3[\,, (0.1 \times 0.3) \times \tfrac{1}{3} + (0.1 \times 0.2) \times \tfrac{1}{5};$$
$$[3,4[\,, 0.1907; [4,5[\,, 0.1880; [5,6[\,, 0.2480; [6,7[\,, 0.0980;$$
$$[7,9[\,, 0.1880; [9,12]\,, 0.0300\}.$$

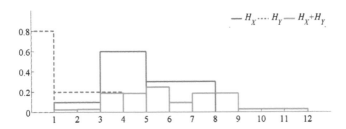

FIGURE 1.3
Representation of histograms H_X, H_Y and $H_X + H_Y$ in Example 5.

The addition of the histogram H_X with the real numbers 2 and -2 are the histograms (see Figure 1.4):

$$H_X + 2 = \{\, [3,5[\,, 0.1; [5,7[\,, 0.6; [7,10]\,, 0.3\}$$

$$H_X - 2 = \{\, [-1,1[\,, 0.1; [1,3[\,, 0.6; [3,6]\,, 0.3\}$$

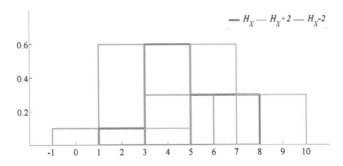

FIGURE 1.4
Representation of histograms H_X, $H_X + 2$ and $H_X - 2$ in Example 5.

Multiplication:
The product of the histogram H_X by the real number 2 and the symmetric of the histogram H_X, $-H_X$, are the histograms given below and represented in Figure 1.5.

$$2H_X = \{\, [2,6[\,, 0.1; [6,10[\,, 0.6; [10,16]\,, 0.3\}$$

$$-H_X = \left\{\, [-8, -5[\, , 0.3; [-5, -3[\, , 0.6; [-3, -1]\, , 0.1 \right\}$$

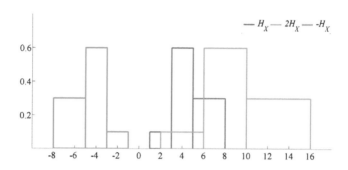

FIGURE 1.5

Representation of histograms H_X, $2H_X$ and $-H_X$ in Example 5.

Considering the histogram H_X, we will illustrate that the addition of a histogram with its symmetric is a symmetric histogram relatively to the yy-axis, as observed in Figure 1.6.

$$H_X - H_X = \left\{\, [-7, -5[\, , 0.012; [-5, -4[\, , 0.042; [-4, -3[\, , 0.057; [-3, -2[\, , 0.072; \right.$$
$$[-2, 0[\, , 0.317; [0, 2[\, , 0.317; [2, 3[\, , 0.072; [3, 4[\, , 0.057;$$
$$\left. [4, 5[\, , 0.042; [5, 7]\, , 0.012 \right\}.$$

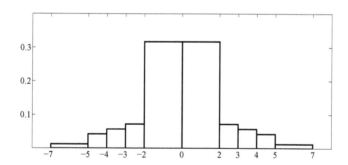

FIGURE 1.6

Representation of the histogram $H_X - H_X$ in Example 5.

When we work with histogram data represented by histograms, the arithmetics proposed above is not in general a good option. The arithmetic is complex and we obtain results that are unexpected. As we illustrated in the previous example, the addition of a histogram with its symmetric is not null but a symmetric histogram, in relation to the yy-axis. Furthermore, the mean of two equal histograms H is not the histogram H.

1.3.2 Operations with Quantile Functions

In more recent studies, the "symbolic observed values" for numerical distributional data have been represented as quantile functions [3, 10, 15]. If we represent the empirical distribution that each unit takes on a histogram-valued variable by a quantile function, then the operations are simplified because, as quantile functions are piecewise functions, the adequate arithmetic for them is a function arithmetic. However, this representation raises, in general, other issues.

Firstly, to operate with quantile functions, it is necessary to define all involved functions with an equal number of pieces. Furthermore, the domain of each piece has to be the same for all functions (the set of cumulative weights is the same in all functions) but not the same in all pieces of each quantile function, i.e. the histograms represented by the quantile functions are not necessarily equiprobable histograms (histograms with equal probability subintervals). To represent the distributions as quantile functions under these conditions, it may be necessary to apply the procedure defined by Irpino and Verde [14], as described in Section 1.2.2.

It is important to avoid that the number of subintervals for each histogram becomes "too" large (which could happen as a consequence of the application of the Irpino and Verde process), in which case the distributions that represent the data would be meaningless. To prevent the situation mentioned above and when the microdata are known, we may consider the option of Colombo and Jaarsma [8] and work with equiprobable histograms. These authors encountered similar problems when operating with histograms using the histogram arithmetic. Colombo and Jaarsma [8] considered advantageous to work with equiprobable histograms because the distributions are reasonably well approximated by equiprobable histograms, the subintervals into which a distribution is subdivided are small when the frequency is high and large when the frequency is low; operations/combinations of equal frequency subintervals form again equal frequency subintervals.

The operations between the quantile functions are the usual operations with functions. However, as the quantile functions present specific characteristics, it is important to explore some operations with this kind of functions.

Consider the quantile functions $\Psi_{Y_1}(t)$ and $\Psi_{Y_2}(t)$, that represent two distributions, defined according to expression (1.4), both with fixed m subintervals. The addition of these quantile functions leads to the function:

$$\Psi_{Y_1}(t) + \Psi_{Y_2}(t) = \begin{cases} l_{11} + l_{21} + \frac{t}{w_1}(a_{11} + a_{21}) & \text{if} \quad 0 \leq t < w_1 \\ l_{12} + l_{22} + \frac{t - w_1}{w_2 - w_1}(a_{12} + a_{22}) & \text{if} \quad w_1 \leq t < w_2 \\ \vdots \\ l_{1m} + l_{2m} + \frac{t - w_{m-1}}{1 - w_{m-1}}(a_{1m} + a_{2m}) & \text{if} \quad w_{m-1} \leq t \leq 1 \end{cases}.$$

When we add two quantile functions we obtain a non-decreasing function.

In this case, both the slopes and the y-intercepts of the resulting function are influenced by the two functions.

The particular case of the addition of a quantile function $\Psi_Y(t)$ with a real number α is the function:

$$
(\Psi_Y + \alpha)(t) = \begin{cases} l_1 + \alpha + \frac{t}{w_1} a_1 & \text{if} \quad 0 \leq t < w_1 \\ l_2 + \alpha + \frac{t - w_1}{w_2 - w_1} a_2 & \text{if} \quad w_1 \leq t < w_2 \\ \vdots & \\ l_m + \alpha + \frac{t - w_{m-1}}{1 - w_{m-1}} a_m & \text{if} \quad w_{m-1} \leq t \leq 1 \end{cases}.
$$

In this case, only the y-intercepts are affected by the operation. We have a translation up when adding a real positive number α and a translation down when the real number α is negative.

The multiplication of the quantile function $\Psi_Y(t)$ by a real number α leads to the function:

$$
\alpha\Psi_Y(t) = \begin{cases} \alpha l_1 + \frac{t}{w_1}(\alpha a_1) & \text{if} \quad 0 \leq t < w_1 \\ \alpha l_2 + \frac{t - w_1}{w_2 - w_1}(\alpha a_2) & \text{if} \quad w_1 \leq t < w_2 \\ \vdots & \\ \alpha l_m + \frac{t - w_{m-1}}{1 - w_{m-1}}(\alpha a_m) & \text{if} \quad w_{m-1} \leq t \leq 1 \end{cases}.
$$

In this case, both the slopes and the y-intercepts are affected by α. If α is positive we will have a non-decreasing function but if α is negative we will obtain a decreasing function that cannot be a quantile function, since quantile functions must always be non-decreasing functions.

The fact that the result of the multiplication of a quantile function by -1 is not a quantile function is a limitation for this representation of the distributions. This issue will have consequences in the development of statistical methods for numerical distributional data [10, 15]. The option to work with distributions represented by histograms is difficult due to the complexity of the histogram arithmetic. However, the alternative strategy of representing the distributions by quantile functions implies that the space where the elements are quantile functions is only a semi-vector space [18].

Quantile functions are a particular kind of functions. Consider the set of the functions defined from $\mathbb{R}$ in $\mathbb{R}$, $\mathcal{F}(\mathbb{R}, \mathbb{R})$, and the usual operations defined in $\mathcal{F}$ as follows:

- *Addition:* $(f + g)(x) = f(x) + g(x), \forall x \in \mathbb{R}$;

- *Product of a function by a real number:* $(\lambda f)(x) = \lambda f(x), \forall x \in \mathbb{R}$, and $\lambda \in \mathbb{R}$.

The $(\mathcal{F}, +, .)$ is a vector space. However, the space $(\mathcal{E}, +, .)$ where $\mathcal{E}([0, 1], \mathbb{R})$ is the set of the quantile functions (piecewise linear functions under the uniformity hypothesis) defined from $[0, 1]$ in $\mathbb{R}$, with the same operations of addition and product by a real number of $(\mathcal{F}, +, .)$, is not a subspace of the vector space $(\mathcal{F}, +, .)$ because if we multiply a quantile function by a negative number we will obtain a decreasing function and consequently not a quantile function. For this reason $(\mathcal{E}, +, .)$ is only a semi-vector space.

In spite of the $-\Psi(t)$ not being a quantile function, it is possible to define a quantile function that represents the symmetric of a given distribution. If we consider a distribution represented by a histogram H and using the histograms arithmetic, it is possible to obtain the symmetric of the histogram H multiplying this histogram H by -1 (Definition 5 in Section 1.3.1). Consequently, it is then possible to define the quantile function that represents the histogram $-H$. However, this function is not obtained by multiplying the quantile function $\Psi(t)$ by -1. Instead, it is obtained by performing the transformations of $\Psi(t)$ to $-\Psi(1 - t)$ with $t \in [0, 1]$.

Definition 6 Consider a histogram H and the respective symmetric $-H$, according to Definition 5. If $\Psi(t)$ is the quantile function that represents the histogram H, $-\Psi(1 - t)$ is the quantile function that represents its symmetric $-H$.

In Figure 1.7(a) [10] are represented the histogram H_X in Example 5 and the respective symmetric histogram. Figure 1.7(b) [10] shows that the function $-\Psi_X(t)$ is different from the quantile function $-\Psi_X(1 - t)$ that corresponds to the histogram $-H_X$.

It is important to underline some conclusions about the function $-\Psi(1-t)$, $t \in [0, 1]$:

- As it is required for quantile functions, $-\Psi(1 - t)$ is a non-decreasing function;

- $\Psi(t) - \Psi(1 - t)$ is not a null function, as expected, but is a quantile function with null symbolic mean, as defined in [4];

- the functions $-\Psi(1 - t)$ and $\Psi(t)$ are linearly independent, providing that $-\Psi(1 - t) \neq \Psi(t)$;

- $-\Psi(1 - t) = \Psi(t)$ only when the histogram is symmetric with respect to the yy-axis.

The following example illustrates the operations with quantile functions and the consequences of their results.

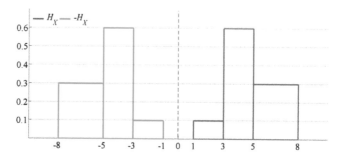

(a) Histogram H_X and the respective symmetric $-H_X$.

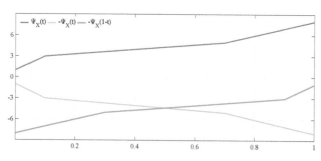

(b) Quantile functions $\Psi_X(t)$, $-\Psi_X(t)$, $-\Psi_X(1-t)$

FIGURE 1.7
Representations of a distribution and the respective symmetric.

Example 6 Consider the histograms H_X and H_Y in Example 5. Now these histograms will be represented by their quantile functions (see Figure 1.8):

$$
\Psi_X(t) = \begin{cases} 1 + \frac{t}{0.1} \times 2 & \text{if} \quad 0 \leq t < 0.1 \\ 3 + \frac{t-0.1}{0.6} \times 2 & \text{if} \quad 0.1 \leq t < 0.7 \\ 5 + \frac{t-0.7}{0.3} \times 3 & \text{if} \quad 0.7 \leq t \leq 1 \end{cases}
$$

$$
\Psi_Y(t) = \begin{cases} \frac{t}{0.8} & \text{if} \quad 0 \leq t < 0.8 \\ 1 + \frac{t-0.8}{0.2} \times 3 & \text{if} \quad 0.8 \leq t \leq 1 \end{cases}
$$

To operate with quantile functions, first the functions have to be rewritten with the same number of pieces and the same set of cumulative weights, using the process proposed by Irpino and Verde [14] and described in Section 1.2.2. Rewriting the quantile functions $\Psi_X(t)$ and $\Psi_Y(t)$ we obtain the following quantile functions:

$$\Psi_X(t) = \begin{cases} 1 + \frac{t}{0.1} \times 2 & \text{if} \quad 0 \leq t < 0.1 \\ 3 + \frac{t-0.1}{0.6} \times 2 & \text{if} \quad 0.1 \leq t < 0.7 \\ 5 + \frac{t-0.7}{0.1} & \text{if} \quad 0.7 \leq t < 0.8 \\ 6 + \frac{t-0.8}{0.2} \times 2 & \text{if} \quad 0.8 \leq t \leq 1 \end{cases}$$

$$\Psi_Y(t) = \begin{cases} \frac{t}{0.1} \times \frac{1}{8} & \text{if} \quad 0 \leq t < 0.1 \\ \frac{1}{8} + \frac{t-0.1}{0.6} \times \frac{6}{8} & \text{if} \quad 0.1 \leq t < 0.7 \\ \frac{7}{8} + \frac{t-0.7}{0.1} \times \frac{1}{8} & \text{if} \quad 0.7 \leq t < 0.8 \\ 1 + \frac{t-0.8}{0.2} \times 3 & \text{if} \quad 0.8 \leq t \leq 1 \end{cases}$$

Addition

From the latter representation of the quantile functions $\Psi_X(t)$ and $\Psi_Y(t)$, the result of its addition is the non-decreasing function represented in Figure 1.8 and defined as follows:

$$\Psi_X(t) + \Psi_Y(t) = \begin{cases} 1 + \frac{t}{0.1} \times \frac{17}{8} & \text{if} \quad 0 \leq t < 0.1 \\ \frac{25}{8} + \frac{t-0.1}{0.6} \times \frac{22}{8} & \text{if} \quad 0.1 \leq t < 0.7 \\ \frac{47}{8} + \frac{t-0.7}{0.1} \times \frac{9}{8} & \text{if} \quad 0.7 \leq t < 0.8 \\ 7 + \frac{t-0.8}{0.2} \times 5 & \text{if} \quad 0.8 \leq t \leq 1 \end{cases}$$

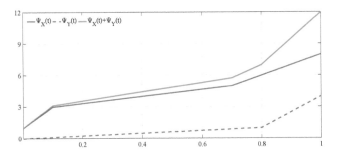

FIGURE 1.8

Representation of the quantile functions Ψ_X, Ψ_Y, $\Psi_X + \Psi_Y$ in Example 6.

When we add the quantile function Ψ_X with a positive real number, *e.g.* 2 and with a negative real number, *e.g.* -2, we obtain non-decreasing functions.

In these cases, we obtain the functions represented above in Figure 1.9:

$$\Psi_X(t) + 2 = \begin{cases} 3 + \frac{t}{0.1} \times 2 & if \quad 0 \leq t < 0.1 \\ 5 + \frac{t-0.1}{0.6} \times 2 & if \quad 0.1 \leq t < 0.7 \\ 7 + \frac{t-0.7}{0.3} \times 3 & if \quad 0.7 \leq t \leq 1 \end{cases}$$

$$\Psi_X(t) - 2 = \begin{cases} -1 + \frac{t}{0.1} \times 2 & if \quad 0 \leq t < 0.1 \\ 1 + \frac{t-0.1}{0.6} \times 2 & if \quad 0.1 \leq t < 0.7 \\ 3 + \frac{t-0.7}{0.3} \times 3 & if \quad 0.7 \leq t \leq 1 \end{cases}$$

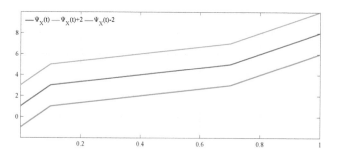

FIGURE 1.9
Representation of the quantile functions Ψ_X, $\Psi_X + 2$, $\Psi_X - 2$ in Example 6.

Multiplication

If we multiply the quantile function $\Psi_X(t)$ by the positive real number 2, we obtain a non-decreasing function but if we multiply the quantile function $\Psi_X(t)$ by the negative real number -1 the resulting function is not a non-decreasing function. The following functions and representations in Figure 1.10 illustrate this situation:

$$2\Psi_X(t) = \begin{cases} 2 + \frac{t}{0.1} \times 4 & if \quad 0 \leq t < 0.1 \\ 6 + \frac{t-0.1}{0.6} \times 4 & if \quad 0.1 \leq t < 0.7 \\ 10 + \frac{t-0.7}{0.3} \times 6 & if \quad 0.7 \leq t \leq 1 \end{cases}$$

$$-\Psi_X(t) = \begin{cases} -1 + \frac{t}{0.1} \times (-2) & if \quad 0 \leq t < 0.1 \\ -3 + \frac{t-0.1}{0.6} \times (-2) & if \quad 0.1 \leq t < 0.7 \\ -5 + \frac{t-0.7}{0.3} \times (-3) & if \quad 0.7 \leq t \leq 1 \end{cases}$$

According to the previous study, the quantile function that represents the

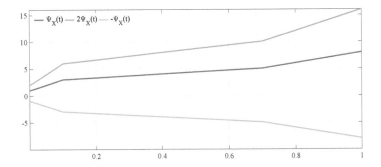

FIGURE 1.10
Representation of the functions $\Psi_X(t)$, $2\Psi_X(t)$, $-\Psi_X(t)$ in Example 6.

symmetric of the histogram H_X, i.e. the histogram $-H_X$, is the piecewise function defined below and shown in Figure 1.7(b).

$$-\Psi_X(1-t) = \begin{cases} -8 + \frac{t}{0.1} \times 3 & \text{if} \quad 0 \leq t < 0.3 \\ -5 + \frac{t-0.3}{0.6} \times 2 & \text{if} \quad 0.3 \leq t < 0.9 \\ -3 + \frac{t-0.9}{0.1} \times 2 & \text{if} \quad 0.9 \leq t \leq 1 \end{cases}$$

An interesting behavior emerges from the use of the Irpino and Verde process (see Section 1.2.2) when it is applied to a set of couples of quantile functions $\Psi(t)$ and $-\Psi(1-t)$, with $t \in [0,1]$. In this case, the distributions are defined with an equal number of subintervals, each of which with associated weights p_ℓ and that verify the condition $p_\ell = p_{m-\ell+1}$, with $\ell \in \{1, 2, \ldots, m\}$.

Considering again the quantile functions $\Psi_X(t)$ and $-\Psi_X(1-t)$, the set of the cumulative weights for these functions is $W = \{0, 0.1, 0.7, 1; 0, 0.3, 0.9, 1\}$. Selecting the weights without repetition, we have $Z = \{0, 0.1, 0.3, 0.7, 0.9, 1\}$. From this set of cumulative weights we have $p_1 = p_5 = 0.1$; $p_2 = p_4 = 0.2$; and $p_3 = 0.4$. So, the quantile functions $\Psi_X(t)$ and $-\Psi_X(1-t)$ may be rewritten as follows:

$$\Psi_X(t) = \begin{cases} 1 + \frac{t}{0.1} \times 2 & \text{if} \quad 0 \leq t < 0.1 \\ 3 + \frac{t-0.1}{0.2} \times \frac{2}{3} & \text{if} \quad 0.1 \leq t < 0.3 \\ \frac{11}{3} + \frac{t-0.3}{0.4} \times \frac{4}{3} & \text{if} \quad 0.3 \leq t < 0.7 \\ 5 + \frac{t-0.7}{0.2} \times 2 & \text{if} \quad 0.7 \leq t < 0.9 \\ 7 + \frac{t-0.9}{0.1} & \text{if} \quad 0.9 \leq t \leq 1 \end{cases}$$

$$-\Psi_X(1-t) = \begin{cases} -8 + \frac{t}{0.1} & if \quad 0 \le t < 0.1 \\ -7 + \frac{t-0.1}{0.2} \times 2 & if \quad 0.1 \le t < 0.3 \\ -5 + \frac{t-0.3}{0.4} \times \frac{4}{3} & if \quad 0.3 \le t < 0.7 \\ -\frac{11}{3} + \frac{t-0.7}{0.2} \times \frac{2}{3} & if \quad 0.7 \le t < 0.9 \\ -3 + \frac{t-0.9}{0.1} \times 2 & if \quad 0.9 \le t \le 1 \end{cases}$$

As we may observe, both functions represent histograms (represented in Figure 1.11) that verify the condition $p_\ell = p_{5-\ell+1}$, with $\ell \in \{1,2,3,4,5\}$.

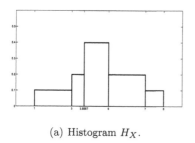

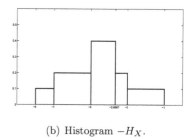

(a) Histogram H_X.   (b) Histogram $-H_X$.

FIGURE 1.11
Representations of the histograms H_X and $-H_X$ rewritten with the same number of subintervals and the same set of cumulative weights.

1.4   Distances between distributions

In classical statistics, the Euclidean distance is often used to compare real values. However, the comparison between distributions is much more complex. Similarly to the classical setup, one possible approach in *SDA* could be computing the difference between two numerical distributions represented by their respective histograms using the histograms' arithmetic.

In their work about forecasting time series applied to histogram-valued variables, Arroyo and Maté [2, 3] also needed to measure the error between the observed and forecasted distributions. Firstly, they considered the possibility of measuring the error between two distributions represented by their respective histograms using the histograms' arithmetic. However, this option turned out to be of little use. As we have seen, it is not easy to operate with the histograms arithmetic and some results are not as expected. For example, as we mentioned above, the difference between two equal histograms is not zero. According to Arroyo [2], this happens because the goal of these arithmetics is to provide a distribution that includes all results that are possible to be obtained by each one of the possible values in each of the terms. So, it is not adequate to analyze the similarity between distributions applying the histogram arithmetic. In some of the statistical methods, such as for example, in linear regression methods proposed for histogram-valued variables [10, 15], the complexity of the elements histograms/quantile functions requires a different approach to measure the dissimilarity between observed and predicted distributions.

As the observations of the histogram-valued variables are distributions, the dissimilarity between these elements will be measured using distances between distributions. According to the selection of divergency measures between distributions, several alternatives to quantify the error of a forecast/prediction [2, 6, 13] were studied. In Table 1.6 [2] the analyzed measures are presented, where $f(x)$ and $g(x)$ are density functions defined for $x \in \mathbb{R}$, $F(x)$ and $G(x)$ the distribution functions and $\Psi_F(t)$ and $\Psi_G(t)$ the respective quantile functions.

TABLE 1.6

Divergency measures between distributions.

Divergency measures	Definitions
Kullback-Leibler	$D_{KL}(f,g) = \int_{\mathbb{R}} log\left(\dfrac{f(x)}{g(x)}\right) f(x) dx$
Jeffrey	$D_J(f,g) = D_{KL}(f,g) + D_{KL}(g,f)$
χ^2	$D_{\chi^2}(f,g) = \int_{\mathbb{R}} \dfrac{|f(x) - g(x)|^2}{g(x)} dx$
Hellinger	$D_H(f,g) = \left[\int_{\mathbb{R}} \left(\sqrt{f(x)} - \sqrt{g(x)}\right) dx\right]^{\frac{1}{2}}$
Total variation	$D_{var}(f,g) = \int_{\mathbb{R}} |f(x) - g(x)| dx$
Wasserstein	$D_W(f,g) = \int_0^1 |\Psi_F(t) - \Psi_G(t)| dt$
Mallows	$D_M(f,g) = \sqrt{\int_0^1 (\Psi_F(t) - \Psi_G(t))^2 dt}$
Kolmogorov	$D_K(f,g) = \max_{\mathbb{R}} |F(x) - G(x)|$

According to Arroyo [2], only the *Wasserstein* and *Mallows* (also known as L_2 Wasserstein) divergency measures are considered adequate to measure the dissimilarity between observed and forecasted distributions (for more details see Chapter 5 in [2]). These measures present interesting properties for error measurement that allow them to be considered as distances: positive definiteness, symmetry and triangle inequality condition. For Arroyo and Maté [2, 3], the *Wasserstein* and *Mallows distance* have intuitive interpretations related to the *Earth Mover's Distance* and are the ones that better adjust to the concept of distance as assessed by the human eye. The measures are defined in terms of the quantile functions and the more further apart are these functions the larger the distance between them. This distance was also used in other works such as those of Irpino and Verde [14, 19], where the L_2 *Wasserstein distance* is successfully applied to cluster histogram data. These authors proved interesting properties of these measures such as, for example, that they allow for the Huygens theorem of decomposition of inertia for clustered data [14]. The same authors used this distance to define basic statistics for numerical distributional data [15]. A linear regression model for histogram-valued variables recently proposed in [10, 15] is also based on this distance. The *Wasserstein* and *Mallows distances* are defined as follows:

Definition 7 Given two quantile functions $\Psi_{Y_1}(t)$ and $\Psi_{Y_2}(t)$ that represent distributions or intervals, the Wasserstein distance is defined as:

$$D_W(\Psi_{Y_1}(t), \Psi_{Y_2}(t)) = \int_0^1 |\Psi_{Y_1}(t) - \Psi_{Y_2}(t)| \, dt \tag{1.8}$$

and the Mallows distance:

$$D_M(\Psi_{Y_1}(t), \Psi_{Y_2}(t)) = \sqrt{\int_0^1 \left(\Psi_{Y_1}(t) - \Psi_{Y_2}(t)\right)^2 dt}. \tag{1.9}$$

If we consider the more general expression:

$$D(\Psi_{Y_1}(t), \Psi_{Y_2}(t)) = \left(\int_0^1 |\Psi_X(t) - \Psi_Y(t)|^p \, dt\right)^{\frac{1}{p}} \tag{1.10}$$

we obtain for the particular cases of $p = 1$ and $p = 2$ the *Wasserstein* and *Mallows distance*, respectively. This is similar to the case of the *Minkowski metric*, that is the *Manhattan distance* when $p = 1$ and the *Euclidean distance* when $p = 2$.

Assuming the Uniform distribution within each of the subintervals of the histograms, the expressions that define the *Mallows* and *Wasserstein distances* may be rewritten using the centers and half ranges of the subintervals for histograms. The simplification of the expression for the *Mallows distance* was proposed by Irpino and Verde [14]; later Arroyo [2] presented a similar simplification for the expression of the *Wasserstein distance*.

Proposition 1 [14] Consider two empirical distributions Y_1 and Y_2 that may be represented by the quantile functions $\Psi_{Y_1}(t)$ and $\Psi_{Y_2}(t)$. Both functions are written with m pieces, and the same set of cumulative weights and the Uniform distribution within subintervals is assumed. The square of the Mallows distance between these distributions is given by

$$D_M^2(\Psi_{Y_1}(t), \Psi_{Y_2}(t)) = \sum_{\ell=1}^{m} p_\ell \left[(c_{1\ell} - c_{2\ell})^2 + \frac{1}{3}(r_{1\ell} - r_{2\ell})^2 \right]$$

where, $c_{1\ell}, c_{2\ell}$ and $r_{1\ell}, r_{2\ell}$ with $\ell \in \{1, \ldots, m\}$ are the centers and half ranges of the subinterval ℓ of the distributions Y_1 and Y_2, respectively.

Proof: Consider the quantile functions $\Psi_{Y_1}(t)$ and $\Psi_{Y_2}(t)$ as in expression (1.5) written with equal numbers of pieces and the same set of cumulative weights. According to expression (1.9) in Definition 7, we have

$$D_M^2(\Psi_{Y_1}(t), \Psi_{Y_2}(t)) = \int_0^1 (\Psi_{Y_1}(t) - \Psi_{Y_2}(t))^2 dt$$

$$= \sum_{\ell=m}^{n} \int_{w_{\ell-1}}^{w_\ell} \left[\left(c_{1\ell} + \left(\frac{2(t-w_\ell)}{w_\ell - w_{\ell-1}} - 1 \right) r_{1\ell} \right) - \left(c_{2\ell} + \left(\frac{2(t-w_\ell)}{w_\ell - w_{\ell-1}} - 1 \right) r_{2\ell} \right) \right]^2 dt$$

$$= \sum_{\ell=1}^{m} \int_{w_{\ell-1}}^{w_\ell} \left[(c_{1\ell} - c_{2\ell}) + \left(\frac{2(t-w_\ell)}{w_\ell - w_{\ell-1}} - 1 \right)(r_{2\ell} - r_{1\ell}) \right]^2 dt.$$

Making the change of variable $v = \frac{t - w_{\ell-1}}{w_\ell - w_{\ell-1}}$ we can rewrite the integral as follows:

$$\sum_{\ell=1}^{m} \int_0^1 p_\ell \left[(c_{1\ell} - c_{2\ell}) + (2v - 1)(r_{2\ell} - r_{1\ell}) \right]^2 dv$$

$$= \sum_{\ell=1}^{m} p_\ell \left[(c_{1\ell} - c_{2\ell})^2 + \frac{1}{3}(r_{2\ell} - r_{1\ell})^2 \right].$$

So, we have

$$D_M^2(\Psi_{Y_1}(t), \Psi_{Y_2}(t)) = \sum_{\ell=1}^{m} p_\ell \left[(c_{1\ell} - c_{2\ell})^2 + \frac{1}{3}(r_{1\ell} - r_{2\ell})^2 \right].$$

$\square$

Proposition 2 [2] Consider two empirical distributions Y_1 and Y_2 that may be represented by the quantile functions $\Psi_{Y_1}(t)$ and $\Psi_{Y_2}(t)$, both written with m pieces and the same set of cumulative weights. Assuming the Uniform distribution within the subintervals, the Wasserstein distance between these distributions is given by

$$D_W(\Psi_{Y_1}(t), \Psi_{Y_2}(t)) = \sum_{\ell=1}^{m} p_\ell |(c_{1\ell} - c_{2\ell})|$$

where, $c_{1\ell}$ and $c_{2\ell}$ are the centers of the subintervals ℓ, with $\ell \in \{1, \ldots, m\}$.

Proof: Consider the quantile functions $\Psi_{Y_1}(t)$ and $\Psi_{Y_2}(t)$ as in expression (1.5) written with equal numbers of pieces and the same set of cumulative weights.

According to expression (1.8) in Definition 7, we have

$$D_W\left(\Psi_{Y_1}(t), \Psi_{Y_2}(t)\right) = \int_0^1 |\Psi_{Y_1}(t) - \Psi_{Y_2}(t)| dt$$

$$= \sum_{\ell=1}^m \int_{w_{\ell-1}}^{w_\ell} \left| \left(c_{1\ell} + \left(\frac{2(t-w_\ell)}{w_\ell - w_{\ell-1}} - 1\right) r_{1\ell}\right) - \left(c_{2\ell} + \left(\frac{2(t-w_\ell)}{w_\ell - w_{\ell-1}} - 1\right) r_{2\ell}\right) \right| dt$$

$$= \sum_{\ell=1}^m \int_{w_{\ell-1}}^{w_\ell} \left| (c_{1\ell} - c_{2\ell}) + \left(\frac{2(t - w_\ell)}{w_\ell - w_{\ell-1}} - 1\right)(r_{2\ell} - r_{1\ell}) \right| dt.$$

Making the change of variable $v = \frac{t - w_{\ell-1}}{w_\ell - w_{\ell-1}}$ we can rewrite the integral as follows:

$$\sum_{\ell=1}^m \int_0^1 p_\ell |(c_{1\ell} - c_{2\ell}) + (2v - 1)(r_{2\ell} - r_{1\ell})| dv.$$

- If $(c_{1\ell} - c_{2\ell}) + (2v - 1)(r_{2\ell} - r_{1\ell}) \geq 0$ we have

$$\sum_{\ell=1}^m \int_0^1 p_\ell |(c_{1\ell} - c_{2\ell}) + (2v - 1)(r_{2\ell} - r_{1\ell})| dv$$

$$= \sum_{\ell=1}^m p_\ell \left[(c_{1\ell} - c_{2\ell}) v + (r_{2\ell} - r_{1\ell}) v^2 - (r_{2\ell} - r_{1\ell}) v \right]_0^1$$

$$= \sum_{\ell=1}^m p_\ell (c_{1\ell} - c_{2\ell}).$$

- If $(c_{1\ell} - c_{2\ell}) + (2v - 1)(r_{2\ell} - r_{1\ell}) < 0$ we have

$$\sum_{\ell=1}^m \int_0^1 p_\ell |(c_{1\ell} - c_{2\ell}) + (2v - 1)(r_{2\ell} - r_{1\ell})| dv$$

$$= - \sum_{\ell=1}^m p_\ell (c_{1\ell} - c_{2\ell}).$$

Therefore,

$$D_W\left(\Psi_{Y_1}(t), \Psi_{Y_2}(t)\right) = \sum_{\ell=1}^m p_\ell |c_{1\ell} - c_{2\ell}|.$$

$\square$

In Example 7, we show that the *Mallows* and *Wasserstein distances* adjust to the concept of distance as assessed by the human eye.

Example 7 Consider again the histogram-valued variable Y_5, "Waiting time for a medical appointment" in Table 1.5. The quantile functions that represent the observations of each healthcare center written as histograms with six subintervals and with the same weight associated to each subinterval of the three variables are defined in Example 4.

To analyze which healthcare centers are more similar, we calculate the square of the *Mallows* and *Wasserstein distances* between the distributions and the values are presented in Table 1.7. The results are in accordance to the graphical behavior of the quantile functions in Figure 1.2 of the Example 4. The healthcare centers B and C are the most similar and the A and B the most different.

TABLE 1.7
Mallows and *Wasserstein distances* between the observations of the histogram-valued variable "Waiting time for a medical appointment" in Table 1.5.

	Mallows distance	Wasserstein distance
$D\big(\Psi_{Y_5(A)}(t), \Psi_{Y_5(B)}(t)\big)$	33.81	33.00
$D\big(\Psi_{Y_5(A)}(t), \Psi_{Y_5(C)}(t)\big)$	23.51	22.50
$D\big(\Psi_{Y_5(B)}(t), \Psi_{Y_5(C)}(t)\big)$	15.12	10.50

1.5 Conclusion

In this chapter, we have introduced the main concepts underlying the representation and analysis of distributional data, under the framework of Symbolic Data Analysis, providing the basic notions for subsequent chapters in this book. The need to consider variability intrinsic to an observation lead to the introduction of new variable types. Distributional data are represented by modal or histogram-valued variables; we have particularly focused on the latter, for which alternative representation models and arithmetics have been discussed. We note that modal-valued variables are also addressed by *Compositional Data Analysis* [1, 12, 17].

Bibliography

[1] John Aitchison. The statistical analysis of compositional data. *Journal of the Royal Statistical Society: Series B (Methodological)*, 44(2):139–160, 1982.

[2] J. Arroyo. *Métodos de Predicción para Series Temporales de Intervalos e Histogramas*. PhD thesis, Universidad Pontificia Comillas, Madrid, Spain, 2008.

[3] J. Arroyo and C. Maté. Forecasting histogram time series with k-nearest neighbours methods. *International Journal of Forecasting*, 25(1):192–207, 2009.

[4] L. Billard and E. Diday. From the statistics of data to the statistics of knowledge: Symbolic Data Analysis. *Journal of the American Statistical Association*, 98(462):470–487, 2003.

[5] L. Billard and E. Diday. *Symbolic Data Analysis: Conceptual Statistics and Data Mining*. John Wiley & Sons, Inc. New York, NY, USA, 2006.

[6] H.-H. Bock and E. Diday, editors. *Analysis of Symbolic Data: Exploratory Methods for Extracting Statistical Information from Complex Data*. Springer-Verlag Berlin, 2000.

[7] P. Brito. Symbolic data analysis: another look at the interaction of Data Mining and Statistics. *WIREs Data Mining and Knowledge Discovery*, 4(4):281–295, 2014.

[8] A.G. Colombo and R.J. Jaarsma. A powerful numerical method to combine random variables. *IEEE Transactions on Reliability*, 29(2):126–129, 1980.

[9] S. Dias. *Linear Regression with Empirical Distributions*. PhD thesis, University of Porto, Porto, Portugal, 2014.

[10] S. Dias and P. Brito. Linear regression model with histogram-valued variables. *Statistical Analysis and Data Mining*, 8(2):75–113, 2015.

[11] E. Diday. The symbolic approach in clustering and related methods of data analysis: The basic choices. In *Classification and Related Methods of Data Analysis. Proceedings of the 1st Conference of the International Federation of Classification Societies (IFCS'87)*, pages 673–684. North Holland, Amsterdam, 1988.

[12] Peter Filzmoser, Karel Hron, and Matthias Templ. Applied compositional data analysis. *Switzerland: Springer Nature*, 2018.

[13] A.L. Gibbs and F.E. Su. On choosing and bounding probability metrics. *International Statistical Review*, 70(3):419–435, 2002.

[14] A. Irpino and R. Verde. A new Wasserstein based distance for the hierarchical clustering of histogram symbolic data. In *Data Science and Classification. Proceedings of the 10th Conference of the International Federation of Classification Societies*, pages 185–192. Springer Berlin Heidelberg, 2006.

[15] A. Irpino and R. Verde. Linear regression for numeric symbolic variables: a least squares approach based on Wasserstein Distance. *Advances in Data Analysis and Classification*, 9(1):81–106, 2015.

[16] M. Noirhomme-Fraiture and P. Brito. Far beyond the classical data models: Symbolic Data Analysis. *Statistical Analysis and Data Mining*, 4(2):157–170, 2011.

[17] V. Pawlowsky-Glahn and A. Buccianti. *Compositional data analysis: Theory and applications*. John Wiley & Sons, 2011.

[18] P. Prakash and R.S. Murat. *Semilinear (Topological) Spaces and Aplications*. Massachsetts Institute of Technology, revised edition, 1971.

[19] R. Verde and A. Irpino. Comparing histogram data using a Mahalanobis-Wasserstein distance. In P. Brito, editor, *Proceedings of the COMPSTAT'2008, 18th International Conference on Computational Statistics, Porto, Portugal, August 2008*, pages 77–89. Physica-Verlag HD, 2008.

[20] R. Williamson. *Probabilistic Arithmetic*. PhD thesis, University of Queensland, Queensland, Australia, 1989.

2

Descriptive Statistics based on Frequency Distribution

Sónia Dias

School of Technology and Management, Polytechnic Institute of Viana do Castelo & LIAAD-INESC TEC, Portugal

Paula Brito

Faculty of Economics, University of Porto & LIAAD-INESC TEC, Porto, Portugal

CONTENTS

To work with distributional data, the definitions of univariate and bivariate descriptive statistics are fundamental concepts. The first descriptive concepts for histogram-valued variables have to take into account the variability within data and were derived from the empirical density functions defined for this kind of variables. Concepts such as sample mean, variance and covariance were derived following a similar reasoning as for the classical concepts of mean and variance of a real-valued variable. Because important properties are not always verified, for some of the symbolic descriptive measures, more than one definition has been proposed for the same concept.

DOI: 10.1201/9781315370545-2

2.1 Introduction

Chronologically, the first statistical concepts and methods of *SDA* were developed for interval-valued variables (with the Uniform distribution being assumed within the intervals). When possible, these definitions and methods were later generalized to histogram-valued variables. This happened with the first basic univariate and bivariate statistics for interval-valued variables proposed by Bertrand and Goupil [7], that afterwards were extended by Billard and Diday [5, 6] to histogram-valued variables. Concepts such as frequency histograms, sample mean, variance and covariance were defined for this kind of symbolic variables. As these concepts were defined considering that each variable is a uniform mixture of distributions [1,5], the variability within data had to be considered. Another consequence is that although the symbolic variables' "values" are distributions and not real numbers, the results of the application of basic statistical concepts are real numbers. For example, the mean of n observations of a histogram-valued variable, as proposed by Billard and Diday [5], is a real number.

In this chapter, univariate and bivariate descriptive measures for histogram-valued variables are defined, derived and illustrated. Based on classical statistics, empirical symbolic mean and empirical variance were derived from the empirical density function. Since important properties of the classical concepts were not verified for their symbolic counterparts, more than one definition for the same concept were proposed. This occurred, for example, for the empirical variance and empirical covariance [2,5].

Most concepts and methods developed within *SDA* approach are descriptive, since a probabilistic assumption is not considered. For this reason, the concepts defined in this chapter are all "empirical" concepts. However, to simplify the flow of the text, the word "empirical" is omitted most of the times. The development of non-descriptive methods for *SDA* is still an open research topic for almost all kinds of symbolic variables [8, 12]. In general, the difficulty to work with symbolic elements under a probabilistic context lays in the extension of the concept of randomness to symbolic variables.

Recently Irpino and Verde [10,11] proposed alternative definitions of some descriptive measures that considered the variability within data and between data. One of the proposed concepts is an alternative definition of mean for histogram-valued variables, which produces a mean distribution, termed as *barycentric histogram*. In this case, the "mean" of a set of distributions is already a distribution. This descriptive measures approach will be studied in a subsequent chapter.

2.2 Univariate statistics

Consider the histogram-valued variable $Y : S \to \mathbb{B}$, with $\mathbb{B}$ a set of distributions over $\mathcal{Y}$. The observation associated with each unit s_i, with $i \in \{1, ..., n\}$, is represented by a histogram composed by m_i subintervals with weight $p_{i\ell}$, and $\ell \in \{1, \ldots, m_i\}$. The histogram associated with unit s_i may be represented as follows:

$$Y(s_i) = \left\{ [l_{i1}, u_{i1}[, p_{i1}; [l_{i2}, u_{i2}[, p_{i2}; \ldots, [l_{im_i}, u_{im_i}], p_{im_i} \right\}.$$

Most of the descriptive univariate concepts for histogram-valued variables that we will present below have been proposed by Billard and Diday ([5]; Chapter 3 of the book [6]). These concepts are generalizations of the descriptive univariate statistics proposed by Bertrand and Goupil (in Chapter 6 of the book [7]) for interval-valued variables.

2.2.1 Frequency of ranges of values in distributional data

The first basic statistical concepts and the location and dispersion symbolic measures for numerical distributional data were derived considering that each histogram-valued variable is a uniform mixture of distributions.

Definition 1 [1, 7] The empirical distribution function $F(\xi)$, is the distribution function of a mixture of n histograms. Each of these histograms is decomposed in m_i subintervals $I_{i\ell} = [l_{i\ell}, u_{i\ell}[, \ell \in \{1, \ldots, m_i\}$ where the values are uniformly distributed and which are associated with weight $p_{i\ell}$. The **empirical distribution function** of a histogram-valued variable is then defined as a uniform mixture of n distributions, more specifically:

$$F(\xi) = \frac{1}{n} \sum_{i=1}^{n} \sum_{\ell=1}^{m_i} P\{Y_{i\ell} \leq \xi\} p_{i\ell}$$

with

$$P\{Y_{i\ell} \leq \xi\} = \begin{cases} 0 & \text{if } \xi < l_{i\ell} \\ \frac{\xi - l_{i\ell}}{u_{i\ell} - l_{i\ell}} & \text{if } l_{i\ell} \leq \xi < u_{i\ell} \\ 1 & \text{if } \xi \geq u_{i\ell} \end{cases}$$

Therefore,

$$F(\xi) = \frac{1}{n} \sum_{i=1}^{n} \left(\sum_{\ell: \xi \geq u_{i\ell}} p_{i\ell} + \sum_{\ell: \xi \in I_{i\ell}} p_{i\ell} \left(\frac{\xi - l_{i\ell}}{u_{i\ell} - l_{i\ell}} \right) \right).$$

Hence, by taking the derivative with respect to ξ, we obtain the empirical density function.

Definition 2 [5] For the histogram-valued variable Y, the **empirical density function** is given by

$$f(\xi) \;=\; \frac{1}{n}\sum_{i=1}^{n}\sum_{\ell=1}^{m_i}\frac{\mathbf{1}_{I_{i\ell}}(\xi)}{\|I_{i\ell}\|}p_{i\ell}, \quad \xi \in \mathbb{R}.$$

where $\mathbf{1}_{I_{i\ell}}$ is the indicator function of subinterval $I_{i\ell}$ and $\|I_{i\ell}\|$ is its length, $\ell = 1,\ldots,m_i; i = 1,\ldots,n$.

As in classical descriptive statistics, $\displaystyle\int_{-\infty}^{+\infty} f(\xi)d\xi = 1$.

We may define the histogram that allows visualizing the frequency distribution of the observed "values". The histogram of all symbolic observations may be constructed as follows. Let

$$I = \Big[\min\{l_{i\ell} : \ell \in \{1,\ldots,m_i\}, i \in S\}, \max\{u_{i\ell} : \ell \in \{1,\ldots,m_i\}, i \in S\}\Big]$$

be the interval which spans all observed values, and let I be partitioned into $r \geq 1$ subintervals

$$I_g = [\zeta_{g-1}, \zeta_g[\;\; \text{with } g \in \{1,\ldots,r-1\} \;\; and \;\; I_r = [\zeta_{r-1}, \zeta_r].$$

To define the histogram from the symbolic observations it is necessary not only to select the subintervals I_g but also to calculate the observed frequency or relative frequency associated with each subinterval. These concepts are defined below.

Definition 3 [5] For a histogram-valued variable Y the **observed frequency** of interval I_g with $g \in \{1,\ldots,r\}$ is given by

$$f(I_g) \;=\; \sum_{i=1}^{n}\sum_{\ell:I(g)\subseteq I_{i\ell}}\frac{\|I_{i\ell}\cap I_g\|}{\|I_{i\ell}\|}p_{i\ell}. \tag{2.1}$$

The **relative frequency** is then,

$$f_r(I_g) \;=\; \frac{f(I_g)}{n}.$$

Notice that each term in the expression that defines $f(I_g)$ represents the portion of the subinterval $I_{i\ell}$, which is spanned by I_g. Consequently, it represents the proportion of its observed relative frequency $p_{i\ell}$, which pertains to the overall histogram interval I_g. It follows that $\sum_{g=1}^{r} f(I_g) = n$.

The graphical representation of the set

$$\Big\{(I_g, f_r(I_g)), g \in \{1,\ldots,r\}\Big\}$$

is the **histogram of a histogram-valued variable**.

2.2.2 Location and dispersion symbolic measures

In *SDA*, the concepts of median or quartile and mode may be obtained by a similar process as for grouped data in classical statistics.

Definition 4 [9] The **interval median** of the histogram-valued variable is the interval I_g, where the cumulative relative frequency is 0.5. Similarly, we obtain as interval first quartile the interval I_g, where the cumulative relative frequency is 0.25 and the interval third quartile the interval I_g, where the cumulative relative frequency is 0.75.

Definition 5 [9] The **interval mode** of the histogram-valued variable is the interval I_g, where the relative frequency is higher. Notice that the **interval mode** may not exist, and if existing it may not be unique.

In classical statistics, the concepts of mean and variance of a random variable are derived from the respective density function. Similarly, symbolic descriptive measures such as the symbolic sample mean and the symbolic sample variance were derived from the empirical density function.

Let Y be a histogram-valued variable and f the empirical density function as presented in Definition 2. The symbolic mean (empirical symbolic mean) of Y is obtained as follows [5]:

$$
\begin{aligned}
\overline{Y} &= \int_{-\infty}^{+\infty} \xi f(\xi) d\xi = \frac{1}{n} \int_{-\infty}^{+\infty} \xi \sum_{i=1}^{n} \left(\sum_{\ell=1}^{m_i} \frac{1_{I_{i\ell}}(\xi)}{\|I_{i\ell}\|} p_{i\ell} \right) d\xi \\
&= \frac{1}{n} \sum_{i=1}^{n} \sum_{\ell=1}^{m_i} \int_{l_{i\ell}}^{u_{i\ell}} \frac{\xi}{\|u_{i\ell} - l_{i\ell}\|} p_{i\ell} d\xi \\
&= \frac{1}{2n} \sum_{i=1}^{n} \sum_{\ell=1}^{m_i} \frac{u_{i\ell}^2 - l_{i\ell}^2}{u_{i\ell} - l_{i\ell}} p_{i\ell} \\
&= \frac{1}{n} \sum_{i=1}^{n} \sum_{\ell=1}^{m_i} \frac{u_{i\ell} + l_{i\ell}}{2} p_{i\ell}
\end{aligned}
\tag{2.2}
$$

Definition 6 [5] The **symbolic mean (empirical symbolic mean)** for histogram-valued variables is given by

$$
\overline{Y} = \frac{1}{n} \sum_{i=1}^{n} \sum_{\ell=1}^{m_i} \frac{u_{i\ell} + l_{i\ell}}{2} p_{i\ell} = \frac{1}{n} \sum_{i=1}^{n} \sum_{\ell=1}^{m_i} c_{i\ell} p_{i\ell}
\tag{2.3}
$$

where $l_{i\ell}$, $u_{i\ell}$ and $c_{i\ell}$ is are, respectively, the lower and upper bounds and center of each subinterval $I_{i\ell}$ with associated weight $p_{i\ell}$, with $i \in \{1, ..., n\}$ and $\ell \in \{1, ..., m_i\}$.

As two definitions of symbolic variance for histogram-valued variables were proposed, we will denote the first definition proposed by Billard and Diday

[5], as symbolic variance 1, (s_1^2) and the definition proposed by Billard and Diday [3, 4] as symbolic variance 2, (s_2^2). Similarly as for the symbolic mean, the symbolic variance 1 for histogram-valued variables is derived as follows:

$$
\begin{aligned}
s_1^2(Y) &= \int_{-\infty}^{+\infty} \left(\xi - \overline{Y}\right)^2 f(\xi)d\xi \\
&= \int_{-\infty}^{+\infty} \xi^2 f(\xi)d\xi - 2\overline{Y}\int_{-\infty}^{+\infty} \xi f(\xi)d\xi + \overline{Y}^2 \int_{-\infty}^{+\infty} f(\xi)d\xi \\
&= \int_{-\infty}^{+\infty} \xi^2 f(\xi)d\xi - \overline{Y}^2
\end{aligned}
$$

with

$$
\begin{aligned}
\int_{-\infty}^{+\infty} \xi^2 f(\xi)d\xi &= \int_{-\infty}^{+\infty} \frac{\xi^2}{n} \sum_{i=1}^{n} \sum_{\ell=1}^{m_i} \frac{1_{I_{i\ell}}(\xi)}{\|I_{i\ell}\|} p_{i\ell} d\xi \\
&= \frac{1}{n} \sum_{i=1}^{n} \sum_{\ell=1}^{m_i} \int_{l_{i\ell}}^{u_{i\ell}} \frac{\xi^2}{u_{i\ell} - l_{i\ell}} p_{i\ell} d\xi \\
&= \frac{1}{3n} \sum_{i=1}^{n} \sum_{\ell=1}^{m_i} \frac{u_{i\ell}^3 - l_{i\ell}^3}{u_{i\ell} - l_{i\ell}} p_{i\ell} \\
&= \frac{1}{3n} \sum_{i=1}^{n} \sum_{\ell=1}^{m_i} \left(u_{i\ell}^2 + l_{i\ell}u_{i\ell} + l_{i\ell}^2\right) p_{i\ell}.
\end{aligned}
$$

Definition 7 [5] The **symbolic variance 1 (empirical symbolic variance 1)** for histogram-valued variables is defined by:

$$
s_1^2(Y) = \frac{1}{3n} \sum_{i=1}^{n} \sum_{\ell=1}^{m_i} \left(u_{i\ell}^2 + l_{i\ell}u_{i\ell} + l_{i\ell}^2\right) p_{i\ell} - \overline{Y}^2 \tag{2.4}
$$

that may also be written as

$$
s_1^2(Y) = \frac{1}{3n} \sum_{i=1}^{n} \left(\sum_{\ell=1}^{m_i} \left(l_{i\ell} - \overline{Y}\right)^2 + \left(l_{i\ell} - \overline{Y}\right)\left(u_{i\ell} - \overline{Y}\right) + \left(u_{i\ell} - \overline{Y}\right)^2\right) p_{i\ell}; \tag{2.5}
$$

Definition 8 As $s_1^2(Y)$ is the symbolic variance of the variable Y, the **symbolic standard deviation**, is usually:

$$
s_1(Y) = \sqrt{s_1^2(Y)}
$$

Concerning the formulas obtained for the symbolic mean and the symbolic variance 1, it is important to underline that:

- As in classical statistics, the expressions of the symbolic mean and symbolic variance 1 are derived from the density function and hence they are consistent with the assumption of uniformity within each subinterval.

- The symbolic mean of a histogram-valued variable Y is obtained applying the classical definition of mean to the weighted center values, for each unit s_i, of the histogram-valued variable Y, i. e. $\overline{Y}_i = \sum_{\ell=1}^{m_i} \frac{u_{i\ell}+l_{i\ell}}{2} p_{i\ell} = \sum_{\ell=1}^{m_i} c_{i\ell} p_{i\ell}$ with $i \in \{1, ..., n\}$. The symbolic mean of the histogram-valued variable is then

$$\overline{Y} = \frac{1}{n} \sum_{i=1}^{n} \overline{Y}_i. \tag{2.6}$$

- If we apply the classical definition of variance to the weighted mean values of the distributions, i.e. $\overline{Y}_i$, with $i \in \{1, ..., n\}$, we do not obtain the expression of the symbolic variance 1.

- In the particular case where all observations of a histogram-valued variable consist of one interval with an associated weight of 1, the symbolic variance 1 coincides with the corresponding symbolic variance definition for interval-valued variables, as proposed by Bertrand and Goupil (in Chapter 6 of the book [7]). If all observations of a histogram-valued variable are degenerate (real values), the symbolic variance 1 coincides with the classical definition of variance.

The concept of the symbolic variance 2 was defined as a particular case of the concept of symbolic covariance, that will be presented below.

2.3 Bivariate descriptive statistics

We now consider now the concepts of descriptive statistics for two histogram-valued variables [1, 3–7]. Let Y_1 and Y_2 be two histogram-valued variables, where Y_j, $j \in \{1,2\}$ is composed by m_{ij} subintervals $I_{Y_{ij\ell_j}}$, each of which with a weight $p_{ij\ell_j}$, $\ell_j \in \{1, \ldots, m_{j\ell}\}$. These histograms are written as:

$$Y_1(s_i) = \left\{ [l_{i11}, u_{i11}[, p_{i11}; [l_{i12}, u_{i12}[, p_{i12}; \ldots, [l_{i1m_{i1}}, u_{i1m_{i1}}], p_{i1m_{i1}} \right\};$$

$$Y_2(s_i) = \left\{ [l_{i21}, u_{i21}[, p_{i21}; [l_{i22}, u_{i22}[, p_{i22}; \ldots, [l_{i2m_{i2}}, u_{i2m_{i2}}], p_{i2m_{i2}} \right\}.$$

2.3.1 Empirical joint distribution and density functions

In this section, the concepts of empirical distribution and density function presented in Section 2.2.1 will be generalized to the bivariate case. Analogously to the definition of empirical symbolic variance, one of the definitions of covariance is derived from the empirical joint density function.

Definition 9 [1] The **empirical joint distribution function** $F_{Y_1 \times Y_2}(\xi_1, \xi_2)$ for histogram-valued variables Y_1 and Y_2, assuming the Uniform distribution within the subintervals of the histograms, is given by

$$
\begin{aligned}
F(\xi_1, \xi_2) &= \frac{1}{n} \sum_{i=1}^{n} \sum_{\ell_1=1}^{m_{i1}} \sum_{\ell_2=1}^{m_{i2}} P\{Y_{i1\ell_1} \leq \xi_1, Y_{i2\ell_2} \leq \xi_2\} p_{i1\ell_1} p_{i2\ell_2} \\
&= \frac{1}{n} \sum_{i=1}^{n} \left(\sum_{\substack{\ell_1 : \xi_1 \geq u_{i1\ell_1} \\ \ell_2 : \xi_2 \geq u_{i2\ell_2}}} p_{i1\ell_1} p_{i2\ell_2} \right. \\
&\qquad + \left. \sum_{\ell_1, \ell_2 : (\xi_1, \xi_2) \in I_{i1\ell_1} \times I_{i2\ell_2}} p_{i1\ell_1} p_{i2\ell_2} \left(\frac{\xi_1 - l_{i1\ell_1}}{u_{i1\ell_1} - l_{i1\ell_1}} \right) \left(\frac{\xi_2 - l_{i2\ell_2}}{u_{i2\ell_2} - l_{i2\ell_2}} \right) \right)
\end{aligned}
$$

with $\ell_j \in \{1, \ldots, m_{ij}\}$; $j \in \{1, 2\}$, and m_{ij} the number of subintervals for the unit i of the histogram-valued variable Y_j.

Definition 10 [5] The **empirical joint density function** is defined for histogram-valued variables Y_1, Y_2 as follows:

$$
f(\xi_1, \xi_2) = \frac{1}{n} \sum_{i=1}^{n} \left(\sum_{\ell_1=1}^{m_{i1}} \sum_{\ell_2=1}^{m_{i2}} \frac{\mathbf{1}_{I_{i1\ell_1} \times I_{i2\ell_2}}(\xi_1, \xi_2)}{\|I_{i1\ell_1} \times I_{i2\ell_2}\|} p_{i1\ell_1} p_{i2\ell_2} \right)
$$

where $\mathbf{1}_{I_{i1\ell_1} \times I_{i2\ell_2}}$ is the indicator function and $\|I_{i1\ell_1} \times I_{i2\ell_2}\|$ is the area of the rectangle $\left[l_{i1\ell_1}, u_{i1\ell_1}\right] \times \left[l_{i2\ell_2}, u_{i2\ell_2}\right]$.

As in classical statistics, we have $\displaystyle\int_{-\infty}^{-\infty} \int_{-\infty}^{-\infty} f(\xi_1, \xi_2) d\xi_1 d\xi_2 = 1$.

Analogously to the univariate situation, we may obtain the **joint histogram** for $Y_1 \times Y_2$ by graphically plotting

$$
\left\{ \left(I_{g_1} \times I_{g_2}, f_r\left(I_{g_1} \times I_{g_2}\right)\right), g_1 \in \{1, \ldots, r_1\}, g_2 \in \{1, \ldots, r_2\} \right\}.
$$

In this histogram we have:

- $I_{g_1} = [\zeta_{g_1-1}, \zeta_{g_1}[$ with $g_1 \in \{1, \ldots, r_1 - 1\}$ and $I_{r_1} = [\zeta_{r_1-1}, \zeta_{r_1}]$;

- $I_{g_2} = [\zeta_{g_2-1}, \zeta_{g_2}[$ with $g_2 \in \{1, \ldots, r_2 - 1\}$ and $I_{r_2} = [\zeta_{r_2-1}, \zeta_{r_2}]$.

The intervals I_{g_1} and I_{g_2} result of a partition of the intervals I_1 and I_2 in the r_1 and r_2 subintervals, respectively. The intervals I_j, with $j \in \{1, 2\}$ are the following:

$$
I_j = \left[\min\{l_{ij\ell_j} : \ell_j \in \{1, \ldots, m_{ij}\}, i \in S\}, \max\{u_{ij\ell_j} : \ell_j \in \{1, \ldots, m_{ij}\}, i \in S\} \right].
$$

To build the joint histogram, it is necessary to calculate the relative frequency associated with each rectangle $I_{g_1} \times I_{g_2}$. To define this frequency, we will first define the observed frequency.

Definition 11 [5] For the histogram-valued variables Y_1 and Y_2, the **observed frequency** of the rectangle $I_{g_1} \times I_{g_2}$ with $g_1 \in \{1, \ldots, r_1\}$ and $g_2 \in \{1, \ldots, r_2\}$ is given by

$$f(I_{g_1} \times I_{g_2}) = \sum_{i=1}^{n} \sum_{\ell_1 : I_{g_1} \subseteq I_{i1\ell_1}} \sum_{\ell_2 : I_{g_2} \subseteq I_{i2\ell_2}} \frac{\| (I_{i1\ell_1} \times I_{i2\ell_2}) \cap (I_{g_1} \times I_{g_2}) \|}{\| I_{i1\ell_1} \times I_{i2\ell_2} \|} p_{i1\ell_1} p_{i2\ell_2}.$$

Each term i in this expression defines the proportion of the observed rectangle $I_{i1\ell_1} \times I_{i2\ell_2}$, which overlaps with the rectangle $I_{g_1} \times I_{g_2}$. The **relative frequency** is, in either case,

$$f_r(I_{g_1} \times I_{g_2}) = \frac{f(I_{g_1} \times I_{g_2})}{n}.$$

In addition to the joint histograms, other graphical representations may be considered for two histogram-valued variables, namely the **scatter plot**.

Consider for each unit i, the histograms that represent Y_{i1} and Y_{i2} that must be defined with the same number of subintervals (this may be done according to the Irpino and Verde process [10], see Chapter 1). Each subinterval that composes the histogram is represented by a rectangle in the xx and yy axis, and the respective weight is represented in the zz-axis. Therefore, the scatter plot for two histogram-valued variables is a three-dimensional figure where each observation i of the pair (Y_{i1}, Y_{i2}) is represented by a set of non-overlapping and contiguous hyperrectangles (if the histograms do not have subintervals with null weight) [9]. In the last section, this representation will be illustrated.

2.3.2 Empirical symbolic covariances

As in the univariate case, bivariate concepts are also derived from the empirical joint density function. Following this approach, Billard and Diday [4] obtained an empirical symbolic covariance definition.

For the symbolic covariance, two definitions have already been proposed. To distinguish the two definitions we will denote the first definition by symbolic covariance 1 (cov_1) [3, 4] and the second by symbolic covariance 2 (cov_2) [2, 5].

Consider the histogram-valued variables Y_1 and Y_2, and the empirical joint density function f. We may define the symbolic covariance 1 for histogram-valued variables as follows:

$$\begin{aligned}
cov_1(Y_1, Y_2) &= \int_{-\infty}^{+\infty} \int_{-\infty}^{+\infty} (\xi_1 - \overline{Y_1})(\xi_2 - \overline{Y_2}) f(\xi_1, \xi_2) d\xi_1 d\xi_2 \\
&= \int_{-\infty}^{+\infty} \int_{-\infty}^{+\infty} \xi_1 \xi_2 f(\xi_1, \xi_2) d\xi_1 d\xi_2 - \overline{Y_1} \int_{-\infty}^{+\infty} \int_{-\infty}^{+\infty} \xi_2 f(\xi_1, \xi_2) d\xi_1 d\xi_2 \\
&\quad - \overline{Y_2} \int_{-\infty}^{+\infty} \int_{-\infty}^{+\infty} \xi_1 f(\xi_1, \xi_2) d\xi_1 d\xi_2 + \overline{Y_1}\,\overline{Y_2} \int_{-\infty}^{+\infty} \int_{-\infty}^{+\infty} f(\xi_1, \xi_2) d\xi_1 d\xi_2.
\end{aligned}$$

We may easily prove that,

$$
\int_{-\infty}^{+\infty} \int_{-\infty}^{+\infty} \xi_1 \xi_2 f(\xi_1, \xi_2) d\xi_1 d\xi_2
$$

$$
= \int_{-\infty}^{+\infty} \int_{-\infty}^{+\infty} \xi_1 \xi_2 \frac{1}{n} \sum_{i=1}^{n} \left(\sum_{\ell_1=1}^{m_{i1}} \sum_{\ell_2=1}^{m_{i2}} \frac{1_{I_{i1\ell_1} \times I_{i2\ell_2}}(\xi_1,\xi_2)}{\|I_{Y_{i1\ell_1}} \times I_{Y_{i2\ell_2}}\|} p_{i1\ell_1} p_{i2\ell_2} \right) d\xi_1 d\xi_2
$$

$$
= \frac{1}{n} \sum_{i=1}^{n} \sum_{\ell_1=1}^{m_{i1}} \sum_{\ell_2=1}^{m_{i2}} \int_{l_{i1\ell_1}}^{u_{i1\ell_1}} \int_{l_{i2\ell_2}}^{u_{i2\ell_2}} \frac{\xi_1 \xi_2 p_{i1\ell_1} p_{i2\ell_2}}{(u_{i1\ell_1} - l_{i1\ell_1})(u_{i2\ell_2} - l_{i2\ell_2})} d\xi_1 d\xi_2
$$

$$
= \frac{1}{4n} \sum_{i=1}^{n} \sum_{i_1=1}^{m_{i1}} \sum_{i_2=1}^{m_{i2}} \frac{p_{i1\ell_1} p_{i2\ell_2} \left(u_{i1\ell_1}^2 - l_{i1\ell_1}^2 \right) \left(u_{i2\ell_2}^2 - l_{i2\ell_2}^2 \right)}{(u_{i1\ell_1} - l_{i1\ell_1})(u_{i2\ell_2} - l_{i2\ell_2})}
$$

$$
= \frac{1}{4n} \sum_{i=1}^{n} \sum_{\ell_1=1}^{m_{i1}} \sum_{\ell_2=1}^{m_{i2}} p_{i1\ell_1} p_{i2\ell_2} \left(u_{i1\ell_1} + l_{i1\ell_1} \right) \left(u_{i2\ell_2} + l_{i2\ell_2} \right)
$$

$$
= \frac{1}{n} \sum_{i=1}^{n} \left[\sum_{\ell_1=1}^{m_{i1}} p_{i1\ell_1} \left(\frac{u_{i1\ell_1} + l_{i1\ell_1}}{2} \right) \sum_{\ell_2=1}^{m_{i2}} p_{i2\ell_2} \left(\frac{u_{i2\ell_2} + l_{i2\ell_2}}{2} \right) \right]
$$

$$(2.7)$$

and similarly that,

$$
\int_{-\infty}^{+\infty} \int_{-\infty}^{+\infty} \xi_j f(\xi_1, \xi_2) d\xi_1 d\xi_2 = \overline{Y_j}
$$

with $\overline{Y_j}$ as in Expression (2.6) for variable $Y_j, j \in \{1, 2\}$. From the above, it follows that

$$
cov_1(Y_1, Y_2) = \int_{-\infty}^{+\infty} \int_{-\infty}^{+\infty} \xi_1 \xi_2 f(\xi_1, \xi_2) d\xi_1 d\xi_2 - \overline{Y_1}\,\overline{Y_2}.
$$

From Expression (2.7), we obtain the empirical definition of symbolic covariance proposed by Billard and Diday [4].

Definition 12 [4] For the histogram-valued variables Y_1 and Y_2, the **symbolic covariance 1 (empirical symbolic covariance 1)** is given by

$$
cov_1(Y_1, Y_2) = \frac{1}{n} \sum_{i=1}^{n} \sum_{\ell_1=1}^{m_{i1}} p_{i1\ell_1} \left(\frac{u_{i1\ell_1} + l_{i1\ell_1}}{2} \right) \sum_{\ell_2=1}^{m_{i2}} p_{j2i_2} \left(\frac{u_{i2\ell_2} + l_{i2\ell_2}}{2} \right) - \overline{Y_1}\,\overline{Y_2}; \quad (2.8)
$$

Relatively to the definition of cov_1, we should note the following:

- Similarly as for symbolic variance 1 and symbolic mean, the expression for the symbolic covariance 1 is obtained from the empirical joint density function that considers that within each subinterval associated with each unit i the values are uniformly distributed.

- The expression of the empirical covariance function for histogram-valued variables is the classical definition of covariance applied to the weighted mean values, for each unit $i \in \{1, \ldots, n\}$, of the variables Y_1 and Y_2,

$$cov_1(Y_1, Y_2) \quad = \quad \frac{1}{n} \sum_{i=1}^{n} \overline{Y}_{i1} \overline{Y}_{i2} - \overline{Y_1} \, \overline{Y_2}.$$

- The covariance definition in Definition 7 is the generalization of the covariance definition for interval-valued variables proposed by Billard and Diday [3].

- If we apply the symbolic covariance 1 to degenerate histogram-valued variables (where real values are associated with all units), we obtain the classical definition of covariance.

- When we consider the histogram-valued variables $Y_1 = Y_2 = Y$, the expression of the covariance does not coincide with the Expression (2.4) in Definition 7 of symbolic variance 1 of the histogram-valued variable Y.

Because $cov_1(Y, Y) = var_1(Y)$ is not verified, a new definition of variance was obtained by particularizing the expression of the symbolic covariance 1 to the case where $Y_1 = Y_2 = Y$. We will designate this new definition of variance by empirical symbolic variance 2, despite the fact that chronologically this definition has been the first one to be proposed.

Definition 13 [4] The symbolic variance **symbolic variance 2 (empirical symbolic variance 2)** is defined as follows:

$$s_2^2(Y) = cov_1(Y, Y) \quad = \quad \frac{1}{n} \sum_{i=1}^{n} \left(\sum_{\ell=1}^{m_i} \frac{u_{i\ell} + l_{i\ell}}{2} p_{i\ell} \right)^2 - \overline{Y}^2;$$

About this definition, some considerations are also important:

- The definition of symbolic variance 2 is not derived from the density function of the symbolic variable.

- The new definition emerges only from the particularization of the definition of symbolic covariance 1 when the two variables are the same.

- If we apply the classical definition of variance to the weighted mean value, for each unit $i \in \{1, \ldots, n\}$, of the variable Y, i.e. to $\overline{Y}_i$, we obtain the expression of symbolic variance 2 for histogram-valued variables. So,

$$s_2^2(Y) \quad = \quad \frac{1}{n} \sum_{i=1}^{n} \overline{Y}_i^2 - \overline{Y}^2.$$

This behavior is not verified by the first definition of symbolic variance (variance 1) in Definition 7.

- As for the definition of symbolic variance 1, when the definition of symbolic variance 2 is applied to the particular case where all observations are degenerate, i.e. real values, it corresponds to the classical definition of variance.

Since by particularizing the concept of symbolic covariance 1 the definition of symbolic variance 1 is not obtained, Billard and Diday [6] proposed a new definition of covariance. As for interval-valued variables, similar definitions of variance and covariance were proposed, and consequently the same limitations occurred. The new definition of covariance was firstly derived for interval-valued variables and later generalized to histogram-valued variables.

The new definition of symbolic covariance proposed by Billard and Diday [6], here named covariance 2, emerges from the similarity between the expressions of classical variance and covariance. To understand the proposed definition, it is convenient to compare the classical expressions with the expression of the symbolic covariance 2 defined for interval-valued variables. The definition for histogram-valued variables was then an extension of the deduction proposed below.

Let Y_1 and Y_2 be two classical variables where to each unit i corresponds the real value y_{ij} with $j \in \{1,2\}$. The classical variance of variable Y_1 is defined as

$$s^2 = \frac{1}{n} \sum_{i=1}^{n} (y_{i1} - \overline{y_1})^2 \tag{2.9}$$

and the classical covariance between the variables Y_1 and Y_2, as

$$cov(Y_1, Y_2) = \frac{1}{n} \sum_{i=1}^{n} (y_{i1} - \overline{y_1}) (y_{i2} - \overline{y_2}). \tag{2.10}$$

Considering the definition of variance 1 for interval-valued variables, derived from the density function (see Expression (2.5) in Definition 7) particularized to this kind of variables, we have:

$$s_1^2(Y) = \frac{1}{3n} \sum_{i=1}^{n} (l_i - \overline{Y})^2 + (l_i - \overline{Y}) (u_i - \overline{Y}) + (u_i - \overline{Y})^2.$$

Let $Q = (l_i - \overline{Y})^2 + (l_i - \overline{Y}) (u_i - \overline{Y}) + (u_i - \overline{Y})^2$, we may write

$$s_1^2(Y) = \frac{1}{3n} \sum_{i=1}^{n} \left(Q^{\frac{1}{2}}\right)^2. \tag{2.11}$$

As such, by analogy with the classical definition of covariance in Expression (2.9), the symbolic covariance between two interval-valued variables is defined [6]. Considering the interval-valued variables Y_1 and Y_2, the **symbolic covariance 2** is given by

$$cov_2(Y_1, Y_2) = \frac{1}{3n} \sum_{i=1}^{n} G_1 G_2 [Q_1 Q_2]^{\frac{1}{2}} \tag{2.12}$$

with

$$Q_j = \left(l_{ij} - \overline{Y}_j\right)^2 + \left(l_{ij} - \overline{Y}_j\right)\left(u_{ij} - \overline{Y}_j\right) + \left(u_{ij} - \overline{Y}_j\right)^2,$$

$$G_i = \begin{cases} -1 & \text{if } c_{ij} \leq \overline{Y}_j \\ 1 & \text{if } c_{ij} > \overline{Y}_j \end{cases}$$

for $j \in \{1, 2\}$ where $\overline{Y}_j$ is the symbolic sample mean of variable Y_j and $c_{ij} = \frac{u_{ij} + l_{ij}}{2}$ is the midpoint, for each unit $i \in \{1, \ldots, n\}$, of variable Y_j.

However, Expression (2.12) has two more factors, G_1 and G_2 than the classical definition of covariance, in Expression (2.10). These factors appear in the expression of symbolic covariance because otherwise the covariance between two interval-valued variables would always be non-negative.

Let us show that Q_j, $j \in \{1, 2\}$ is always non-negative. Since

$$Q_j = \left(l_{ij} - \overline{Y}_j\right)^2 + \left(l_{ij} - \overline{Y}_j\right)\left(u_{ij} - \overline{Y}_j\right) + \left(u_{ij} - \overline{Y}_j\right)^2$$

to prove that Q_j is non-negative it is sufficient to prove that

$$\left(l_{ij} - \overline{Y}_j\right)\left(u_{ij} - \overline{Y}_j\right) \geq 0.$$

- If $\overline{Y}_j \leq l_{ij}$ as $l_{ij} \leq u_{ij}$ we have $l_{ij} - \overline{Y}_j \geq 0$ and $u_{ij} - \overline{Y}_j \geq 0$, so $Q_j \geq 0$;

- If $\overline{Y}_j \geq u_{ij}$ as $l_{ij} \leq u_{ij}$ we have $u_{ij} - \overline{Y}_j \leq 0$ and $l_{ij} - \overline{Y}_j \leq 0$, so $Q_j \geq 0$;

- If $l_{ij} \leq \overline{Y}_j \leq u_{ij}$ we have $\left(l_{ij} - \overline{Y}_j\right)\left(u_{ij} - \overline{Y}_j\right) \leq 0$. We may rewrite Q_j as $Q_j = \left(\left(l_{ij} - \overline{Y}_j\right) + \left(u_{ij} - \overline{Y}_j\right)\right)^2 - \left(l_{ij} - \overline{Y}_j\right)\left(u_{ij} - \overline{Y}_j\right)$. So, in this situation we have $Q_j \geq 0$.

In conclusion, in all situations $Q_j \geq 0$ and consequently, if the expression of covariance did not have the factors G_1 and G_2, it would always be non-negative.

In the Expression (2.12), that defines covariance 2, an analogous behavior is imposed to the centers of the intervals c_{ij}, to $j \in \{1, 2\}$. The considered conditions allow obtaining the definition of the factors G_j.

When all observations of the interval-valued variables are degenerate intervals, Expression (2.12) reduces to the classical expression of variance. On the other hand, the symbolic definition of covariance 2 for interval-valued variables was generalized to histogram-valued variables.

Definition 14 [6] For histogram-valued variables Y_1 and Y_2, the **symbolic covariance 2 (empirical covariance 2)** is given by

$$cov_2(Y_1, Y_2) = \frac{1}{3n} \sum_{i=1}^{n} \sum_{\ell_1=1}^{m_{i1}} \sum_{\ell_2=1}^{m_{i2}} p_{i1\ell_1} p_{i2\ell_2} G_1 G_2 [Q_1 Q_2]^{\frac{1}{2}} \qquad (2.13)$$

with

$$Q_j = \left(l_{ij\ell_j} - \overline{Y}_j\right)^2 + \left(l_{ij\ell_j} - \overline{Y}_j\right)\left(u_{ij\ell_j} - \overline{Y}_j\right) + \left(u_{ij\ell_j} - \overline{Y}_j\right)^2,$$

$$G_j = \begin{cases} -1 & if \quad \overline{Y}_{ij} \leq \overline{Y}_j \\ 1 & if \quad \overline{Y}_{ij} > \overline{Y}_j \end{cases}$$

for $j \in \{1,2\}$ where $\overline{Y}_j$ is the symbolic sample mean of the variable Y_j and $\overline{Y}_{ij} = \sum_{\ell_j=1}^{m_{ij}} \frac{l_{ij\ell_j} + u_{ij\ell_j}}{2} p_{ij\ell_j}$ is the weighted mean value, for each observation $i \in \{1, \ldots, n\}$ of variable Y_j.

Although Definition 14 is only one generalization for histograms, from the case of the intervals, the behavior of this symbolic covariance for interval-valued variables and for histogram-valued variables is different. For histogram-valued variables, if we particularize the definition of symbolic covariance 2 to the case where $Y_1 = Y_2$, we do not obtain the definition of symbolic variance 1 derived from the density function.

Independently of the definitions of variance and covariance used, the empirical correlation coefficient is calculated according to the next definition.

Definition 15 For histogram-valued variables Y_1 and Y_2 the **empirical correlation coefficient** is given by

$$r(Y_1 Y_2) = \frac{cov(Y_1, Y_2)}{\sqrt{s_{Y_1}^2 s_{Y_2}^2}}$$

where $cov(Y_1, Y_2)$ is the empirical covariance function and s_{Y_1}, s_{Y_2} the symbolic variance of the variables Y_1 and Y_2, respectively.

If the variance and covariance between the histogram-valued variables Y_1 and Y_2 are computed using defintions s_2^2 and cov_1, respectively, for which $s_2^2(Y) = cov_1(Y; Y)$ (see Definition 13), then it is trivial to conclude that $-1 \leq r(Y_1, Y_2) \leq 1$, as in the classical case.

2.4   Illustrative Example

Consider the symbolic data table [6], in Table 2.1.

In this table the distributions of the records of the levels of hematocrit and hemoglobin of a set of ten patients attending a healthcare center during one month are presented.

For this symbolic data table we will define the histogram of the symbolic variable, *Level of hemoglobin*. The observation distributions span the interval $I = [9.92, 16.75]$ since

$$\min \{l_{i\ell} : \ell \in \{1, \ldots, m_{10}\}, i \in \{1, \ldots, 10\}\} = 9.92$$

TABLE 2.1

Symbolic data table (macrodata) when the levels of hematocrit and hemoglobin are symbolic variables.

Patients	Level of hematocrit (Y)	Level of hemoglobin (X)
1	$\{[33.29; 37.52[,0.6;[37.52; 39.61],0.4\}$	$\{[11.54; 12.19[,0.4;[12.19; 12.8],0.6\}$
2	$\{[36.69; 39.11[,0.3;[39.11; 45.12],0.7\}$	$\{[12.07; 13.32[,0.5;[13.32; 14.17],0.5\}$
3	$\{[36.70; 42.64[,0.5;[42.64; 48.68],0.5\}$	$\{[12.38; 14.2[,0.3;[14.2; 16.16],0.7\}$
4	$\{[36.38; 40.87[,0.4;[40.87; 47.41],0.6\}$	$\{[12.38; 14.26[,0.5;[14.26; 15.29],0.5\}$
5	$\{[39.19; 50.86],1\}$	$\{[13.58; 14.28[,0.3;[14.28; 16.24],0.7\}$
6	$\{[39.7; 44.32[,0.4;[44.32; 47.24],0.6\}$	$\{[13.81; 14.5[,0.4;[14.5; 15.2],0.6\}$
7	$\{[41.56; 46.65[,0.6;[8.8146.65; 4],0.4\}$	$\{[14.34; 14.81[,0.5;[14.81; 15.55],0.5\}$
8	$\{[38.4; 42.93[,0.7;[42.93; 45.22],0.3\}$	$\{[13.27; 14.0[,0.6;[14.0; 14.6],0.4\}$
9	$\{[28.83; 35.55[,0.5;[35.55; 41.98],0.5\}$	$\{[9.92; 11.98[,0.4;[11.98; 13.8],0.6\}$
10	$\{[44.48; 52.53],1\}$	$\{[15.37; 15.78[,0.3;[15.78; 16.75],0.7\}$

$$\max\{u_{i\ell} : \ell \in \{1,\ldots,m_{10}\}, i \in \{1,\ldots,10\}\} = 16.75$$

Let us build a histogram of the records of hemoglobin with $r = 5$ subintervals all with length 1.37. Associated with each interval is a frequency or relative frequency computed according to Definition 3. The complete information about the five subintervals and respective frequencies is given in Table 2.2 and the histogram is represented in Figure 2.1.

TABLE 2.2

Table of frequencies of the histogram-valued variable *Level of hemoglobin* in Table 2.1.

$\mathbf{I_g}$	$\mathbf{f(I_g)}$	$\mathbf{f_r(I_g)}$	$\mathbf{F_r(I_g)}$
$[9.92, 11.29[$	0.27	0.027	0.027
$[11.29, 12.66[$	1.58	0.158	0.184
$[12.66, 14.03[$	2.73	0.273	0.457
$[14.03, 15.40[$	4.05	0.405	0.862
$[15.40, 16.75]$	1.65	0.165	1

According to the observed results, only 2.7% of the monthly records of the level of hemoglobin in the set of the ten patients range between 9.92 and 11.29. The subinterval $[14.03,15.40[$ is the modal interval because it has the highest relative frequency. The interval median of the histogram-valued variable under study is $[14.03,15.40[$, because the cumulative relative frequency associated with this interval is 0.86. The equation that defines the cumulative frequency polygon in interval $[14.03,15.40[$, where the median is included, is $y = 0.296x - 3.69$. So, the median, obtained for $x = 0.5$, is 14.18. Similarly, we may compute the value of the 1st and 3rd quartiles. In Figure 2.2, we may observe the cumulative frequency polygon of this variable and the respective quartiles.

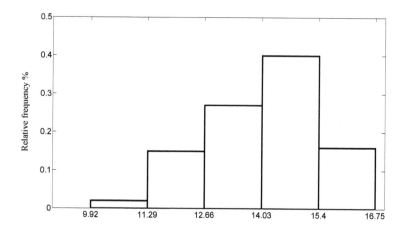

FIGURE 2.1
Histogram of the histogram-valued variable *Level of hemoglobin* in Table 2.1.

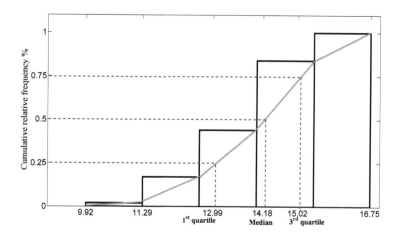

FIGURE 2.2
Cumulative frequency polygon of the histogram-valued variable *Level of hemoglobin* in Table 2.1.

The distributions associated with each of the ten patients of the histogram-valued variables *Level of hemoglobin* and *Level of hematocrit* in Table 2.1 are represented by a scatter plot in Figure 2.3 [10]. The scatter plot was obtained according to the process described in Section 2.3.1. To represent this graph, all observed distributions have to be rewritten with an equal number of subintervals and the same set of cumulative weights. In this case, six subintervals will be considered (see Section 1.2.2, in Chapter 1) and each distribution associated with a patient is represented with a different color.

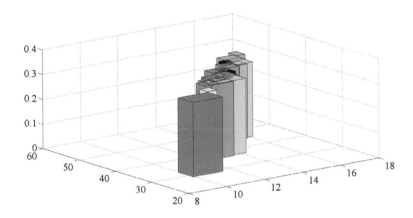

FIGURE 2.3
Scatter Plot of the histogram data in Table 2.1.

From a scatter plot we can verify if the collection of observations corresponding to a given pair of variables displays a linear behavior. For classical variables, this means that the observed values are aligned and also allows knowing whether the linear relation is direct or inverse. For symbolic variables, when the sets of hyperrectangles are well aligned in the scatter plots, we may say that the symbolic variables are in a linear relation. The direction of this alignment also indicates if the relation is direct or inverse. In Chapter 13 we will analyze this behavior in more detail.

The symbolic descriptive statistics defined in this chapter for the variables *Level of hemoglobin* and *Level of hematocrit* in Table 2.1 are presented in Table 2.3. The values of covariance calculated using the proposed definitions may also be compared. The results illustrate the conclusions presented when the definitions of variance and covariance were introduced. Covariance 1, when applied to two identical variables, coincides with the definition of variance 2. On the other hand, when the definition of covariance 2 is applied to two equal histogram-valued variables, the obtained value is not coincident with the value of the variance 1 for this variable.

In Table 2.3, the values obtained for the covariance and variance are presented, considering the histogram-valued variables rewritten either with the same number (Situation I) or with a different number (Situation II) of subintervals. Comparing the obtained results, the conclusion is that the symbolic mean and variance obtained with both definitions are always the same, but the values for the covariance only coincide when the definition of covariance 1 is applied. Moreover, we may conclude that the values obtained when the two definitions of both variance and covariance are applied, are obviously different.

TABLE 2.3

Values of the descriptive measures for the data in Table 2.1.

Descriptive measures	Situation I	Situation II [1]
$\overline{X}$	14.0507	14.0507
$\overline{Y}$	42.2639	42.2639
$s_1^2(X)$	1.8367	1.8367
$s_2^2(X)$	1.3688	1.3688
$s_1^2(Y)$	21.6990	21.6990
$s_2^2(Y)$	14.1394	14.1394
$cov_1(X,Y)$	4.3276	4.3276
$cov_2(X,Y)$	5.993	5.2263
$cov_1(X,X)$	1.3688	1.3688
$cov_2(X,X)$	1.6595	1.5624
$cov_1(Y,Y)$	14.1394	14.1394
$cov_2(Y,Y)$	19.9008	17.7531

2.5   Conclusion

In this chapter, we have introduced univariate and bivariate descriptive statistics for symbolic variables classified as histogram-valued variables. As in classical statistics, the main concepts are defined from the empirical density functions for this type of variables. The concepts introduced are the basis for most statistical methods for the analysis of numerical distributional data. However, as important properties observed in classical statistics measures are not verified, more than one definition has been proposed for some of the symbolic descriptive measures, such as variance and covariance. To illustrate the descriptive measures, an example is presented in the last section comparing the results for univariate and bivariate descriptive statistics using different definitions. Following an alternative approach, Irpino and Verde [11] proposed other definitions for some descriptive measures based on the L_2 Wasserstein distance. These concepts are introduced in the next chapter.

---

[1] The observations associated with histogram-valued variable X and Y are rewritten with the same number of subintervals and the same set of cumulative weight (according to the Irpino and Verde process [10]).

Bibliography

[1] Javier Arroyo. *Métodos de Predicción para Series Temporales de Intervalos e Histogramas*. PhD thesis, Universidad Pontificia Comillas, Madrid, Espanha, 2008.

[2] L. Billard. Dependencies and variation components of symbolic interval-valued data. In P. Brito, G. Cucumel, P. Bertrand, and F. De Carvalho, editors, *Selected Contributions in Data Analysis and Classification*, pages 3–12. Springer-Verlag Berlin, 2007.

[3] L. Billard and E. Diday. Regression analysis for interval-valued data. In *Data Analysis, Classification and Related Methods. Proceedings of the 7th Conference of the International Federation of Classification Societies*, pages 369–374. Springer Berlin Heidelberg, 2000.

[4] L. Billard and E. Diday. Symbolic regression analysis. In *Classification, Clustering, and Data Analysis. Proceedings of the 8th Conference of the International Federation of Classification Societies*, pages 281–288. Springer Berlin Heidelberg, 2002.

[5] L. Billard and E. Diday. From the statistics of data to the statistics of knowledge: Symbolic Data Analysis. *Journal of the American Statistical Association*, 98(462):470–487, 2003.

[6] L. Billard and E. Diday. *Symbolic Data Analysis: Conceptual Statistics and Data Mining*. John Wiley & Sons, Inc. New York, NY, USA, 2006.

[7] H.-H. Bock and E. Diday, editors. *Analysis of Symbolic Data: Exploratory Methods for Extracting Statistical Information from Complex Data*. JSpringer-Verlag Berlin, 2000.

[8] P. Brito and A.P. Duarte Silva. Modelling interval data with Normal and Skew-Normal distributions. *Journal of Applied Statistics*, 39(1):3–20, 2012.

[9] S. Dias. *Linear regression with empirical distributions*. PhD thesis, University of Porto, Porto, Portugal, 2014.

[10] A. Irpino and R. Verde. A new Wasserstein based distance for the hierarchical clustering of histogram symbolic data. In *Data Science and Classification. Proceedings of the 10th Conference of the International Federation of Classification Societies*, pages 185–192. Springer Berlin Heidelberg, 2006.

[11] A. Irpino and R. Verde. Basic statistics for distributional symbolic variables: a new metric-based approach. *Advances in Data Analysis and Classification*, 9(2):143–175, 2015.

[12] E.A. Lima Neto, G.M. Cordeiro, and F.A.T. De Carvalho. Bivariate symbolic regression models for interval-valued variables. *Journal of Statistical Computation and Simulation*, 81(11):1727–1744, 2011.

3

Descriptive Statistics for Numeric Distributional Data

Antonio Irpino

Department of Mathematics and Physics, University of Campania "L. Vanvitelli", Caserta, Italy

Rosanna Verde

Department of Mathematics and Physics, University of Campania "L. Vanvitelli", Caserta, Italy

CONTENTS

Like other types of multi-valued data, distributional data are characterized by two kinds of variability: internal and between data variability. This chapter gives an overview of descriptive statistics tools for numeric distributional data based on the (2-norm) Wasserstein metric. The descriptive statistics based on the Wasserstein metric allow interpreting the role of the two variability sources compared to the other descriptive statistics proposed for such data. The presented statistics extend some properties of classic descriptive measures and tools to numeric distributional data. Using a dataset from the CPS survey on

DOI: 10.1201/9781315370545-3

the labor force in the US, we present an application of the proposed statistics. The results provide evidence of the interpretative properties of the proposed descriptive statistics.

3.1   Introduction

This chapter provides an overview of the tools of descriptive statistics for numeric distributional data developed in [11], which are based on the (2-norm) Wasserstein metric or L_2 Wasserstein distance (also known as Mallows distance). Symbolic Data Analysis (SDA) [4] formalized data described by multi-valued variables, where a *numeric distributional* variable is a particular case of a *modal-numeric* variable. Such data result from an aggregation of single-valued data. In the numeric case, they are usually represented by empirical frequency distributions on numeric support in terms, for example, of histograms (in this case, they are called histogram data). For this kind of data, new statistics and methods have been proposed. Although several proposals have been made to analyze histogram data, in this chapter, we present a generalization of some central tendency, dispersion and bivariate measures for the descriptive analysis of data described by density distribution-valued variables, where histogram one is considered as a particular case.

A first proposal to treat such type of data was proposed in [1]. The authors introduced a set of descriptive univariate and bivariate measures for numeric distributional variables that were integrated and extended by [3]. Further developments for the quantification of the variability and the dependence relationships between variables of a set of multi-valued data can be found in [2] and [5]. Statistics proposed by [2] and [1] especially for interval data consider, for example, an interval-valued data $[a, b]$ as the support of a uniformly distributed variable $U \sim (a, b)$. By considering histograms as weighted sets of contiguous intervals, [1] and [3] extended the univariate measures (mean, variance and standard deviation) and bivariate statistics (covariance and correlation) defined for interval data to histogram data.

Distributional data, like other types of multi-valued data, are characterized by two kinds of variability: *internal data variability* and *between data variability*. The first is related to the multiplicity of values that describe a single observation: for example, an interval of real values has a proper variability related to its width. The second type of variability is related to the different multi-valued observations: two intervals can differ in terms of position (w.r.t. the midpoints), width, or both. The descriptive tools proposed by [1] and [3] are not sensible to express the role of the two sources of variability (for example, the variance of a set of identical multi-valued data is, in general, not null despite the classical property of the variance for constant observations). In this chapter, we consider a novel set of univariate and bivariate statistics that better consider the two sources of variability and extend some properties of classic descriptive measures and tools to numeric distributional data.

The chapter is organized as follows: in Section 3.2 we recall the definition of numeric distributional data, as a particular case of modal numerical data, according to the classical definition given in SDA literature. In Section 3.3, we show state-of-the-art descriptive tools and their consistency with the central tendency and dispersion measures of a finite mixture of distributions. We discuss their use and illustrate new descriptive tools that solve some problems of the former approach. The new univariate tools are defined by considering a measure of variability related to a distance between distributions. Among the different distances presented in the literature, we have chosen the L_2 Wasserstein distance [13] because it allows keeping most properties of the descriptive tools for classical data as well as providing consistency with the double source of variability of a set of multi-valued data.

In Section 3.4, we propose an extension of the classical covariance and correlation measures to numeric distributional variables. Also in such a case, the different sources of variability in the data are taken into account.

Using a dataset from the CPS survey on labor force in the US, in Section 14.3 we present an application of the proposed statistics. The results provide evidence of the interpretative properties of the proposed descritive statistics.

3.2 Modal-numeric data and numeric distributional data

The definition of numeric distributional data given in this chapter is consistent with the modal-numeric data definition given in the framework of SDA literature [3, 4].

A modal variable Y on a set S of objects s_i with domain $\mathcal{Y}$ is defined by a mapping:

$$Y(s_i) = (I(s_i), p_i), \forall s_i \in S \qquad (3.1)$$

where p_i is a measure or a (frequency, probability or weight) distribution on the domain $\mathcal{Y}$ of possible observation values (completed by a σ-field), and $I(s_i) \subseteq \mathcal{Y}$ is the support of p_i in the domain $\mathcal{Y}$. In this chapter, we assume that $I(s_i)$ is numeric.

A variable Y that assumes a distribution of probability (or of relative frequency) f_i (with respective cumulative distribution function F_i) for each object s_i, is termed *modal-numeric* variable or *numeric distributional variable*.

Histogram valued data. A particular type of numeric distributional variable is the histogram one. Let us consider the support of the description of the object s_i partitioned into a set of m_i intervals (or bins):

$$I(s_i) = \{I_{i1}, \dots, I_{im_i}\}, \text{ where } I_{i\ell} = [l_{i\ell}, u_{i\ell}) \text{ and } \ell = 1, \dots, m_i.$$

Each interval $I_{i\ell}$ is supposed to be the support of a uniform distribution and is associated with a weight (frequency) $p_{i\ell}$.

The cumulative distribution function F_i of $Y(s_i)$ is then:

$$F_i(y) = \sum_{h<\ell} p_{ih} + p_{i\ell} \cdot \frac{y - l_{i\ell}}{u_{i\ell} - l_{i\ell}} \ where \ (\ell = 1, \ldots, m_i : l_{i\ell} \leq y \leq u_{i\ell}).$$

Thus, it is possible to express the histogram data description as follows:

$$Y(s_i) = \{(I_{i\ell}, p_{i\ell}) \mid \forall I_{i\ell} \in I(s_i); \ p_{i\ell} = F_i(u_{i\ell}) - F_i(l_{i\ell}) \geq 0\}.$$

3.3 Univariate statistics

In this section, we introduce central tendency and dispersion tools for numeric distributional variables. A few proposals have been introduced for data described by numeric distributions whereas, some proposals for interval or histogram-valued data have been given in [4] and [3]. When data are described by numeric distributions, two main approaches have been proposed: the Bertrand and Goupil approach [1], further developed by Billard and Diday [3] for interval and histogram-valued data, and the approach of Irpino and Verde [11]. In a classical framework, let us think of individuals data coming from different sub-populations described by a numeric variable; both approaches assume that numerical distributional data are summaries of sub-populations (or groups in the population). The first approach [1, 3] aims at computing scalar measures of central tendency and dispersion (mean, standard deviation, and so on) related to the whole population, such that they correspond to the central tendency and dispersion (scalar) measures of a mixture of the sub-population distributions having the same mixing weights.

The latter approach [11], starting from a probabilistic distance between distributions (or probability measures), introduced a set of descriptive tools statistics where, for example, the mean of a distributional variable is an average distribution such that a distance criterion is minimized, and the variance is a scalar value computed as the average of the squared distances between the observations and the mean distribution.

3.3.1 Fréchet and Chisini mean for numeric distributional variables

While in probability theory the mean corresponds to the expected value of a random variable, in descriptive statistics the mean of a variable describing a set of objects can assume several definitions. For example, starting from proximity relations among data, it is possible to define a so-called *Fréchet* mean, while starting from the definition of a function of the observed data, it is possible to define the so called *Chisini* mean. More formally:

Fréchet (or Karcher) mean [10] a Fréchet type mean (barycenter) M is the object x which leads to the solution of the following minimization problem:

$$M = \arg \min_x \sum_{i=1}^n w_i d^2(y_i, x) \tag{3.2}$$

provided that a unique minimizer exists.

Chisini mean according to [6], given a set of n objects described by the single-valued numeric variable Y and a function $f(y_1, \ldots, y_n)$, M is a mean in the sense of *Chisini* if:

$$F(y_1, \ldots, y_i, \ldots, y_n) = f(\underbrace{M, \ldots, M}_{n \; times}). \tag{3.3}$$

For example, if f is the sum of the n observed values, then the arithmetic mean is a *Chisini* mean which is invariant with respect to the sum function, i.e.:

$$\sum_{i=1}^n y_i = \sum_{i=1}^n M = nM \; \Rightarrow \; M = \frac{1}{n} \sum_{i=1}^n y_i.$$

To extend *Chisini* means to numeric distributional variables, functions and operators for numeric distributional data must be defined.

The definition of a *Fréchet* and *Chisini* compatible mean for numeric distributional variables requires two conditions: the definition of a distance between distributions (or random variables) and the definition of, at least, the sum of distributions and the product of a distribution by a scalar.

In [4, Chap. 8], some dissimilarities for multi-valued data are discussed. For numeric distributional variables, starting from the study in [8], the authors of [14] investigated a set of dissimilarities and distances for probability distributions. However, the authors observed that the considered distances or dissimilarities are not always acceptable for identifying a unique distribution as a *center* (or representative) of a collection of distributions; moreover, *centers* could not be expressed as a distribution for all the considered dissimilarities or distances. As a result, the authors noticed that only two distances allow for the definition of a single center as a distribution and are compatible with the *Fréchet* and *Chisini* concepts of mean: the Euclidean and the L_2 Wasserstein distance between distributions.

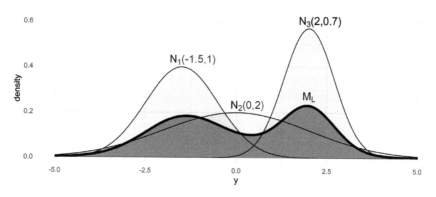

FIGURE 3.1
The *pdf* (in bold) of the *Fréchet* mean of three Normal density functions according to L_2 distance.

3.3.1.1 Euclidean distance-based mean

The L_2 distance between two density functions f_i and $f_{i'}$ (for continuous distributions) is defined as:

$$d_L(f_i, f_{i'}) = \sqrt{\int_{-\infty}^{+\infty} |f_i(y) - f_{i'}(y)|^2 \, dy}. \qquad (3.4)$$

It is straightforward to prove that the *Fréchet* mean g, which is a density function, of the distributional variable Y observed on n objects described by density functions $f_1, \ldots, f_n$ based on d_L (assuming equal weights w_i) is given by their finite mixture as follows:

$$M_L(Y) \sim \arg\min_g \sum_{i=1}^{n} d_L^2(f_i, g) = \frac{1}{n} \sum_{i=1}^{n} f_i \qquad (3.5)$$

and, thus, the mean and the variance of $M_L(Y)$ correspond to those presented in [7] for the finite mixtures of distributions, and correspond to those developed by [1]. Generally, the density function describing $M_L(Y)$ has a different shape with respect to the set of summarized densities: for example, as shown in Figure 3.1, usually, a density function of the mixture of Normal distributions is not Normal.

3.3.1.2 L_2 Wasserstein distance-based mean

As stated above and in [14], the L_2 version of the L_p Wasserstein distance was considered as a good candidate for the definition of a *Fréchet*-type mean of a numeric distributional variable. The literature provides different formulations of the Wasserstein distance but we use the formalization used in [13] which expresses the distance between two probability distributions having density functions f_i and $f_{i'}$ using the respective quantile functions Ψ_i and $\Psi_{i'}$, namely, the inverse of the *cdfs* F_i and $F_{i'}$ functions respectively, as follows:

$$d_{W_p}(f_i, f_{i'}) = \left(\int_0^1 |\Psi_i(t) - \Psi_{i'}(t)|^p \, dt \right)^{\frac{1}{p}}. \tag{3.6}$$

The formulation proposed in [13] shows that d_{W_p} distance can be considered as an extension of the classical L_p Minkowski distance between quantile functions. The quantile functions (*qfs*) are in a one-to-one correspondence with the density functions; *qfs* are defined on a finite domain ($t \in [0; 1]$), and *qfs* are non-decreasing functions.

For the sake of simplicity, hereafter we denote with d_W the L_2 Wasserstein distance between two continuous probability distributions:

$$d_W(f_i, f_{i'}) = \sqrt{\int_0^1 [\Psi_i(t) - \Psi_{i'}(t)]^2 \, dt}. \tag{3.7}$$

The *Fréchet* mean of the numeric distributional variable Y, based on d_W and assuming equal weights w_i, has a *pdf* g which solves the following optimization problem:

$$M_W(Y) \sim \arg \min_g \sum_{i=1}^n d_W^2(f_i, g). \tag{3.8}$$

Proposition 1 The minimum value of the function in Equation (3.8) is found for the *pdf* f_x associated with the quantile function $\bar{\Psi}(t)$, that is:

$$\bar{\Psi}(t) = \frac{1}{n} \sum_{i=1}^n \Psi_i(t), \quad \forall t \in [0, 1].$$

Proof. Assuming that g is a *pdf*, $\bar{G}$ is its *cdf* and $\bar{\Psi}$ is the corresponding quantile function, the optimal solution of Equation (3.8) is obtained for each $t \in [0, 1]$ according to the classical first order condition, as follows:

$$\frac{d}{dt} \left[\sum_{i=1}^n (\Psi_i(t) - \bar{\Psi}(t))^2 \right] = 0 \Rightarrow \bar{\Psi}(t) = \frac{1}{n} \sum_{i=1}^n \Psi_i(t), \quad \forall t \in [0, 1]. \tag{3.9}$$

$\square$

The *Fréchet* mean distribution corresponds to the *pdf* that is into a one-to-one correspondence with $\bar{\Psi}(t)$, that is:.

$$M_W(Y) \sim g = \frac{d}{dt} \bar{\Psi}^{-1} = \frac{d}{dy} \bar{F}. \tag{3.10}$$

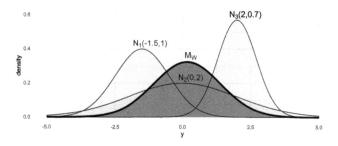

FIGURE 3.2
The *pdf* (in bold) of the *Fréchet* mean according to L_2 Wasserstein distance among three Normal distributions.

Also in this case, it is possible to obtain a mean distribution from a set of distribution functions. If all the distributions have a *pdf* of the same type (for example, they are all Normal distributions), an interesting property of this approach is that the *pdf* of the mean distribution has the same shape too. For example, in Figure 3.2, it is interesting to note that the average distribution $M_W(Y)$ (in bold) of three Normal distributions is a Normal one too, differently from the mean distribution in Figure 3.1 obtained by considering the Euclidean distance between *df*'s.

To discuss the *centrality* properties of $M_W(Y)$, we need to introduce a measure of correlation between two quantile functions. Given two observations y_i and $y_{i'}$ described by the respective *pdfs* f_i and $f_{i'}$, we denote the respective means by μ_i and $\mu_{i'}$, standard deviations by σ_i and $\sigma_{i'}$. With each *pdf* is associated its *cdf* F_i (resp. $F_{i'}$) and the corresponding *qf* Ψ_i (resp. $\Psi_{i'}$). We recall that the mean, the standard deviation, the skewness and the kurtosis index of y_i can be computed using the corresponding quantile function [9] as follows:

$$\mu_i = \int_{-\infty}^{\infty} y\, f_i(y)dy = \int_0^1 \Psi_i(t)dt, \qquad (3.11)$$

$$\sigma_i = \sqrt{\int_0^1 [\Psi_i(t)]^2\, dt - \mu_i^2}, \qquad (3.12)$$

$$sk_i = \sqrt[3]{\int_0^1 \left[\frac{\Psi_i(t) - \mu_i}{\sigma_i}\right]^3 dt}, \qquad (3.13)$$

$$ku_i = \sqrt[4]{\int_0^1 \left[\frac{\Psi_i(t) - \mu_i}{\sigma_i}\right]^4 dt}. \qquad (3.14)$$

The Pearson correlation coefficient between the two quantile functions, denoted by $\rho_{i,i'}$ is defined as follows:

$$\rho_{i,i'} = \int_0^1 \frac{(\Psi_i(t) - \mu_i)}{\sigma_i} \frac{(\Psi_{i'}(t) - \mu_{i'})}{\sigma_{i'}} dt = \frac{\int_0^1 \Psi_i(t)\Psi_{i'}(t)dt - \mu_i\mu_{i'}}{\sigma_i\sigma_{i'}}. \qquad (3.15)$$

Being a correlation measure between two not decreasing functions, $\rho_{i,i'}$ is always positive, and it is null when at least one of the two distributions has no variability (namely, it is a degenerate distribution or single-valued observation).

According to a classical Q-Q (Quantile-Quantile) plot, where the quantiles of two distributions are plotted against each other, $\rho_{i,i'}$ is the correlation coefficient associated with the points scattered on the plot. Moreover, it can be considered as a measure of the similarity between the *shapes* of two distribution functions. In fact, $\rho_{i,i'} = 1$ only if the two distributions have the same standardized quantiles, which occurs when the two distributions have the same shape.

Using Equation (3.7) the squared L_2 Wasserstein distance can be decomposed in three components as showed here below.

Proposition 2 The squared L_2 Wasserstein distance between $y_i \sim f_i$ and $y_{i'} \sim f_{i'}$ with quantile functions, respectively, Ψ_i and $\Psi_{i'}$ is

$$d_W^2(f_i, f_{i'}) = \int_0^1 [\Psi_i(t) - \Psi_{i'}(t)]^2 \, dt$$

and it can be expressed as

$$d_W^2(f_i, f_{i'}) = \underbrace{(\mu_i - \mu_{i'})^2}_{Location} + \underbrace{\underbrace{(\sigma_i - \sigma_{i'})^2}_{Size} + \underbrace{2\sigma_i\sigma_{i'}(1 - \rho_{i,i'})}_{Shape}}_{Variability}. \qquad (3.16)$$

Proof. The proof is given in [11].

□

The decomposition allows for the interpretation of the (squared) distance between two distributions according to two *additive* components: one related to the *Location* of the distributions that emphasizes the difference in position of the respective means; the second is related to the different *Variability* structure of the two distributions due to the different standard deviations (the *Size* component) and to the different shapes of the density functions (the *Shape* component). While the *Size* component is expressed by the (squared Euclidean) distance between the standard deviations, the *Shape* component depends on the value of $\rho_{i,i'}$.

Mean and variance of $M_W(Y)$ mean distribution.

Recalling that $M_W(Y)$ is a distribution, we can compute its main moments according to the L_2 Wasserstein distance-based approach using its quantile function.

Proposition 3 The mean of $M_W(Y)$ is the arithmetic mean of the means of the y_i ($\forall i \in \{1, \ldots, n\}$):

$$\mu_{M_W(Y)} = \frac{1}{n} \sum_{i=1}^{n} \mu_i. \tag{3.17}$$

Proof. The proof is given in [11].

□

Definition 1 Given two quantile functions Ψ_i and $\Psi_{i'}$, associated with two pdf's f_i and $f_{i'}$ with means μ_i and $\mu_{i'}$ and standard deviations σ_i and $\sigma_{i'}$, and considering Equation (3.7), the product of two quantile functions is defined as follows:

$$\langle \Psi_i, \Psi_{i'} \rangle = \int_0^1 \Psi_i(t)\Psi_{i'}(t)dt = \rho_{i,i'}\sigma_i\sigma_{i'} + \mu_i\mu_{i'}. \tag{3.18}$$

Using the definition of variance through quantile functions given in [9, pag. 71][1] and according to Equation (3.18), we obtain the following result for the variance of $M_W(Y)$.

Proposition 4 The variance of $M_W(Y)$ can be expressed as a function of the standard deviations of y_i ($\forall i \in \{1, \ldots, n\}$) and of the correlation terms between all the pairs of the respective quantile functions, as follows:

$$\sigma^2_{M_W(Y)} = \sum_{i=1}^{n} \left[\frac{\sigma_i}{n}\right]^2 + \frac{2}{n^2} \sum_{i=1}^{n-1} \sum_{i'>i} [\rho_{i,i'}\sigma_i\sigma_{i'}].$$

Proof. The proof is given in [11].

□

It is worth noting that if all the distributions have the same shape then $\rho_{i,i'} = 1$ for each couple of distributions and the variance of $M_W(Y)$ reaches its maximum value. The minimum value is obtained when all the observed data are single valued (i.e. $\sigma_i = 0$ for each $i = 1, \ldots, n$), thus:

$$0 \leq \sigma^2_{M_W(Y)} \leq \left(\frac{1}{n} \sum_{i=1}^{n} \sigma_i\right)^2. \tag{3.19}$$

---

[1]where $\sigma_i^2 = \int_0^1 [\Psi_i(t) - \mu_i]^2 \, dt$.

3.3.2 The variance of a distributional variable

After defining the average, borrowing from the concept of *Fréchet* variance, we derive the variance of a numeric distributional variable Y. The variance of a numeric distributional variable can be considered as the average of the squared distances from the mean distribution. We propose the L_2 Wasserstein distance, however, a Euclidean distance-based variance is also discussed in order to highlight some advantages of the Wasserstein based approach for the variance measure.

3.3.2.1 Euclidean distance-based variance

Considering the distance in Equation (3.4), the mean distribution is the mixture of the observed distributions. In this case, the variance of a numeric distributional variable Y is

$$S_L^2(Y) = \frac{1}{n} \sum_{i=1}^{n} d_L^2(f_i, f_{M_L(Y)}) = \frac{1}{n} \sum_{i=1}^{n} \int_{-\infty}^{+\infty} \left[f_i(y) - f_{M_L(Y)}(y) \right]^2 dy. \quad (3.20)$$

However, to the authors' best knowledge, $S_L^2(Y)$ cannot be generally expressed in a closed form.

3.3.2.2 L_2 Wasserstein distance based variance

By analogy with the classical population variance, the variance of a numeric distributional variable Y can be considered as the average of the squared L_2 Wasserstein distances between y_i ($\forall i \in \{1, \ldots, n\}$) and $M_W(Y)$.

Proposition 5 The variance of a numeric distributional variable Y is defined as follows:

$$
\begin{aligned}
S_W^2(Y) &= \frac{1}{n} \sum_{i=1}^{n} d_W^2(f_i, f_{M_W(Y)}) = \frac{1}{n} \sum_{i=1}^{n} \int_0^1 \left[\Psi_i(t) - \bar{\Psi}(t) \right]^2 dt \\
&= \underbrace{\left[\frac{1}{n} \sum_{i=1}^{n} \mu_i^2 - \mu_{M_W(Y)}^2 \right]}_{SM_W^2(Y)} + \underbrace{\left[\frac{1}{n} \sum_{i=1}^{n} \sigma_i^2 - \frac{1}{n^2} \sum_{i=1}^{n} \sum_{i'=1}^{n} \rho_{i,i'} \sigma_i \sigma_{i'} \right]}_{SV_W^2(Y)}.
\end{aligned}
$$

$$(3.21)$$

We note that $S_W^2(Y)$ is the sum of two kinds of variability: $SM_W^2(Y)$ is the variance of the means of the distributions, while $SV_W^2(Y)$ is a measure of variance related to the (squared) differences of the internal variability of the n distributions.

When for each couple of distributions, $\rho_{i,j} = 1$, (i.e., all the distributions have the same shape), and if $\sigma_i > 0$ $i = 1, \ldots, n$, $\sum_{i'=1}^{n} \rho_{i,i'} \sigma_i \sigma_{i'}$ reaches its maximum value. Consequently, we observe that the minimum value of $SV_W^2(Y)$

is obtained when all the distributions have the same shape (namely, when $\rho_{i,j} = 1 \ \forall (i, i')$ such that $i \in 1, \ldots, n$ and $i' \in 1, \ldots, n$):

$$SV_W^2(Y) = \frac{1}{n} \sum_{i=1}^{n} \sigma_i^2 - \left[\frac{1}{n} \sum_{i=1}^{n} \sigma_i \right]^2.$$

In this case, it corresponds to the variance of the standard deviations of the n distributions. Finally, $SV_W^2(Y)$ is equal to zero in two cases:

i. all the distributions are identical except for their means;

ii. all the data are single valued.

In Tables 3.1 and 3.2 summarize the basic descriptive univariate tools for numeric distributional variables.

TABLE 3.1

Descriptive indices and distributions of a numeric distributional variable based on the L_2 Euclidean distance.

Name	Symbol	Formula	Type
Mean	$M_L(Y)$	$f_{M_L(Y)} = \frac{1}{n} \sum_{i=1}^{n} f_i$	*pdf*
Mean of $M_L(Y)$	$\mu_{M_L(Y)}$	$\frac{1}{n} \sum_{i=1}^{n} \mu_i$	Scalar
Variance of $M_L(Y)$	$\sigma_{M_L(Y)}^2$	$\frac{1}{n} \sum_{i=1}^{n} \left(\mu_i^2 + \sigma_i^2 \right) - \mu_{M_L(Y)}^2$	Scalar
Variance	$S_L^2(Y)$	$\frac{1}{n} \sum_{i=1}^{n} d_L^2 \left(f_i, f_{M_L(Y)} \right)$	Scalar

The generalization of the standard deviation of the numeric distributional variable Y is straightforward as follows:

$$S_W(Y) = \sqrt{S_W^2(Y)} \tag{3.22}$$

3.4 Bivariate descriptive indices for numeric distributional variables

In this section, we present an extension of the classical indices of association, like the covariance and the correlation measures between numeric distributional variables.

TABLE 3.2

Descriptive indices and distributions of a numeric distributional variable based on the L_2 Wasserstein distance.

Name	Symbol	Formula	Type
Correlation between qf's	$\rho_{i,i'}$	$\dfrac{\int_0^1 \Psi_i(t)\Psi_{i'}(t)dt - \mu_i\mu_{i'}}{\sigma_i\sigma_{i'}}$	Scalar
Mean	$M_W(Y)$	$f_{M_W(Y)} = \frac{d}{dy}(\bar{\Psi}^{-1}) = \frac{d}{dy}(\bar{F})$ where $\bar{\Psi} = \frac{1}{n}\sum_{i=1}^{n}\Psi_i$	pdf
Mean of $M_W(Y)$	$\mu_{M_W(Y)}$	$\frac{1}{n}\sum_{i=1}^{n}\mu_i$	Scalar
Variance of $M_W(Y)$	$\sigma^2_{M_W(Y)}$	$\sum_{i=1}^{n}\left[\frac{\sigma_i}{n}\right]^2 + \frac{2}{n^2}\sum_{i=1}^{n-1}\sum_{i'>i}[\rho_{i,i'}\sigma_i\sigma_{i'}]$	Scalar
Variance	$S^2_W(Y)$	$\left[\frac{1}{n}\sum_{i=1}^{n}\mu_i^2 - \mu^2_{M_W(Y)}\right] +$ $+\left[\frac{1}{n}\sum_{i=1}^{n}\sigma_i^2 - \frac{1}{n^2}\sum_{i=1}^{n}\sum_{i'=1}^{n}\rho_{i,i'}\sigma_i\sigma_i'\right]$	Scalar

Bertand and Goupil [1] proposed a first definition of the covariance measure between numeric distributional data. Given two numeric distributional variables Y_1 and Y_2, we denote by $COV_{BG}(Y_1, Y_2)$ the covariance index introduced by the authors for two numeric-symbolic variables as follows:

$$COV_{BG}(Y_1, Y_2) = \frac{1}{n}\sum_{i=1}^{n}\mu_{i1}\cdot\mu_{i2} - \bar{Y}_1\bar{Y}_2, \tag{3.23}$$

where $\bar{Y}_1 = \frac{1}{n}\sum_{i=1}^{n}\mu_{i1}$ and $\bar{Y}_2 = \frac{1}{n}\sum_{i=1}^{n}\mu_{i2}$.

It is immediately clear that $C\hat{O}V_{BG}(Y_1, Y_2)$ corresponds to the covariance of the means of the numeric distributional observations. Reading Equation (3.23), one can note that it depends only on the means of the numeric distributional observations. Therefore, the internal variability of each observation does not affect its final value. Actually, an essential part of information related to the nature of data is irremediably lost.

3.4.1 L_2 Wasserstein distance-based covariance

Using the L_2 Wasserstein metric and the product of qfs defined in Equation (3.18), a measure of covariance, and thus of correlation, is extended to evaluate the association between two numeric distributional variables. Let Y_1 and Y_2 be two numeric distributional variables observed on a set of n objects, the generic i object is described by the ordered pair $Y(i) = \{y_{i1}, y_{i2}\}$, where

$y_{i1} \sim f_{i1}$ and $y_{i2} \sim f_{i2}$, with respective means equal to μ_{i1} and μ_{i2}, and standard deviations σ_{i1} and σ_{i2}. With each f_{i1} (resp. f_{i2}) is associated the corresponding *cdf* denoted by F_{i1} (resp. F_{i2}) and the respective *qf* are denoted by Ψ_{i1} (resp. Ψ_{i2}). Further, we denote by $M_1 \sim \bar{f}_{Mw(Y_1)}$ and $M_2 \sim \bar{f}_{Mw(Y_2)}$ the mean distributions of the numeric distributional variables Y_1 and Y_2. The *pdf* $\bar{f}_{Mw(Y_1)}$ (resp. $\bar{f}_{Mw(Y_2)}$) is associated with the corresponding *cdf*, that, for simplifying the notation, we denote by $\bar{F}_1$ (resp. $\bar{F}_2$) and the corresponding *qf* denoted by $\bar{\Psi}_1$ (resp. $\bar{\Psi}_2$). Finally, we denote by μ_{M_1} (resp. μ_{M_2}), σ_{M_1} (resp. σ_{M_2}) the mean and the standard deviation of M_1 (resp. M_2).

Definition 2 We define $COV_W(Y_1, Y_2)$ as the covariance between Y_1 and Y_2 based on the L_2 Wasserstein metric as follows:

$$COV_W(Y_1, Y_2) = \frac{1}{n} \sum_{i=1}^{n} \int_0^1 \left[\Psi_{i1}(t) - \bar{\Psi}_1(t) \right] \cdot \left[\Psi_{i2}(t) - \bar{\Psi}_2(t) \right] dt \qquad (3.24)$$

Given the i and the i' generic objects, the $\rho_{i1,i'2}$ index, consistently with Equation (3.7), is the correlation measure between Ψ_{i1} and $\Psi_{i'2}$, while $\rho_{.1,.2}$ is the correlation between $\bar{\Psi}_1$ and $\bar{\Psi}_2$.

Proposition 6 Using the proposed notation and the product between two *qfs* defined in Equation (3.18), $COV_W(Y_1, Y_2)$ is:

$$COV_W(Y_1, Y_2) = \underbrace{\left(\frac{1}{n} \sum_{i=1}^{n} \mu_{i1}\mu_{i2} - \mu_{M_1}\mu_{M_2} \right)}_{CM_W(Y_1, Y_2)} \\ + \underbrace{\left(\frac{1}{n} \sum_{i=1}^{n} \rho_{i1,i2}\sigma_{i1}\sigma_{i2} - \frac{1}{n^2} \sum_{i=1}^{n} \sum_{i'=1}^{n} \rho_{i1,i'2}\sigma_{i1}\sigma_{i'2} \right)}_{CV_W(Y_1, Y_2)} \qquad (3.25)$$

Similarly to the variance in Equation (3.24), it is worth noting that the index $COV_W(Y_1, Y_2)$ is the sum of two kinds of covariance: the first denoted by $CM_W(Y_1, Y_2)$ is the covariance of the means (which corresponds to COV_{BG}), while the latter, denoted by $CV_W(Y_1, Y_2)$, is related to the internal variability of the numeric distributional-valued observations. In this case, it is possible that the two components have different signs. On the other hand, it allows us to consider different aspects when numeric distributional variables are compared. For example, if all the distributions have the same shape (e.g., they are all normally distributed) then all the ρ's are equal to 1 and $COV_W(Y_1, Y_2)$ can be simplified as follows:

$$COV_W(Y_1, Y_2) = \left(\frac{1}{n} \sum_{i=1}^{n} \mu_{i1}\mu_{i2} - \mu_{M_1}\mu_{M_2} \right) + \left(\frac{1}{n} \sum_{i=1}^{n} \sigma_{i1}\sigma_{i2} - \sigma_{M_1}\sigma_{M_2} \right).$$

$$(3.26)$$

Further, if all the distributions are equally-shaped and have the same standard deviations, then the second term of Equation (3.26) goes to zero, and only in this case $COV_W(Y_1, Y_2)$ coincides with COV_{BG}. Finally, as in the classical case, it is noteworthy that if identical numeric distributional variables are observed, then $COV_W(Y_1, Y_2) = S_W^2(Y_1) = S_W^2(Y_2)$.

3.4.2 The correlation coefficient

Once defined the covariance and the standard deviation of two numeric distributional variables, an extension of the classical Pearson correlation coefficient for numeric distributional variables can be derived. Given the covariance $COV_W(Y_1, Y_2)$ between two numeric distributional variables Y_1 and Y_2 and the respective standard deviations $S_W(Y_1)$ and $S_W(Y_2)$, the correlation index $R_W(Y_1, Y_2)$ is defined as follows:

$$R_W(Y_1, Y_2) = \frac{C_W(Y_1, Y_2)}{S_W(Y_1) \cdot S_W(Y_2)}. \tag{3.27}$$

Even if $R_W(Y_1, Y_2)$ depends on $COV_W(Y_1, Y_2)$, the ratio does not allow us to decompose the contribution of the means and of the variability of the distributions as a sum of two distinct and summable correlation indices. However, it is possible to consider $R_W(Y_1, Y_2)$ as the sum of two components: the first, denoted $RM_W(Y_1, Y_2)$, is related to the means, while the latter, denoted $RV_W(Y_1, Y_2)$, is related to the variability of the data as follows:

$$
\begin{aligned}
R_W(Y_1, Y_2) &= RM_W(Y_1, Y_2) + RV_W(Y_1, Y_2) \\
&= \frac{CM_W(Y_1, Y_2)}{S_W(Y_1) \cdot S_W(Y_2)} + \frac{CV_W(Y_1, Y_2)}{S_W(Y_1) \cdot S_W(Y_2)}.
\end{aligned} \tag{3.28}
$$

Finally, $R_W(Y_1, Y_2) = 1$ if, and only if, Y_1 is equal to Y_2 (namely, y_{i1} and y_{i2} are identically distributed for all $i = 1, \ldots, n$), while this is not generally true for the correlation measure proposed by Bertrand and Goupil in [1], because the $COV_{BG}(Y_1, Y_2)$ does not consider the internal variability of the data.

3.5 Example

The descriptive statistics presented in this chapter have been implemented in the R package HistDAWass[2] to analyze data described by histogram variables. We recall that a histogram-valued variable is a particular type of distributional variable that allows observations to be described through histograms.

---

[2]HistDAWass is freely available and an open-source package for the descriptive analysis of data described by histogram-valued variables. It can be downloaded from the official CRAN repository following the URL: https://CRAN.R-project.org/package=HistDAWass.

The descriptive statistics here introduced are generalized to data described by histograms with a different number of bins and with bins of variable width. The use of histograms allows the descriptive statistics introduced in this chapter to be calculated precisely and in a computationally efficient manner. It is straightforward to express the *cdf* and the corresponding *qf* associated with a histogram through a piecewise linear function, which greatly simplifies the computations. Other ways of representing empirical frequency distributions (such as through Kernel Density Estimators or Gaussian mixtures) can be used. Although they are estimators which are more efficient statistics than histograms, it is not always possible to express in closed form their quantile functions. Consequently, the calculations of the necessary integrals for quantifying the descriptive statistics are generally approximated by using a histogram representation of the estimated density functions. For more details about histogram-valued data, we suggest to refer to [12].

In this section, we show an application of the proposed descriptive statistics to the CPS Survey. Each month, the US Department of Labor's Bureau of Labor Statistics conducts the Current Population Survey (CPS), which collects statistics on the population's labor force characteristics, such as employment, unemployment, and wages. Each month, a random sample of around 65,000 homes in the United States of America is polled. A random selection of addresses from a database is used to generate the sample. The Handbook of Labor Statistics has details, and the Bureau of Labor Statistics' website (http://www.bls.gov/) provides further information.

Each March, a more extensive poll is undertaken, in which respondents are asked about their wages throughout the previous year. The data sets include information for 2004 (based on the March 2005 poll), as well as some older years (up to 1992).

If education is provided, it is for full-time employees, defined as those who work more than 35 hours per week for a period of at least 48 weeks in the preceding year. The data are for employees with a high school diploma and a bachelor's degree as their greatest educational attainment.

Earnings before 2004 were inflation-adjusted using the Consumer Price Index (CPI). The CPI market basket price increased by 34.6 percent between 1992 and 2004. 1992 earnings are inflated by the level of total CPI price inflation to make them comparable to 2004 earnings, by multiplying 1992 earnings by 1.346 to convert to 2004 dollars.

Samples of the data are collected in the AER R package[3]. The package contains a set of CPS related datasets. CPSSW9204 dataset contains a sample of 61,395 college-educated full-time employees in the United States between 1992 and 2004. Hourly wages of working college graduates aged 25–34 in the United States from 1992 to 2004 are considered (in 2004 USD).

---

[3]AER: Applied Econometrics with R is a free and open-source package available at the URL: https://CRAN.R-project.org/package=AER

Histogram data definition

From the CPSSW9204 dataset, we considered 20 sub-groups of employees that are classified according to gender ($m = male$, $f = female$) and age. For each group, we extracted the histogram (using the classic **hist** function in the base R) of the Hourly wages for those who declared to hold a *bachelor* or a *high-school* degree in 1992 and 2004. Finally, we obtain a 20 × 4 histogram-valued data table[4].

In Table 3.3, we show the data table using, for the sake of space, the first two and the last objects of the dataset, while in Figure 3.3, we plot the entire dataset using their representation through histograms. We remark that histograms have different bins and bin-widths.

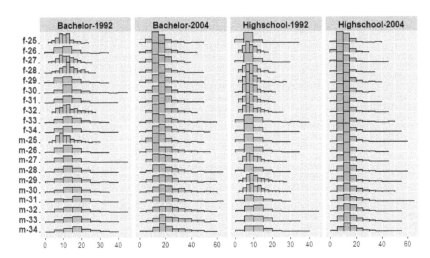

FIGURE 3.3
Histogram representation of the CPS data.

Mean histograms

In Figure 3.4, the histograms associated with the *Fréchet* means M_W and M_L are represented for each variable.

It is simple to observe the various shapes of the mean distributions by inspecting Figure 3.4. M_L appears as a mixture of the twenty distributions for each variable, but M_W appears as a distribution with intermediate position, variability, and shape characteristics. Due to the absence of details on the gender and age composition of employees at the population level, M_L may

---

[4]The code used for generating the outputs is freely available at the URL: https://github.com/Airpino/Clustering_DD_app/blob/main/Descriptive_example/ Desc_stat.md

TABLE 3.3

CPS data: first two and last objects of the histogram-valued dataset. Objects are made using m=male and f=female and the age class. Histograms are considered as the relative frequency (p) observed for each bin of the corresponding variable.

Objects	Bachelor-1992		Bachelor-2004		Highschool-1992		Highschool-2004	
	bins	p	bins	p	bins	p	bins	p
	[2;4)	0.03	[0;5)	0.02	[0;5)	0.2	[0;5)	0.04
	[4;6)	0.07	[5;10)	0.13	[5;10)	0.62	[5;10)	0.42
	[6;8)	0.12	[10;15)	0.36	[10;15)	0.16	[10;15)	0.36
	[8;10)	0.22	[15;20)	0.27	[15;20)	0.01	[15;20)	0.14
	[10;12)	0.2	[20;25)	0.1	[20;25)	0.01	[20;25)	0.01
f-25	[12;14)	0.22	[25;30)	0.06	[25;30)	0.001	[25;30)	0.001
	[14;16)	0.08	[30;35)	0.03	[30;35]	0.01	[30;35)	0.001
	[16;18)	0.03	[35;40)	0.01			[35;40]	0.01
	[18;20)	0.03	[40;45)	0.01				
	[20;22)	0.01	[45;50]	0.01				
	[22;24]	0.01						
	bins	p	bins	p	bins	p	bins	p
	[0;5)	0.05	[0;5)	0.03	[2;4)	0.06	[0;5)	0.11
	[5;10)	0.33	[5;10)	0.08	[4;6)	0.19	[5;10)	0.37
	[10;15)	0.4	[10;15)	0.32	[6;8)	0.27	[10;15)	0.33
	[15;20)	0.17	[15;20)	0.31	[8;10)	0.19	[15;20)	0.15
f-26	[20;25)	0.04	[20;25)	0.12	[10;12)	0.16	[20;25)	0.03
	[25;30)	0.001	[25;30)	0.08	[12;14)	0.06	[25;30]	0.01
	[30;35]	0.01	[30;35)	0.04	[14;16)	0.04		
			[35;40)	0.01	[16;18]	0.01		
			[40;45]	0.02				
	⋯	⋯	⋯	⋯	⋯	⋯	⋯	⋯
	bins	p	bins	p	bins	p	bins	p
	[0;5)	0.02	[0;5)	0.02	[0;5)	0.07	[0;5)	0.02
	[5;10)	0.13	[5;10)	0.03	[5;10)	0.35	[5;10)	0.19
	[10;15)	0.21	[10;15)	0.15	[10;15)	0.34	[10;15)	0.33
	[15;20)	0.33	[15;20)	0.26	[15;20)	0.17	[15;20)	0.23
	[20;25)	0.17	[20;25)	0.18	[20;25)	0.05	[20;25)	0.12
m-34	[25;30)	0.1	[25;30)	0.09	[25;30)	0.01	[25;30)	0.06
	[30;35)	0.03	[30;35)	0.1	[30;35)	0.001	[30;35)	0.03
	[35;40]	0.01	[35;40)	0.06	[35;40]	0.01	[35;40)	0.01
			[40;45)	0.04			[40;45)	0.01
			[45;50)	0.04			[45;50)	0.001
			[50;55)	0.01			[50;55]	0.01
			[55;60]	0.01				

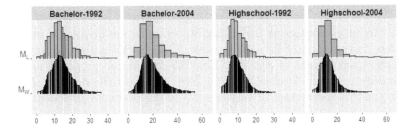

FIGURE 3.4
CPS dataset: M_L and the M_W mean distributions of wages for each variable.

provide an erroneous approximation of the population distribution and so induce erroneous judgments about the observed phenomena.

In Table 3.4, for each histogram variable, we reported the mean, standard deviation, skewness (the third standardized centered moment), and kurtosis (the fourth standardized centered moment) indices of the distributions of M_L and M_W compared with the mean and standard deviation of the same indices calculated over the 20 objects. From the table, it is evident that the descriptive statistics of M_W approximate the average of the indices of the observed objects for each histogram variable more than those of M_L. This result suggests that M_W can provide a better interpretation of the observed phenomenon. From comparing the M_Ws for each variable, we can observe an increase in average hourly wage and dispersion, right skewness, and kurtosis in 2004 with respect to 1992.

TABLE 3.4
CPS data: average (aver.) of the mean, standard deviation, skewness (third standardized moment) and kurtosis (fourth standardized moment) for the 20 histogram-valued observations of the dataset compared with the same indices for M_L and M_W histograms for each variable.

Y	Mean			Standard deviation		
	aver.	$\mu_{M_L(Y)}$	$\mu_{M_W(Y)}$	aver.	$\sigma_{M_L(Y)}$	$\sigma_{M_W(Y)}$
Bachelor-1992	13.96	13.96	13.96	5.79	6.12	5.76
Bachelor-2004	19.92	19.92	19.92	9.20	9.67	9.16
Highschool-1992	9.70	9.70	9.70	4.54	4.76	4.52
Highschool-2004	13.22	13.22	13.22	6.52	6.81	6.49

Y	Skewness			Kurtosis		
	aver.	$sk_{M_L(Y)}$	$sk_{M_W(Y)}$	aver.	$ku_{M_L(Y)}$	$ku_{M_W(Y)}$
Bachelor-1992	0.64	0.82	0.62	3.96	4.34	3.71
Bachelor-2004	0.96	1.09	0.93	4.22	4.47	4.01
Highschool-1992	0.94	1.08	0.88	5.03	5.59	4.52
Highschool-2004	1.35	1.51	1.30	6.90	7.66	6.33

Variability measures

For a careful assessment of variability, we examine the dispersion measures proposed in this chapter for each variable. Since $S_L^2(Y)$ is not decomposable, we restrict our results to $S_W^2(Y)$. Table 3.5 reports the proposed measures of variance for numeric distributional data. We show the decomposition of $S_W^2(Y)$ and the percentage of variance due to the *position* component.

TABLE 3.5

CPS data: variance of histogram variables of the hourly wages according to the L_2 Wasserstein distance between distributions. Decomposition of the variance and standard deviation for each histogram variables.

Y	Bachelor-1992	Bachelor-2004	Highschool-1992	Highschool-2004
$S_W^2(Y)$	4.26	9.58	2.22	4.26
$SM_W^2(Y)$	3.01	6.95	1.46	2.82
$SV_W^2(Y)$	1.24	2.63	0.76	1.44
$\frac{SM_W^2(Y)}{S_W^2(Y)}\%$	70.8%	72.5%	65.8%	66.2%
$S_W(Y)$	2.06	3.10	1.49	2.06

We note an increase in variability over time for both bachelor employees and high-school ones. This result suggests that the wage distributions have reflected the mutation, related to the increase in heterogeneity, of the market labor between 1992 and 2004 in the US. However, in general, the variability in salaries has been consistently higher for bachelor employees than for high-school ones. Considering the components of the variance measures, we can see that the variability due to the average wages of the gender-age groups accounted for about two-thirds of the total variability. Interestingly, although the salaries of bachelor's degree employees exhibit a more significant variability, the variability of the wages of high-school employees has a more prominent SV_W^2 component. This result suggests that, in general, the histograms of high-school employees are more different in spread and shape than those of bachelor employees.

Covariance and correlation measures

We conclude the application with an analysis of the covariance and correlation measures derived from the L_2 Wasserstein distance between distributions. In Table 3.6, we show the values of the covariance index and its decomposition for the part related to the position component and the one related to the variability component. Because the variables are expressed in the same unit of measure, the analysis of covariance measures can be done by considering that the values we observe are in squared dollars. However, when analyzing histogram variables or distributional variables, the interpretation of covariance may be generally more complicated. Starting with the sign and the decomposition of $COV_W(Y_i, Y_j)$, we can observe that, in general, observing that all the $CM_W(Y_i, Y_j)$ components are positive, there is a concordance between the

TABLE 3.6

CPS data: covariance between histogram variables of the hourly wages according to the L_2 Wasserstein distance between distributions and decomposition of the covariance.

Y_i vs. Y_j	$COV_W(Y_i, Y_j)$	$CM_W(Y_i, Y_j)$	$CV_W(Y_i, Y_j)$
Bach. 1992 vs. Bach. 2004	5.342	4.267	1.074
Bach. 1992 vs. High. 1992	2.238	1.820	0.417
Bach. 1992 vs. High. 2004	2.984	2.306	0.677
Bach. 2004 vs. High. 1992	3.636	2.860	0.776
Bach. 2004 vs. High. 2004	4.961	3.764	1.197
High. 1992 vs. High. 2004	2.548	1.932	0.615

average wage levels across variables. A special interpretative effort must be made for the $CV_W(Y_i, Y_j)$ component since it encloses the variability due to differences in the size and shape of the histograms. From Figure 3.3, we observe that the histograms are unimodal and right-skewed, and we can assume that the histograms differ in their standard deviations mainly. Therefore, observing a positive $CV_W(Y_i, Y_j)$ suggests that there is some agreement between the standard deviations of the histograms belonging to two different variables too. This result is interesting because we are dealing with phenomenon which is concordant by position (i.e., the average wages of the considered sex-age employees have grown over time) and internal variability (i.e., the average spreads around the mean wages for the observed groups have also grown over time).

TABLE 3.7

CPS data: correlation between histogram variables of the hourly wages according to the L_2 Wasserstein distance between distributions and decomposition of the correlation. The last column contains the correlation between the means of the histograms of the variables, $R_c(Y_i, Y_j)$.

Y_i vs. Y_j	$R_W(Y_i, Y_j)$	$RM_W(Y_i, Y_j)$	$RV_W(Y_i, Y_j)$	$R_c(Y_i, Y_j)$
Bach. 1992 vs. Bach. 2004	0.837	0.668	0.168	*0.932*
Bach. 1992 vs. High. 1992	0.728	0.592	0.136	*0.867*
Bach. 1992 vs. High. 2004	0.701	0.542	0.159	*0.791*
Bach. 2004 vs. High. 1992	0.788	0.620	0.168	*0.898*
Bach. 2004 vs. High. 2004	0.776	0.589	0.187	*0.850*
High. 1992 vs. High. 2004	0.828	0.628	0.120	*0.951*

In Table 3.7, we show the correlations. It is interesting to compare the measures discussed in this chapter with the correlations of the means of the observed histograms of two histogram variables. We denote it by $R_c(Y_i, Y_j)$ and it can be expressed as follows:

$$R_c(Y_i, Y_j) = \frac{CM_W(Y_i, Y_j)}{\sqrt{SM_W^2(Y_i)SM_W^2(Y_j)}}.$$

We recall that it is possible to prove that $R_c(Y_i, Y_j)$ corresponds to the correlation measure between two numeric distributional variables according to the Bertrand and Goupil approach [1]. Looking at the $R_c(Y_i, Y_j)$ only, all the variables appear to be very highly correlated, especially when observing the wages of those employees with the same degree over time.

Instead, analyzing the correlations between distributional variables, the approach derived from the L_2 Wasserstein distance between distributions shows slightly less pronounced correlations.

Again, it is interesting to note that the highest correlation values are between the wages observed for employees with the same educational level in 1992 and 2004. The decomposition of the correlation measure allows us to capture the different impacts of the position component RM_W and the variability one RV_W. We observe how the conclusions described after the analysis of the covariance measures are confirmed.

3.6 Conclusion

In this chapter, we have addressed how to conduct a descriptive analysis of numerical distributional variables. We have discussed the descriptive statistics developed in [11] that allow for the definition of measures and distributions of central tendency, measures of dispersion for a single distributional variable, and measures of association between distributional variables. The descriptive tools based on the L_2 Wasserstein distance have proven to be particularly useful in interpreting the effect of the different components of the variability of a distributional variable. The application conducted on the data coming from the CPS survey about the wages of US employees has shown the interpretative advantages of using the proposed tools.

Bibliography

[1] P. Bertrand and F. Goupil. Descriptive statistics for symbolic data. In H.-H. Bock and E. Diday, editors, *Analysis of Symbolic Data: Exploratory Methods for Extracting Statistical Information from Complex Data*, pages 106–124. Springer Berlin Heidelberg, 2000.

[2] L. Billard. Dependencies and variation components of symbolic interval-valued data. In P. Brito et al., editors, *Selected Contributions in Data Analysis and Classification*, Studies in Classification, Data Analysis, and Knowledge Organization, pages 3–12. Springer Berlin Heidelberg, 2007.

[3] L. Billard and E. Diday. *Symbolic Data Analysis: Conceptual Statistics and Data Mining*. Wiley, Chichester, Hoboken (N.J.), 2006.

[4] H.-H. Bock and E. Diday, editors. *Analysis of Symbolic Data: Exploratory Methods for Extracting Statistical Information from Complex Data*. Springer-Verlag Berlin, 2000.

[5] P. Brito. On the Analysis of Symbolic Data. In P. Brito et al., editors, *Selected Contributions in Data Analysis and Classification*, pages 13–22. Springer Berlin Heidelberg, 2007.

[6] O. Chisini. Sul concetto di media. *Periodico di Matematiche*, 4:106–116, 1929.

[7] S. Frühwirth-Schnatter. *Finite Mixture and Markov Switching Models*. Springer, 2006.

[8] A.L. Gibbs and F.E. Su. On choosing and bounding probability metrics. *International Statistical Review*, 70(3):419–435, 2002.

[9] W. Gilchrist. *Statistical Modelling with Quantile Functions*. CRC Press, Abingdon, 2000.

[10] C.E. Ginestet, A. Simmons, and E.D. Kolaczyk. Weighted Fréchet means as convex combinations in metric spaces: Properties and generalized median inequalities. *Statistics and Probability Letters*, 82(10):1859–1863, 2012.

[11] A. Irpino and R. Verde. Basic statistics for distributional symbolic variables: a new metric-based approach. *Advances in Data Analysis and Classification*, 9(2):143–175, 2015.

[12] A. Irpino, R. Verde, and Y. Lechevallier. Dynamic clustering of histograms using Wasserstein metric. In *COMPSTAT 2006*, pages 869–876. Physica-Verlag, 2006.

[13] L. Rüshendorff. Wasserstein metric. In *Encyclopedia of Mathematics*. Springer, 2001.

[14] R. Verde and A. Irpino. Dynamic clustering of histogram data: using the right metric. In P. Brito et al., editors, *Selected Contributions in Data Analysis and Classification*, Studies in Classification, Data Analysis, and Knowledge Organization, pages 123–134. Springer, Berlin, Heidelberg, 2007.

4

The Quantile Methods to Analyze Distributional Data

Manabu Ichino

Tokyo Denki University, Japan

Paula Brito
Faculty of Economics, University of Porto & LIAAD-INESC TEC, Porto, Portugal

CONTENTS

This chapter presents the quantile approach for the analysis of distributional data. This consists in representing each distributional observation by a given set of quantiles, thereby transforming the distributional data array to a standard numerical data array. The number m of quantiles used controls the granularity, therefore, a set of $m+1$ numerical vectors, called the quantile vectors, represents each object. The methods developed then explore the monotonicity property of quantile vectors. Visualization of each object is made by parallel monotone line graphs. Principal Component Analysis is applied to the quantile vectors based on the rank order correlation coefficients. Hierarchical conceptual clustering is performed based on the quantile vectors and on the concept sizes of the p-dimensional hyper-rectangles spanned by the quantile vectors. An application illustrates the proposed approaches.

4.1   Introduction

This chapter describes the quantile methods to manipulate distributional data. The quantile method transforms the given (n objects) $\times$ (p variables) distributional data array to a standard ($n \times (m + 1)$ sub-objects) $\times$ (p variables) numerical data array [4], [7], [8], [9], [10], [11], [16], where m is a preselected integer number that controls the granularity to represent the given objects. Therefore, a set of ($m + 1$) p-dimensional numerical vectors, called the quantile vectors, represents each object. According to the monotonicity of quantile vectors, we present the following three methods to analyze distributional data.

1. Visualization: We visualize each object by parallel monotone line graphs, called the data accumulation graph (DAG) [9], [11]. Each line graph is composed of line segments accumulating the zero-one normalized variable values. We present three types of the DAG for distributional data based on the quantile vectors.

2. PCA: When the given objects have a monotone structure in the representation space, the structure confines the corresponding quantile vectors to a similar geometrical shape. We apply the PCA to the quantile vectors based on the rank order correlation coefficients. We reproduce each object as m series of arrow lines that connect from the minimum quantile vector to the maximum quantile vector in the factor planes [7], [8].

3. Clustering: We present a hierarchical conceptual clustering based on the quantile vectors. We define the concept sizes of p-dimensional hyper-rectangles spanned by quantile vectors. The concept size may be used as similarity measure between sub-objects, i.e., quantile vectors, and it plays also the role of the measure for cluster quality [4], [10], [16].

4.2 Common representation of distributional data by the quantiles

This section describes the quantile representations for distributional data. The quantile method transforms each object represented by different types of distributional data to a common number of quantile values.

4.2.1 Histogram-valued data

For an object $s_i, i = 1, 2, \ldots, n$, let a histogram variable Y_j take the value H_{ij} given by

$$H_{ij} = \{I_{ij1}, p_{ij1}; I_{ij2}, p_{ij2}; \ldots; I_{ijm_j}, p_{ijm_j}\} \tag{4.1}$$

where $\sum_{\ell=1}^{m_j} p_{ij\ell} = 1$ and $I_{ij\ell} = [a_{ij\ell}, b_{ij\ell}], \ell = 1, 2, \ldots, m_j$.

We assume that the bins $I_{ij\ell}$ are mutually disjoint and that, for each histogram-valued variable Y_j and for each object s_i, within each bin the variable follows a uniform distribution.

We define the distribution function as follows:

$$\begin{cases} F_{ij}(x) = 0 & \text{for} \quad x < a_{ij1} \\ F_{ij}(x) = p_{ij1} \times \dfrac{(x - a_{ij1})}{(b_{ij1} - a_{ij1})} & \text{for} \quad a_{ij1} \le x < b_{ij1} \\ F_{ij}(x) = F_{ij}(b_{ij1}) + p_{ij2} \times \dfrac{(x - a_{ij2})}{(b_{ij2} - a_{ij2})} & \text{for} \quad a_{ij2} \le x < b_{ij2} \\ \ldots \\ F_{ij}(x) = F_{ij}(b_{ij(m_j-1)}) + p_{ijm_j} \times \dfrac{(x - a_{ijm_j})}{(b_{ijm_j} - a_{ijm_j})} & \text{for} \quad a_{ijm_j} \le x < b_{ijm_j} \\ F_{ij}(x) = 1 & \text{for} \quad b_{ijm_j} \le x \end{cases} \tag{4.2}$$

For a preselected integer m, we can easily find $(m + 1)$ quantile values including the minimum and the maximum values by solving the following equations:

$$\begin{cases} F_{ij}(\min_{ij}) = 0 \ (i.e., \min_{ij} = a_{ij1}) \\ F_{ij}(Q_{ij2}) = 2/m \\ \vdots \\ F_{ij}(Q_{ij(m_j-1)}) = (m-1)/m \\ F_{ij}(\max_{ij}) = 1 \ (i.e., \max_{ij} = b_{ijm_j}) \end{cases} \tag{4.3}$$

Therefore, we have a set of $(m + 1)$ quantile values:
$(\min_{ij}, Q_{ij1}, Q_{ij2}, \ldots, Q_{ij(m_j-1)}, \max_{ij})$.
In the quartile case, we select $m = 4$, and we have five quantile values:
$(\min_{ij}, Q_{ij1}, Q_{ij2}, Q_{ij3}, \max_{ij})$.

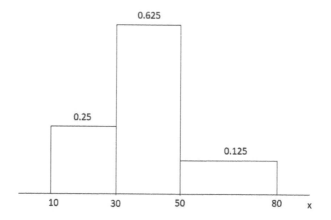

FIGURE 4.1
A three-bin histogram.

Example 1 Figure 4.1 illustrates a three-bin histogram. According to (4.2), we have the following distribution function $F(x)$:

$$\begin{cases} F(x) = 0 & for \quad x < 10 \\ F(x) = 0.25 \times \dfrac{x - 10}{30 - 10} & for \quad 10 \leq x < 30 \\ F(x) = 0.25 + 0.625 \times \dfrac{x - 30}{50 - 30} & for \quad 30 \leq x < 50 \quad (4.4) \\ F(x) = 0.875 + 0.125 \times \dfrac{x - 50}{80 - 50} & for \quad 50 \leq x < 80 \\ F(x) = 1 & for \quad 80 \leq x \end{cases}$$

In the quartile case, we have the following five quantile values, $\min, Q_1, Q_2, Q_3,$ and $\max$, by (4.3). It is clear that $\min = 10, Q_1 = 30,$ and $\max = 80$. We have $Q_2 = 38$ by solving the equation: $0.5 = 0.25 + 0.625 \times \frac{x-30}{50-30}$. Similarly, we have $Q_3 = 46$ by solving the equation: $0.75 = 0.25 + 0.625 \times \frac{x-30}{50-30}$.

4.2.2   Other types of variables

4.2.2.1   Interval-valued data

We should note that an interval-valued variable Y_j is a special type of histogram-valued variable, where the histogram (4.1) is reduced to only a single interval I_{ij1} and its probability p_{ij1} is one. We can easily obtain the desired $(m + 1)$ quantile values for a selected integer m by assuming e.g. a Uniform distribution.

4.2.2.2 Categorical modal data

For an object s_i, let variable Y_j take as value a subset of the possible ordered categorical values $\{c_{j1}, c_{j2}, \ldots, c_{jm_j}\}$ with positive weights whose sum is one. For example, the set of possible categorical values of Y_j is $\{a, b, c, d, e\}$ and an object s_i takes b with weight 0.8 and c with weight 0.2. Then, for this value of Y_j, we may use a representation similar to the histogram one:

$$H_{ij} = \{a, 0; b, 0.8; c, 0.2; d, 0; e, 0\}. \tag{4.5}$$

We replace each categorical value by a unit interval as follows:

$$H_{ij} = \{[0, 1), 0; [1, 2), 0.8; [2, 3), 0.2; [3, 4), 0; [4, 5), 0\}. \tag{4.6}$$

Then, we apply the same procedure as for histogram-valued data in the preceding sub-section. For the quartile case, we have min $= 1$, $Q_1 = 1.3125$, $Q_2 = 1.625$, $Q_3 = 1.9375$, and max $= 4$.

We assume the general representation of the categorical values of Y_j for s_i:

$$H_{ij} = \{c_{j1}, p_{ij1}; c_{j2}, p_{ij2}; \ldots; c_{jm_j}, p_{ijm_j}\} \tag{4.7}$$

where several probabilities may take the value zero and their total sum is one. We replace each categorical value in (4.7) by a unit interval as follows.

$$H_{ij} = \{[0, 1), p_{ij1}; [1, 2), p_{ij2}; \ldots; [m_j - 1, m_j], p_{ijm_j}\}. \tag{4.8}$$

Then, we apply the same procedure as for histogram-valued data to obtain the common quantile representation.

4.2.3 Quantile vectors and monotone property

By the quantile representation in the preceding section, we can transform each object s_i of $(n$ objects$) \times (p$ variables$)$ mixed variable-type distributional data to p sets of $(m + 1)$ quantile values for preselected integer number m:

$$(\min_{ij}, Q_{ij1}, Q_{ij2}, \ldots, Q_{ij(m_j-1)}, \max_{ij}), j = 1, \ldots, p. \tag{4.9}$$

We define $(m+1)$ p-dimensional numerical vectors, called the quantile vectors, as follows.

$$
\begin{aligned}
x_{i1} &= (\min_{i1}, \min_{i2}, \ldots, \min_{ip}) \\
x_{i2} &= (Q_{i11}, Q_{i12}, \ldots, Q_{i1p}) \\
x_{i3} &= (Q_{i21}, Q_{i22}, \ldots, Q_{i2p}) \\
&\vdots \\
x_{im} &= (Q_{i(m_j-1)1}, Q_{i(m_j-1)2}, \ldots, Q_{i(m_j-1)p} \\
x_{i(m+1)} &= (\max_{i1}, \max_{i2}, \ldots, \max_{ip}).
\end{aligned}
\tag{4.10}
$$

From the monotone property of quantile values for each variable, the quantile vectors in (4.10) satisfy the following monotone property:

$$x_{i1} \leq x_{i2} \leq \ldots \leq x_{i(m+1)}. \tag{4.11}$$

4.2.4   Hardwood data

We use the Hardwood data in North America throughout this chapter in order to illustrate the quantile methods. From the database in [15], we selected five species of hardwoods: ACER WEST, ALSUS WEST, FRAXINUS WEST, JAGLANS WEST, and QUERCUS WEST. The following eight histogram-valued variables describe these hardwoods.

1. Annual Temperature: ANNT (°C);

2. January Temperature: JANT (°C);

3. July Temperature: JULT (°C);

4. Annual Precipitation: ANNP (mm);

5. January Precipitation: JANP (mm);

6. July Precipitation: JULP (mm);

7. Growing Degree Days on 5°C base × 1000: GDC5; and

8. Moisture Index: MITM.

In this dataset, seven quantile values for 0%, 10%, 25%, 50%, 75%, 90%, and 100% describe each histogram variable. In order to simplify our data table, we select the integer m as $m = 3$. We select only four variables Y1: ANNT (°C), Y2: JULT (°C), Y3: ANNP (mm), and Y4: MITM, and among seven quantile values, we select only four quantile values for 0%, 50%, 90%, and 100%. Therefore, four four-dimensional quantile vectors describe each hardwood in the simplified data array, represented in Table 4.1. In this data table, ACER WEST, for example, is composed of four sub-objects Ac1, Ac2, Ac3, and Ac4, and each of them is described by a four-dimensional quantile vector, respectively. These four sub-objects satisfy the monotone property, since the quantile values of each variable satisfy this monotone property. Therefore, the sub-objects Ac1 and Ac4 determine the macroscopic region of ACER WEST in the four-dimensional representation space, and the region contains the other sub-objects Ac2 and Ac3 (see (4.11)). We use the monotone property of sub-objects, i.e., quantile vectors, in the subsequent quantile methods.

TABLE 4.1

Hardwood data.

		ANNT	JULT	ANNP	MITM
ACER	Ac1	-3.9	7.1	105	0.14
	Ac2	4.2	14.9	750	0.75
	Ac3	10.3	19.9	1860	0.98
	Ac4	20.6	29.2	4370	1
ALNUS	Al1	-12.2	7.1	170	0.22
	Al2	0.3	14.4	510	0.72
	Al3	7.6	17.5	1385	0.97
	Al4	18.7	28.3	4685	1
FRAXINUS	Fr1	2.6	12.5	85	0.09
	Fr2	17.2	24.3	485	0.49
	Fr3	22.7	30.4	1155	0.78
	Fr4	14.4	33.1	2555	0.97
JUGLANS	Ju1	7.3	17.1	235	0.2
	Ju2	16.3	22.7	625	0.69
	Ju3	22.7	27.7	905	0.89
	Ju4	26.6	31.3	1245	0.94
QUERCUS	Qu1	-1.5	9.7	85	0.08
	Qu2	14.6	21.1	540	0.63
	Qu3	19.1	27.4	1160	0.88
	Qu4	27.2	33.8	2555	0.99

4.3 Visualization: The data accumulation graph (DAG)

The data accumulation graph is a parallel line graph obtained by accumulating normalized variable values for each object [9], [11]. For each variable in Table 4.1, we find the minimum and the maximum values, and then apply the zero-one normalization to each variable. Table 4.2 is the result.

4.3.1 Quantile vectors data accumulation graph (QVDAG)

Figure 4.2 is the QVDAG for the normalized Hardwood data in Table 4.2. In Figure 4.2, ACER WEST, for example, is composed of four monotone line graphs corresponding to quantile vectors for sub-objects Ac1, Ac2, Ac3, and Ac4, respectively. We plot the line graph for Ac1 by accumulating the values for Ac1 in Table 4.2 as follows.

Horizontal axis:	1	2	3	4
Vertical axis:	0.2107	0.2107 + 0	0.2107 + 0.0043	0.2107 + 0.0043 + 0.0652

$$(4.12)$$

We plot subsequent line graphs after suitable spacing. We should note the following properties of the QVDAG:

TABLE 4.2

Normalized Hardwood data.

		ANNT	JULT	ANNP	MITM
	Ac1	0.2107	0	0.0043	0.0652
ACER	Ac2	0.4162	0.2921	0.1446	0.7283
	Ac3	0.5711	0.4794	0.3859	0.9783
	Ac4	0.8325	0.8277	0.9315	1
	Al1	0	0	0.0185	0.1522
ALNUS	Al2	0.3173	0.2734	0.0924	0.6957
	Al3	0.5025	0.3895	0.2826	0.9674
	Al4	0.7843	0.7940	1	1
	Fr1	0.3756	0.2022	0	0.0109
FRAXINUS	Fr2	0.7462	0.6442	0.0870	0.4457
	Fr3	0.8858	0.8727	0.2326	0.7609
	Fr4	0.9289	0.9738	0.5370	0.9674
	Ju1	0.4949	0.3745	0.0326	0.1304
JUGLANS	Ju2	0.7234	0.5843	0.1174	0.6630
	Ju3	0.8858	0.7715	0.1783	0.8804
	Ju4	0.9848	0.9064	0.2522	0.9348
	Qu1	0.2716	0.0974	0	0
QUESCUS	Qu2	0.6802	0.5243	0.0989	0.5978
	Qu3	0.8147	0.7603	0.2337	0.8696
	Qu4	1	1	0.537	0.9891

1. Each line graph owes to the sum of normalized unit-less values. Therefore, these line graphs show abstract sizes of the sub-objects.

2. Since the quantile vectors satisfy the monotone property (4.11), the size of the first line graph is minimum and that of the last one is maximum, for each hardwood.

3. The shapes of line graphs owe to the order of variables whose values are accumulated. However, the final sizes are independent from the order.

4. We are able to find various similar and dissimilar properties between hardwoods by comparing line graphs for sub-objects. For example, Ac4 and Al4 are very similar with respect to the selected four variables. We can also find similar properties between Fr3, Ju3, and Qu3, and others. We should note that the QVDAG figures out the distributional differences of hardwoods, i.e., the differences between the corresponding quantile vectors.

4.3.2   Variable-wise data accumulation graph (VWDAG)

As a different DAG, Figure 4.3 illustrates the VWDAG for Table 4.2. In this figure, we generate four line graphs for a hardwood by accumulating the data

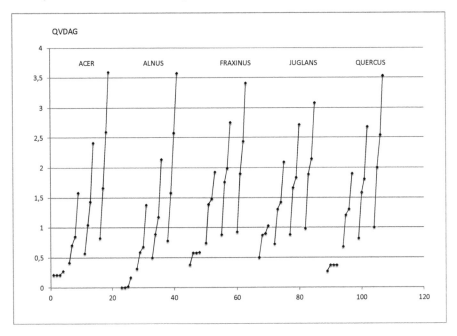

FIGURE 4.2
Quantile vectors data accumulation graph.

values for each variable separately. For example, according to (4.12), the first line graph for ACER is obtained by accumulating the values of ANNT as follows.

Horizontal axis:	1	2	3
Vertical axis:	0.2107	0.2107 + 0.4162	0.2162 + 0.4162 + 0.5711
Horizontal axis:			4
Vertical axis:			0.2162 + 0.4162 + 0.5711 + 0.8325

$$(4.13)$$

In Figure 4.3, the obtained four line graphs for each hardwood do not satisfy the monotone property as in Figure 4.2. However, we are able to find various properties depending on the selected variables among five hardwoods. For example, as a macroscopic property of the set of four line graphs, ACER and ALNUS are similar, while the remaining three hardwoods are also similar among them. From a microscopic viewpoint, ACER and ALNUS are strongly dependent on the MITM. On the other hand, FRAXINUS, JUGLANS, and QUERCUS are strongly dependent on variables ANNT and JULT.

4.3.3   Total data accumulation graph (TDAG)

By a further accumulation of the VWDAG, we can obtain the total data accumulation graph (TDAG) as in Figure 4.4. In this graph, each line graph (for

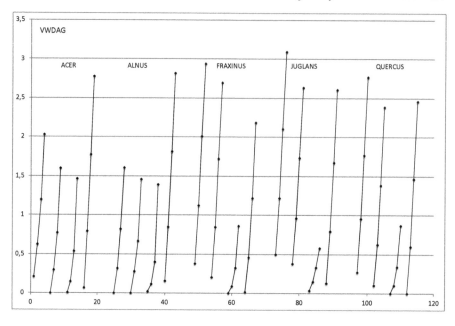

FIGURE 4.3
Variable-wise data accumulation graph.

each hardwood) is obtained by accumulating four feature-wise data accumulation line graphs in the given order. We should note the following properties for the TDAG.

1. The TDAG captures the macroscopic size of each object. Therefore, it may be useful to detect outliers in the given data in general.

2. The local shapes of the line graphs show the properties of the hardwoods associated with the variables. For example, the differences of dependency for the ANNP and the MITM are well figured out in each line graph.

4.4 The quantile method Principal Component Analysis (PCA)

Several different methods of PCA have been developed [1], [2], [5]. This section presents the quantile method PCA for distributional data [7], [8].

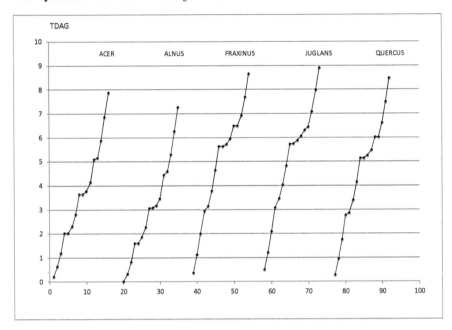

FIGURE 4.4
Total data accumulation graph.

4.4.1   Quantile method PCA

Suppose that we have $(n$ objects$) \times (p$ variables$)$ distributional data. For a preselected integer number m, we transform the given data to a standard numerical data array of the size $(n \times (m + 1)$ sub-objects$) \times (p$ variables$)$. We should note that a set of $(m + 1)$ sub-objects, described by p-dimensional quantile vectors, represents each object, and that the integer m controls the resolution of the descriptions for the given distributional data. For an object s_i, the minimum quantile vector x_{i1} and the maximum quantile vector $x_{i(m+1)}$ defined in (4.10) span a p-dimensional hyper-rectangle $E_i = E_{i1} \times E_{i2} \times \ldots \times E_{ip}$ in $\mathbb{R}^p$ by the intervals:

$$E_{ij} = [\min_{ij}, \max_{ij}], j = 1, 2, \ldots, p. \tag{4.14}$$

The hyper-rectangle E_i contains other $(m - 1)$ quantile vectors from the relations in (4.11).

Now we define a monotone property for objects based on the hyper-rectangles in the p-dimensional Euclidean space $\mathbb{R}^p$. Let E_i and $E_{i'}$ be hyper-rectangles in $\mathbb{R}^p$ for objects s_i and $s_{i'}$, respectively. Let $E_{ii'}$ denote the minimum hyper-rectangle spanned by the two hyper-rectangles E_i and $E_{i'}$ defined by:

$$E_{ii'} = E_i \boxplus E_{i'} = (E_{i1} \boxplus E_{i'1}) \times (E_{i2} \boxplus E_{i'2}) \times \ldots \times (E_{ip} \boxplus E_{i'p}) \tag{4.15}$$

where $(E_{ij} \boxplus E_{i'j})$ is the minimum interval that includes two intervals E_{ij} and $E_{i'j}$ and is called the Cartesian join (region) of E_{ij} and $E_{i'j}$, and where $E_i \boxplus E_{i'}$ is called the Cartesian join of E_i and $E_{i'}$ [7], [8], [12] (see Figure 4.5).

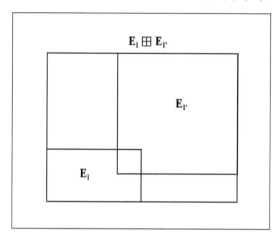

FIGURE 4.5
A two-dimensional Cartesian join region.

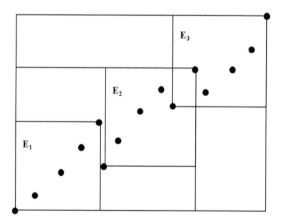

FIGURE 4.6
A nesting structure by three two-dimensional objects.

We say that the sequence of objects $s_1, s_2, \ldots, s_n$ are monotone if their hyper-rectangles satisfy the following nesting structure:

$$E_1 \subseteq E_1 \boxplus E_i \subseteq E_1 \boxplus E_{(i+1)}, \quad i = 2, 3, \ldots, n-1. \qquad (4.16)$$

Figure 4.6 illustrates a nesting structure by three two-dimensional rectangles for the quartile case. Each rectangle includes five quantile vectors with the minimum and the maximum vertices.

If the given set of objects satisfies monotone property (4.16), the nesting structure confines all quantile vectors, including quantile vectors contained in each hyper-rectangle, to have the same or similar nesting structures along the axes of p variables (see Figure 4.6).

Therefore, we can evaluate the similarity of nesting structures of quantile vectors on the axes of $\mathbb{R}^p$, i.e., the correlation between variables, by using the Kendall or the Spearman rank order correlation coefficients. We summarize the procedure for the quantile method of PCA as follows.

1. Select an integer number m, and transform the given (n objects) $\times$ (p variables) distributional data to a numerical data array of size ($n \times (m+1)$ sub-objects) $\times$ (p variables).

2. Apply traditional PCA based on the correlation matrix by the Kendall or the Spearman's rank correlation coefficient.

3. On the factor planes, reproduce each object as a series of m arrow lines connecting the minimum quantile vector to the maximum quantile vector.

4.4.2 Application to the Hardwood data

We apply the quantile method of PCA to the Hardwood data in Table 4.1 assuming $m = 3$. Table 4.3 is the correlation matrix by the Spearman's rank order correlation coefficient. Two pairs of variables (ANNT, JULT) and (ANNP, MITM) are highly correlated. The contribution ratios of the first two principal components Pc1 and Pc2 are 86.89% and 12.53%, respectively. Therefore, the first factor plane by the Pc1 and Pc2 reproduces almost the whole structural information of the quantile vectors. Table 4.4 gathers the eigenvectors for the first two principal components. Pc1 is the size factor, opposing vectors with low values to vectors with high values in all variables; Pc2 has negative weights for ANNT and JULT and positive weights for ANNP and MITM, therefore it opposes vectors with high values in ANNP and MITM and low values in ANNT and JULT to vectors in the inverse situation.

TABLE 4.3

Spearman's correlation matrix for Hardwood data.

	ANNT	JULT	ANNP	MITM
ANNT	1.00	0.99	0.74	0.70
JULT	0.99	1.00	0.79	0.74
ANNP	0.74	0.79	1.00	0.98
MITM	0.70	0.74	0.98	1.00

Figure 4.7 shows the results of PCA. ACER and ALNUS are mutually similar hardwoods, and the other three hardwoods organize a separate cluster in this factor plane. The shapes and positions of connected arrow lines tell

TABLE 4.4
Eigenvectors for the first two
principal components.

Eigenvectors	Pc1	Pc2
ANNT	0.49	-0.55
JULT	0.51	-0.45
ANNP	0.51	0.45
MITM	0.49	0.54

us various properties of hardwoods. ACER and ALNUS have a strong dependency on the precipitation (ANNP) and the moisture index (MITM). On the other hand, the other three objects have a dependency on the temperatures. Among them, JUGLANS takes a middle position. In the first 50% portion of quantiles, JUGLANS shows a dependency on the precipitation and the moisture index. Then, the dependency changes toward the temperatures. We may obtain more detailed understandings by combining the result of Figure 4.7 with various DAGs described in the preceding section. We should note that this type of possibility is strongly depending on the fact that the quantile methods use mainly the same notion of the "monotone property".

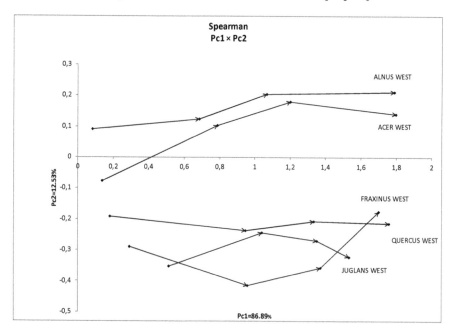

FIGURE 4.7
The result of PCA for Hardwood data.

4.5 The quantile method of hierarchical conceptual clustering

Many methods for clustering have been developed [6], [13]. One category of these is conceptual clustering [14], where conjunctive logical expressions describe clusters. There are k-means type methods and hierarchical methods. This section describes the quantile method for agglomerative hierarchical conceptual clustering [4], [10], [16].

4.5.1 Compactness as the measures of similarity and cluster quality

Let each object $s_i, i = 1, 2, \ldots, n$, be divided into $(m + 1)$ sub-objects, i.e., the minimum sub-object s_{i1}, $(m - 1)$ quantile sub-objects, s_{iq}, $q = 2, 3, \ldots, m$, and the maximum sub-object $s_{i(m+1)}$ for a preselected integer number m. These sub-objects are described by p-dimensional quantile vectors $x_{i1}, x_{i2}, \ldots, x_{i(m+1)}$ defined in (4.10), respectively. For variables $Y_j, j = 1, 2, \ldots, p$, their domains D_j are the following intervals:

$$D_j = [x_{j\min}, x_{j\max}], j = 1, 2, \ldots, p, \qquad (4.17)$$

where $x_{j\min} = \min\{\min_{1j}, \min_{2j}, \ldots, \min_{nj}\}$ and $x_{j\max} = \max\{\max_{1j}, \max_{2j}, \ldots, \max_{nj}\}$.

We define the (whole) concept space by:

$$\mathbf{D}^p = D_1 \times D_2 \times \ldots \times D_p. \qquad (4.18)$$

Then, the following minimum vertex $x_{\min}$ and the maximum vertex $x_{\max}$ span the concept space $\mathbf{D}^p$:

$$x_{\min} = (x_{1\min}, x_{2\min}, \ldots, x_{p\min}) \text{ and } x_{\max} = (x_{1\max}, x_{2\max}, \ldots, x_{p\max}). \qquad (4.19)$$

Example 2 From Table 4.1, we have the minimum and the maximum vertices as follows: $x_{\min} = (-12.2, 7.1, 85, 0.08)$ and $x_{\max} = (27.2, 33.8, 4685, 1)$.

Let $\mathbf{E} = E_1 \times \ldots \times E_p$ be a hyper-rectangle, i.e., a concept, spanned by quantile vectors in the concept space $\mathbf{D}^p$. We define the concept size of an interval E_j in terms of variable Y_j as

$$P(E_j) = \frac{|E_j|}{|D_j|}, \ j = 1, 2, \ldots, p, \qquad (4.20)$$

where $|*|$ denotes the length of interval $*$.

Then, we define the concept size $P(\mathbf{E})$ of hyper-rectangle $\mathbf{E}$ by the arithmetic mean:

$$P(\mathbf{E}) = \frac{P(E_1) + P(E_2) + \ldots + P(E_p)}{p}. \qquad (4.21)$$

Example 3 From Example 2, we have $|D_1| = (27.2 - (-12.2)) = 39.4$, $|D_2| = (33.8 - 7.1) = 26.7$, $|D_3| = (4685 - 85) = 4600$, and $|D_4| = (1 - 0.08) = 0.92$. On the other hand, for example, the sub-objects Ac1 and Ac4 span ACER from the monotone property (6.13). Therefore, from Table 4.1, we have $|E_1| = (20.6 - (-3.9)) = 24.5$, $|E_2| = (29.2 - 7.1) = 22.1$, $|E_3| = (4370 - 105) = 4265$, and $|E_4| = (1 - 0.14) = 0.86$. As a result, we have $P(E_1) = 24.5/39.4 = 0.6218$, $P(E_2) = 22.1/26.7 = 0.8277$, $P(E_3) = 4265/4600 = 0.9272$, and $P(E_4) = 0.86/0.92 = 0.9348$. Then, we have the concept size of ACER as

$$P(\mathbf{E}) = \frac{P(E_1) + P(E_2) + P(E_3) + P(E_4)}{4} = \frac{0.6218 + 0.8277 + 0.9272 + 0.9348}{4} = 0.8279.$$

The normalized data in Table 4.2 simplifies these calculations.

We should note that:

$$0 \le P(E_j) \le 1, j = 1, 2, \dots, p, \tag{4.22}$$

$$0 \le P(\mathbf{E}) \le 1, \tag{4.23}$$

and

$$P(\mathbf{D}^p) = 1. \tag{4.24}$$

Note that the "generality degree" measure proposed by Brito for hierarchical and pyramidal clustering [3] (see also Chapter 10 in [5]) corresponds to the geometric mean of the variable-wise concept sizes.

Now, we define the compactness $C(s_i, s_{i'})$ of the concept generated by two objects s_i and $s_{i'}$ by using the definitions in (4.15) and (4.21) as follows:

$$C(s_i, s_{i'}) = P(E_i \boxplus E_{i'}) = \frac{(P(E_{i1} \boxplus E_{i'1}) + P(E_{i2} \boxplus E_{i'2}) + \dots + P(E_{ip} \boxplus E_{i'p})}{p}. \tag{4.25}$$

The compactness satisfies the following properties.

1. $0 \le C(s_i, s_i), C(s_{i'}, s_{i'}) \le C(s_i, s_{i'}) \le 1$
2. $C(s_i, s_{i'}) = C(s_{i'}, s_i)$

The triangle law may not hold in general. The compactness plays a role of similarity measure between objects although it is not a metric.

In Figure 4.8, the Cartesian join regions generated by hyper-rectangles E_i and $E_{i'}$ are the same for both figures. However, the values of the similarity (or the dissimilarity) between these hyper-rectangle pairs are different in general. Therefore, the similarity (or the dissimilarity) of objects (and clusters) is a different aspect from the size of the generalized concept (i.e., the Cartesian join region) that is obtained by the objects (and clusters). In our hierarchical conceptual clustering method, we generate clusters so as to minimize the concept sizes of the clusters. To minimize the concept size of a cluster corresponds to maximize a dissimilarity between the concept associated with the cluster and the whole concept. In this context, the compactness plays also the role of a measure of cluster quality.

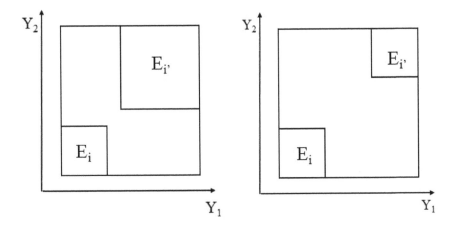

FIGURE 4.8
Compactness and similarity.

4.5.2 Algorithm of hierarchical conceptual clustering

Let $S = \{s_1, s_2, \ldots, s_n\}$, and let the goal of our algorithm be to build a hierarchy H.

Step 0) Initialization: Set $H \equiv S$.

Step 1) For each pair of objects s and s' in S, calculate the compactness $C(s, s')$ as in (4.25) and find the pair s_i and $s_{i'}$ that minimizes the compactness C.

Step 2) Generate the merged concept $s_{ii'}$ of s_i and $s_{i'}$ in S, and add $s_{ii'}$ to H and S. Then, delete s_i and $s_{i'}$ from S. The new object $s_{ii'}$ (a cluster) is described by the Cartesian join $E_{ii'} = E_i \boxplus E_{i'}$ in the concept space $\mathbf{D}^p$.

Step 3) Repeat Steps 1 and 2, until S contains only the whole cluster.

The algorithm generates a hierarchy H, i.e., a dendrogram. At each node of the dendrogram, the generated cluster has the smallest concept size, i.e., it is compact.

4.5.3 A sufficient condition for mutually disjoint concepts, and examples

We apply our algorithm to the five hardwoods described by four-dimensional hyper-rectangles spanned by the minimum and the maximum quantile vectors. Figure 4.9 is the dendrogram for the Hardwood data. The structure of the dendrogram agrees well with the result of PCA in Figure 4.7. In hierarchical clustering we determine the number of clusters k based on the obtained

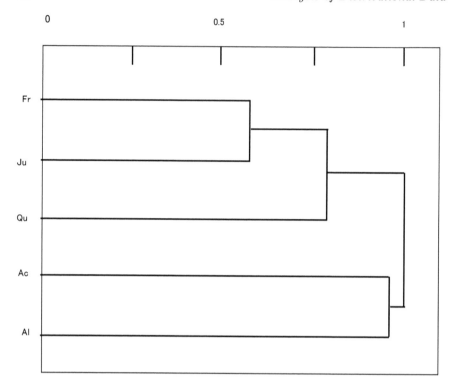

FIGURE 4.9
Dendrogram by the support regions of Hardwood data.

dendrogram with some degree of arbitrariness. However, in our hierarchical conceptual clustering, we can find a simple sufficient condition to determine mutually disjoint concepts as follows. If two hyper-rectangles E_i and $E_{i'}$ span the concept space, i.e., $\mathbf{D}^p = E_i \boxplus E_{i'}$, we have

$$P(\mathbf{D}^p) = P(E_i \boxplus E_{i'}) = \frac{P(E_{i1} \boxplus E_{i'1}) + P(E_{i2} \boxplus E_{i'2}) + \ldots + P(E_{ip} \boxplus E_{i'p})}{p} = 1.$$
$$(4.26)$$

Therefore, if the concept sizes of E_i and $E_{i'}$ satisfy the inequality:

$$P(E_i) + P(E_{i'}) < 1, \qquad (4.27)$$

there exists at least one variable Y_j that satisfies the inequality:

$$P(E_{ij}) + P(E_{i'j}) < P(E_{ij} \boxplus E_{i'j}) = 1. \qquad (4.28)$$

This asserts the fact that the two hyper-rectangles E_i and $E_{i'}$ are mutually disjoint with respect to the variable Y_j . Thus, if we cut the dendrogram at the concept size 0.5, we can find desirable candidate clusters that satisfy (4.27).

In Figure 4.9, the minimum conceptual cluster exceeds the concept size 0.5. In Table 4.2, we can easily check the fact that the support intervals of hardwoods are mutually overlapping with respect to each of the four variables.

We then apply our clustering method to the twenty sub-objects in Table 4.2, the result is represented in Figure 4.10. If we cut the dendrogram at the concept size 0.5, we can find the four clusters shown in Table 4.5. The concept size of each cluster is small compared to the cutting value 0.5. Table 4.6 summarizes the descriptions for each conceptual cluster based on the normalized values. We can easily check the fact that the four concepts are mutually disjoint. We should note that the sub-object based result in Figure 4.10 describes local properties of the hardwoods against the result in Figure 4.9. We should

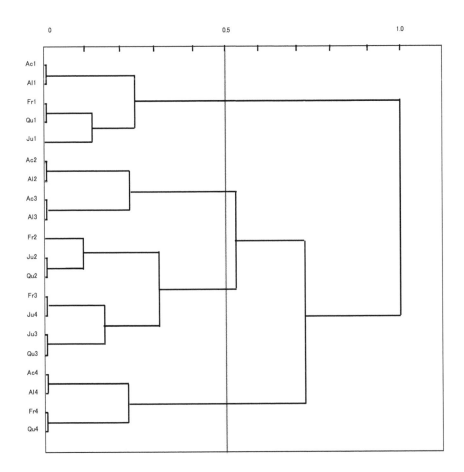

FIGURE 4.10
Dendrogram by the sub-object based clustering.

TABLE 4.5
Clusters of sub-objects and their concept sizes.

Cluster 1	Ac1, Al1, Fr1, Ju1, Qu1	0.242
Cluster 2	Ac2, Al2, Ac3, Al3	0.259
Cluster 3	Fr2, Ju2, Qu2, Fr3, Ju3, Qu3, Ju4	0.355
Cluster 4	Ac4, Al4, Fr4, Qu4	0.224

TABLE 4.6
Description of clusters.

	ANNT	**JULT**	**ANNP**	**MITM**
Cluster 1	[0, 0.495]	[0, 0.375]	[0, 0.033]	[0, 0.065]
Cluster 2	[0.317, 0.571]	[0.273, 0.479]	[0.092, 0.386]	[0.696, 0.978]
Cluster 3	[0.680, 0.985]	[0.524, 0.986]	[0.087, 0.252]	[0.446, 0.935]
Cluster 4	[0.784, 1]	[0.794, 1]	[0.537, 1]	[0.989, 1]

4.6   Summary of the quantile methods

This chapter presented three quantile methods to analyze distributional data: the data accumulation graphs (DAG's), principal component analysis (PCA), and hierarchical conceptual clustering. The key idea of the quantile methods is the "monotone property". The three quantile methods show a good agreement among themselves based on the examples by the Hardwood data that constitutes typical distributional data.

Bibliography

[1] L. Billard and E. Diday. *Symbolic Data Analysis: Conceptual Statistics and Data Mining.* John Wiley & Sons, Inc. New York, NY, USA, 2006.

[2] H.-H. Bock and E. Diday, editors. *Analysis of Symbolic Data: Exploratory Methods for Extracting Statistical Information from Complex Data.* JSpringer-Verlag Berlin, 2000.

[3] P. Brito. Symbolic objects: order structure and pyramidal clustering. *Annals of Operations Research*, 55(2):277–297, 1995.

[4] P. Brito and M. Ichino. Symbolic clustering based on quantile representation. In *Proc. COMPSTAT 2010*, Paris, France, 2010.

[5] E. Diday and M. Noirhomme-Fraiture. *Symbolic Data Analysis and the SODAS Software.* John Wiley & Sons, 2008.

[6] L. Hubert. Some extensions of Johnson's hierarchical clustering algorithms. *Psychometrika*, 37(3):261–274, 1972.

[7] M. Ichino. Symbolic PCA for histogram-valued data. *Proc. IASC 2008*, pages 5–8, 2008.

[8] M. Ichino. The quantile method for symbolic principal component analysis. *Statistical Analysis and Data Mining: The ASA Data Science Journal*, 4(2):184–198, 2011.

[9] M. Ichino. Quantile method for symbolic data analysis. In *Tutorial at the V Workshop on Symbolic Data Analysis*, Orléans, France, 2015.

[10] M. Ichino and P. Brito. A hierarchical conceptual clustering based on the quantile method for mixed feature-type data. In *Proceedings of World Statistics Congress of the International Statistical Institute*, 2013.

[11] M. Ichino and P. Brito. The data accumulation graph (DAG) to visualize multi-dimensional symbolic data. In *Proc. IV Workshop on Symbolic Data Analysis*, Taipei, Taiwan, 2014.

[12] M. Ichino and H. Yaguchi. Generalized Minkowski metrics for mixed feature-type data analysis. *IEEE Transactions on Systems, Man, and Cybernetics*, 24(4):698–708, 1994.

[13] S.C. Johnson. Hierarchical clustering schemes. *Psychometrika*, 32(3):241–254, 1967.

[14] R.S. Michalski and R.E. Stepp. Learning from observation: Conceptual clustering. In Mitchell T.M., Michalski R.S., Carbonell J.G., editors, *Machine Learning*, pages 331–363. Springer, Berlin, Heidelberg, 1983.

[15] Climate-Vegetation Atlas of North America. Histogram data by the U.S. geological survey. `http://pubs.usgs.gov/pp/p1650-b/`.

[16] K. Umbleja, M. Ichino, and H. Yaguchi. Hierarchical conceptual clustering based on quantile method for identifying microscopic details in distributional data. *Advances in Data Analysis and Classification*, 15(2):407–436, 2021.

Part II

Clustering and Classification

5

Partitive and Hierarchical Clustering of Distributional Data using the Wasserstein Distance

Rosanna Verde

Department of Mathematics and Physics, University of Campania "L. Vanvitelli", Caserta, Italy

Antonio Irpino

Department of Mathematics and Physics, University of Campania "L. Vanvitelli", Caserta, Italy

CONTENTS

This chapter presents partitive and hierarchical clustering algorithms to analyze a set of observations described by numeric distributional variables. The two distance-based algorithms use the 2-norm Wasserstein distance as a building block, exploiting its interpretative advantages. The first is a dynamic clustering algorithm that generalizes the well-known k-means. The second is an extension of the hierarchical agglomerative clustering algorithm. Clustering outputs are evaluated according to internal indices, which can be interpreted through the type and the size of the variability explained by the obtained clusters. This chapter presents an application of the clustering algorithms on a temperature dataset.

DOI: 10.1201/9781315370545-5

5.1 Introduction

This chapter deals with some clustering methods for distributional data. Clustering algorithms look for a partition of a set of data into homogeneous clusters, such that similar objects belong to the same cluster, while different objects belong to different ones. Clustering methods can be classified into hierarchical and partitive ones. Hierarchical methods aim at grouping data into clusters by joining objects and clusters progressively according to their minimum distance into more and more numerous groups. Usually, agglomeration paths of the clusters are visualized by hierarchical structures, like dendrograms. Some variants of the agglomerative methods allow objects to belong to more than one cluster, such as pyramidal clustering algorithm [2].

Partitive clustering algorithms look for clusters optimizing a criterion of homogeneity of the clusters. Generally, this is obtained by an iterative procedure that moves objects from one group to another to obtain more homogeneous clusters. A general algorithm of this type is the Dynamical Clustering Algorithm (DCA) [8]. DCA is an iterative algorithm looking for the best partition of a set data and the best representation of the clusters according to a criterion of best fitting between the clusters' objects and the clusters' representative elements. The well-known *k-means* algorithm is a particular case of DCA, where the minimized criterion is based on the Euclidean distance between the objects to allocate to the clusters and the barycenters of clusters (means). Other types of partitive clustering algorithms are based on a density criterion that requires the definition of a distance, or a dissimilarity, implicitly. Indeed, they are based on the definition of *closeness* measures between data which depend on suitable similarity/dissimilarity/distance measures between data. Finally, the third kind of partitive algorithm is based on criteria based on models of probability distributions, where the problem is the estimation of the parameters of a mixture of distributions (sub-populations generating clusters) that best fits the observed data. A well-known method of this kind of partitive algorithm is the Gaussian Mixture Model [14].

In the framework of Symbolic Data Analysis, several extensions of classical clustering algorithms have been proposed, both for hierarchical and partitive algorithms. All the proposed methods are based on dissimilarity or distance measures defined for set-valued (or symbolic) data [5], [6]. In the framework of model-based partitive clustering, the authors in [15] proposed a new algorithm for distributional data.

Dealing with distributional data, distance-based partitive and hierarchical clustering are extensions of classical algorithms using suitable dissimilarities or distances. Such dissimilarities/distances between set-valued descriptions are introduced into the optimized criteria. The convergence to a stable solution into a finite number of steps is provided [4].

In Chapter 3, L_2 Wasserstein distance is proposed for comparing distributional data. The same chapter shows some properties of the distance and a generalization of classical variability measures for distributional variables. Using the properties of the L_2 Wasserstein distance, two clustering methods for distributional data are presented:

- the first algorithm is the Dynamical Clustering Algorithm (DCA) [8] adapted to distributional data. DCA is a prototype-based partitive algorithm that can be considered a generalization of *k-means*;

- the second algorithm is a hierarchical agglomerative clustering using the Single, Complete, Average and Ward's linkages for joining clusters.

5.2 Dynamic Clustering Algorithm for distributional data

In several contexts, the interest in analyzing distributional-valued data and looking for the most representative kind of distribution among the observed ones is increasing. In such a way, a synthesis of the set of data can be performed by clustering distributions in a predefined number of homogeneous clusters. The representative element of each cluster, expressed as the mean of the cluster's distributional data, can be a proper representative element of the typology that describes the distributional data. An example can be provided by air pollution control sensors located in different zones of an urban area: different distributions of air pollutants recorded in a day for each sensor can be synthesized by grouping the different controlled zones into homogeneous clusters.

Clustering algorithms require the definition of dissimilarity or distance measures to compare pairs of objects, and, for comparing two density functions having the first two moments finite, the Wasserstein metric [9] [18] is a very suitable measure. This choice is motivated by the possibility of finding a centroid, expressed as an average distribution, which is a linear combination of quantile functions (see [19], [11] for a comparison with other probability metrics). In particular, this can be obtained by using the L_2 Wasserstein distance (that is a Euclidean distance between quantile functions).

We briefly recall the L_2 Wasserstein distance. Given two distributional-valued units described by two density functions f_i and $f_{i'}$ associated with the respective quantile functions Ψ_i and $\Psi_{i'}$ (namely, the inverse functions of the *cdf*s F_i and $F_{i'}$), the L_2 Wasserstein distance is defined as follows:

$$d_W(f_i, f_{i'}) = \sqrt{\int_0^1 |\Psi_i(t) - \Psi_{i'}(t)|^2 \, dt}. \tag{5.1}$$

When objects are described by p distributional variables, the L_2 Wasserstein distance is obtained by combining the distances computed for each variable. Considering two objects s_i and $s_{i'}$, described respectively by two vectors of density functions $y(i) = [f_{i1}, \ldots, f_{ip}]$ and $y(i') = [f_{i'1}, \ldots, f_{i'p}]$, the L_2 Wasserstein distance between the two distributional data is expressed as follows:

$$D_W\left(y(i), y(i')\right) = \sqrt{\sum_{j=1}^{p} d_W^2\left(f_{ij}, f_{i'j}\right)}. \qquad (5.2)$$

Dynamic Clustering Algorithm (DCA) [8] is a general class of centroid (or prototype)-based partitive clustering algorithms. Once defined a fixed number k of clusters, DCA looks for a partition of a dataset alternating two steps, a *representation* and an *allocation* step, until a convergence to a *local* optimum of a criterion function is obtained. The two steps are performed according to a fitting criterion function, i.e., the best *representation* of a cluster and the best *allocation* of an object to a cluster are obtained by optimizing the same criterion function. Generally, the representation of a cluster is expressed by a *prototype*, which depends on the criterion function and can be a point, an object, a set of objects, a probability law, or a factorial axis. The *allocation* of the objects is based on a minimum distance between objects and clusters' prototypes, such that the criterion function is minimized. The *k-means* algorithm can be considered a particular case of DCA, where the criterion function is a cluster homogeneity measure based on the within-cluster sum of squares (WSS). Indeed, a (local) minimum of the criterion is obtained when the clusters' prototypes are the cluster averages (the means). The allocation is performed by using the Squared Euclidean distance between data and prototypes.

Dynamical Clustering Algorithm

Formally, given a dissimilarity/distance measure δ between a pair of objects s_i and $s_{i'}$ belonging to a set S of size n and described by p distributional variables Y_j $(j = 1, \ldots, p)$, and a fixed number k of desired clusters, DCA looks for the partition $P^* \in P_k$ of S in k clusters, among all the possible partitions P_k, and the vector $L^* \in L_k$ of k prototypes representing the clusters in P, such that, they minimize a fitting criterion Δ between L and P:

$$\Delta(P^*, L^*) = \min\{\Delta(P, L) \mid P \in P_k, L \in L_k\}. \qquad (5.3)$$

Such a criterion is defined as the sum of dissimilarity or distance measures $\delta(y(i), G_h)$ between the unit s_i which belongs to a cluster $C_h \in P$ and described by $y(i)$, and the cluster prototype $G_h \in L$:

$$\Delta(P, L) = \sum_{h=1}^{k} \sum_{s_i \in C_h} \delta(y(i), G_h). \qquad (5.4)$$

Generally, the $\Delta(P, L)$ criterion is based on an additive distance with respect to p descriptors. The algorithm is initialized by generating k random clusters (or, alternatively, k random prototypes). The algorithm alternates two steps until convergence to a stationary value of the criterion function:

Representation step Fixed the partition P, k prototypes are computed such that Equation (5.4) is minimized, assuming that the solution is unique (this depends on the choice of δ).

Allocation step Fixed the set of k prototypes, a partition P is obtained by assigning each object to the cluster having the *closest* (accordingly to δ) prototype.

A suitable choice of δ is fundamental the convergence of the criterion to an optimum. Generally, quadratic distances (for example, Euclidean distance) guarantee that the criterion reaches a local optimum. In the literature, several choices for δ have been proposed according to the nature of data [1]. However, for distributional data, a few dissimilarity/distance measures allow the convergence of the Δ criterion. Further, few distances [19] [11] allow us to obtain a prototype of a cluster of distributions as a distribution. Using the (Squared) L_2 Wasserstein distance, and denoting by f_{ij} the distribution-valued data (a density function) describing the s_i object for the j-th distributional variable, and by G_{hj} the distribution associated with the prototype of the h-th cluster for the j-th distributional variable, the Δ criterion is as follows:

$$\Delta(P,L) = \sum_{h=1}^{k} \sum_{j=1}^{p} \sum_{s_i \in C_h} d_W^2 \left(f_{ij}, G_{hj} \right). \tag{5.5}$$

The criterion corresponds to *within-clusters sum of squared distances*(WSS) to the clusters prototypes, thus it guarantees that the optimal partition is reached minimizing an homogeneity criterion. If we consider the *total sum of squares* (TSS) as the sum of squared distances between each object and the overall prototype as a dispersion measure of a dataset, using the L_2 Wasserstein distance for comparing distributions, we prove (see Appendix of this chapter) that a classical results of decomposition of the TSS holds also when data are distributions. Denoting by BSS, the between-clusters sum of squares of a set of n objects grouped into k clusters as a weighted (by the cluster size n_h) sum of squared distances between each prototype G_{hj} and the general one, denoted by G_j, the following equality holds:

$$\underbrace{\sum_{i=1}^{n} \sum_{j=1}^{p} d_W^2 \left(f_{ij}, G_j \right)}_{TSS} = \underbrace{\sum_{h=1}^{k} \sum_{j=1}^{p} \sum_{s_i \in C_h} d_W^2 \left(f_{ij}, G_{hj} \right)}_{WSS} + \underbrace{\sum_{h=1}^{k} n_h \sum_{j=1}^{p} d_W^2 \left(G_{hj}, G_j \right)}_{BSS}.$$
$$\tag{5.6}$$

This proves that the minimization of the Δ homogeneity criterion corresponds to maximize the separation between clusters (namely, the BSS). Further, let μ_{ij} be the mean of the distributional observations having f_{ij} as *pdf*, and denoting with f_{ij}^c the *centered* w.r.t. μ_{ij} *pdf*, namely f_{ij} shifted by μ_{ij}, having *cdf* denoted by F_{ij} and F_{ij}^c respectively, and *qf* denoted by Ψ_{ij} and Ψ_{ij}^c the squared L_2 Wasserstein distance between two distributional observations can be rewritten as [10]:

$$d_W^2(f_{ij}, f_{i'j}) = \underbrace{(\mu_{ij} - \mu_{i'j})^2}_{Position} + \underbrace{\int_0^1 \left| \Psi_{ij}^c(t) - \Psi_{i'j}^c(t) \right|^2 dt}_{Variability}. \quad (5.7)$$

If it is possible to estimate the $\rho(\Psi_{ij}, \Psi_{i'j})$ correlation index between the quantile functions, [11] showed that:

$$d_W^2(f_{ij}, f_{i'j}) = \underbrace{(\mu_{ij} - \mu_{i'j})^2}_{Position} + \underbrace{(\sigma_{ij} - \sigma_{i'j})^2 + 2\sigma_{ij}\sigma_{i'j} \left[1 - \rho(\Psi_{ij}, \Psi_{i'j})\right]}_{Variability}. \quad (5.8)$$

The (Squared) L_2 Wasserstein distance between two distributional observations can be decomposed into the (squared) Euclidean distance between the means of the two distributions and the (squared) Wasserstein distance between the two *centered* distributions. Therefore, the Total Sum of Squares (TSS) in Equation (5.6), the Within Sum of Squares (WSS) and the Between Sum of Squares (BSS) can be further decomposed into a *position* component, related to the means, and a *variability-shape* component, related to the standard deviations and to the correlation measure between the centered distributions:

$$TSS = TSS_P + TSS_V \quad (5.9)$$

$$TSS_P = \sum_{i=1}^n \sum_{j=1}^p \sum_{s_i \in C_h} (\mu_{ij} - \mu_j)^2 \quad (5.10)$$

$$TSS_V = \sum_{i=1}^n \sum_{j=1}^p \sum_{s_i \in C_h} \left\{ (\sigma_{ij} - \sigma_j)^2 + 2\sigma_{ij}\sigma_j \left[1 - \rho\left(\Psi_{ij}, \bar{\Psi}_j\right)\right] \right\} \quad (5.11)$$

$$WSS = WSS_P + WSS_V \quad (5.12)$$

$$WSS_P = \sum_{j=1}^p \sum_{h=1}^k \sum_{s_i \in C_h} (\mu_{ij} - \bar{\mu}_{hj})^2 \quad (5.13)$$

$$WSS_V = \sum_{j=1}^p \sum_{h=1}^k \sum_{s_i \in C_h} \left\{ (\mu_{ij} - \bar{\mu}_{hj})^2 + 2\sigma_{ij}\bar{\sigma}_{hj} \left[1 - \rho\left(\Psi_{ij}, \bar{\Psi}_{hj}\right)\right] \right\} \quad (5.14)$$

$$BSS = BSS_P + BSS_V \quad (5.15)$$

$$BSS_P = \sum_{j=1}^p \sum_{h=1}^k n_h \left(\bar{\mu}_{hj} - \mu_j\right)^2 \quad (5.16)$$

$$BSS_V = \sum_{j=1}^p \sum_{h=1}^k n_h \left\{ (\bar{\mu}_{hj} - \mu_j)^2 + 2\bar{\sigma}_{hj}\sigma_j \left[1 - \rho\left(\bar{\Psi}_{hj}, \bar{\Psi}_j\right)\right] \right\} \quad (5.17)$$

$$TSS_P = WSS_P + BSS_P \quad (5.18)$$

$$TSS_V = WSS_V + BSS_V \quad (5.19)$$

Such relationships are useful, for example, for the definition of indices for evaluating the quality of the partition obtained by the DCA. For example, it is possible to extend the partition quality index, as defined in [4]:

$$Q(P_k) = 1 - \frac{\sum\limits_{h=1}^{k} \sum\limits_{s_i \in C_h} \delta(s_i, G_h)}{\sum\limits_{s_i \in E} \delta(s_i, G_S)} \qquad (5.20)$$

where G_S is the overall prototype of the S set (containing all the s_i objects). $Q(P_k)$ index is a generalization of the ratio between the Within-clusters Sum of Squares and the Total Sum of Squares. The more the clusters are compact and well separated from each other, the closer to 1 is the index. The more the cluster structure is absent the closer to zero is the index. Using the squared L_2 Wasserstein distance in DCA, the index is:

$$Q(P_k) = 1 - \frac{WSS}{TSS} = \frac{BSS}{TSS}. \qquad (5.21)$$

This relationship generally does not hold when the distance is not quadratic.

5.3 Agglomerative hierarchical clustering using Wasserstein distance

Hierarchical clustering algorithms look for nested groups of objects such that they form a hierarchy of clusters. Hierarchical clustering algorithms are either divisive (top-down), where the whole data set is recursively split into narrower subgroups, or agglomerative (bottom-up), where objects are merged into wider groups. Differently from DCA, hierarchical clustering algorithms do not require to fix the number k of desired clusters in advance. The splitting or the merging of clusters is done using greedy strategies based on a dissimilarity/distance. A dendrogram generally represents the merging or splitting path of clusters.

Agglomerative hierarchical clustering needs the specification of a metric (distance) for the objects and a merging criterion (namely, the linkage criterion) for joining clusters. The choice of the metric and the linkage affects the aggregation and the shape of the obtained clusters. Denoting by $\delta(s_i, s_{i'})$ the metric, C_h and $C_{h'}$ two clusters to merge, G_h and $G_{h'}$ the respective centers, and n_h and $n_{h'}$ their sizes, the most common linkage criteria used for hierarchical clustering algorithms are:

- Single-linkage

$$min\left\{\delta(s_i, s_{i'}) \mid s_i \in C_h, s_{i'} \in C_{h'}\right\}.$$

It suffers from the well-known chain effect.

- Complete-linkage

$$max \left\{ \delta \left(s_i, s_{i'} \right) | s_i \in C_h, s_{i'} \in C_{h'} \right\}.$$

It suffers from the presence of outlier data.

- Average-linkage

$$\frac{1}{n_h n_{h'}} \sum_{s_i \in C_h} \sum_{s_{i'} \in C_{h'}} \delta \left(s_i, s_{i'} \right)$$

It is one of the most chosen linkage because it mitigates the previous two effects.

- Ward-linkage [21]

$$\frac{n_h n_{h'}}{n_h + n_{h'}} \delta^2 \left(G_h, G_{h'} \right)$$

Two clusters are joined if the decreasing of homogeneity is minimum. Choosing the L_2 Wasserstein distance, the within-cluster sum of squared distances $WSS(C_h \cup C_{h'})$ of a new cluster joining C_h and $C_{h'}$, is defined as follows

$$WSS(C_h \cup C_{h'}) = WSS(C_h) + WSS(C_{h'}) + \frac{n_h n_{h'}}{n_h + n_{h'}} D_W^2 \left(G_h, G_{h'} \right).$$

This particular criterion is based on the minimization of the Within-cluster Sum of Squares (see [16] and the references therein for a detailed discussion).

5.4   Example

This section presents an example of application of the DCA and hierarchical clustering to a climatic dataset. The original dataset `drd964x.tmpst.txt`[1] contains the sequential "Time Biased Corrected" State climatic division monthly Average Temperatures recorded in the 48 (Hawaii and Alaska are not present in the dataset) states of US from 1895 to 2014.

In order to obtain distributional data, using the R packages `HistDAWass`[2] and `histogram`[3], for each State and each month, we have estimated a histogram according to the procedure proposed in [17]. Then, the data have been organized into a histogram-valued data table constituted by 48 rows (States) and 12 columns (Monthly temperature in deg. F. as the distributional variables)[4]. The choice of this kind of density estimation has allowed us

---

[1]freely available at the National Climatic Data Center website of US http://www1.ncdc.noaa.gov/pub/data/cirs/

[2]https://cran.r-project.org/web/packages/HistDAWass/

[3]https://cran.r-project.org/web/packages/histogram/

[4]The distributional data tables in R format are freely available at: https://github.com/Airpino/Clustering_DD_app.

to speed up the computing of the L_2 Wasserstein distance-based statistics. In Table 5.1, we report a small part of the data table where, in each cell is described a histogram, while in Figure 5.1, for the sake of space, we have reported only 6 of the 48 States described by the twelve monthly temperature distributional variables.

TABLE 5.1
USA temperature dataset. Part of the data table reporting the histogram-valued data of three States observed on January, February and March.

State	January		February		March		···
	Temp.	p	Temp.	p	Temp.	p	···
	[33.3; 37.8)	0.050	[37.60; 41.05)	0.067	[45.50; 49.30)	0.025	···
	[37.8; 42.3)	0.175	[41.05; 44.50)	0.092	[49.30; 53.10)	0.286	···
ALA	[42.3; 46.8)	0.367	[44.50; 47.95)	0.319	[53.10; 56.90)	0.328	···
	[46.8; 51.3)	0.317	[47.95; 51.40)	0.302	[56.90; 60.70)	0.277	···
	[51.3; 55.8)	0.075	[51.40; 54.85)	0.184	[60.70; 64.50]	0.084	···
	[55.8; 60.3]	0.017	[54.85; 58.3]	0.034			···
	Temp.	p	Temp.	p	Temp.	p	···
	[29.60; 32.93)	0.008	[36.10; 39.20)	0.034	[43.90; 48.73)	0.311	···
	[32.93; 36.27)	0.042	[39.20; 42.30)	0.118	[48.73; 53.57)	0.538	···
ARI	[36.27; 39.60)	0.200	[42.30; 45.40)	0.361	[53.57; 58.40]	0.151	···
	[39.60; 42.93)	0.417	[45.40; 48.50)	0.361			···
	[42.93; 46.27)	0.300	[48.50; 51.60]	0.126			···
	[46.27; 49.60]	0.033					···
	Temp.	p	Temp.	p	Temp.	p	···
	[26.60; 32.05)	0.042	[31.10; 35.22)	0.050	[40.10; 45.65)	0.067	···
	[32.05; 37.50)	0.208	[35.22; 39.34)	0.118	[45.65; 51.20)	0.420	···
ARK	[37.50; 42.95)	0.525	[39.34; 43.46)	0.319	[51.20; 56.75)	0.403	···
	[42.95; 48.40]	0.225	[43.46; 47.58)	0.344	[56.75; 62.30]	0.110	···
			[47.58; 51.70]	0.168			···
···	···	···	···	···	···	···	···

5.4.1 USA temperature dataset: Dynamic Clustering

The analysis[5] consisted in the following three steps:

1. 12 monthly mean temperatures of each state have been represented by distributions using histograms;

2. for each $k = 2, \ldots, 8$, 100 DCAs have been run with a different starting random partition of objects;

3. for each solution, the Calinski and Harabasz (CH) [3] validity index has been computed to identify an optimal number of clusters. CH

---

[5]We used the WH_kmeans() function implemented in the HistDAWass package.

is defined as:

$$CH(k) = \frac{BSS(k)/(k-1)}{WSS(k)/(n-k)}$$

where k denotes the number of clusters, and $BSS(k)$ and $WSS(k)$ denote the between and within-cluster sums of squares of the partition into k clusters, respectively. The highest value of CH index suggests the optimal number of clusters.

Table 5.2 reports the best indices for the 700 runs of the algorithm. The results suggest that a good choice for the number of cluster is obtained for $k = 3$.

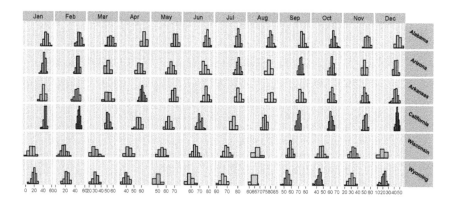

FIGURE 5.1
USA temperature dataset. Part of the data table showing the histogram-valued data of six States observed on the 12 months (distributional variables).

TABLE 5.2
DCA output: criterion values; $Q(P_k)$ and CH validity indices. In bold, the maximum vakue of the $CH(k)$ index, that suggests $k = 3$.

#k	$\Delta(P_k, L_k)$	$Q(P_k)$	$CH(k)$
2	13,368.39	0.66	90.78
3	6,787.43	0.83	**109.27**
4	5,145.23	0.87	98.64
5	3,837.46	0.90	100.60
6	3,374.70	0.92	90.54
7	2,913.48	0.93	86.40
8	2,502.63	0.94	85.05

TABLE 5.3

DCA: composition of the 3 clusters.

Cluster	N.	States
1	13	California, Arizona, Oklahoma, North Carolina, Tennessee, Texas, Alabama, Mississippi, Georgia, South Carolina, Arkansas, Louisiana, Florida
2	21	Washington, Oregon, Massachusetts, Nebraska, Pennsylvania, Connecticut, Rhode Island, New Jersey, Indiana, Nevada, Utah, Ohio, Illinois, Delaware, West Virginia, Maryland, Kentucky, Kansas, Virginia, Missouri, New Mexico
3	14	Montana, Maine, North Dakota, South Dakota, Wyoming, Wisconsin, Idaho, Vermont, Minnesota, New Hampshire, Iowa, New York, Colorado, Michigan

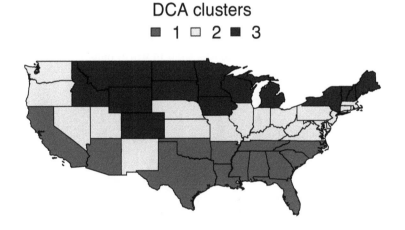

FIGURE 5.2

USA temperature dataset. DCA partition in 3 clusters.

Table 5.3 lists the States belonging to each cluster as resulted from the DCA, with $k = 3$, while Figure 5.2 shows the map of clustered States. The three climatic zones represented by clusters seem consistent with the geomorphology of the USA (except for Alaska and Hawaii states).

Figure 5.3 represents the descriptions of the prototypes of the three clusters obtained from DCA. It is possible to note that clusters are ordered from the warmest (Cluster 1) to the coldest one (Cluster 3). Looking at the plot, and in particular at the sizes of the prototypes' distributions for each month, it is interesting to note that generally Cluster 3 (the cold cluster), for each month, has mean distributions which have a higher variability than the ones of Cluster 1 (the warmest one) and that Cluster 2 prototypes have an intermediate behavior with respect to the variability.

Denoting by $BSS(j)$ the between-cluster sum of squares for each variable, and considering that $\sum_{j=1}^{p} BSS(j) = BSS$, we considered $Q(P_k)(j) = \frac{BSS(j)}{TSS}$, such that $\sum_{j=1}^{p} Q(P_k)(j) = Q(P_k)$, as the part of $Q(P_k)$ related to the j-th variable. Naturally, this measure is sensible to the different scales of the variables. However, in the proposed example, these differences are negligible, thus, it is possible to interpret the value of $Q(P_k)(j)$ as a discriminant power of the variable between clusters. In Table 5.4 reports a decomposition of the $Q(P_k)$ index for each variable. We observe that winter months appear as more discriminant with respect to summer ones. Therefore, we can interpret that greater differences may be observed in cold months than in warm ones.

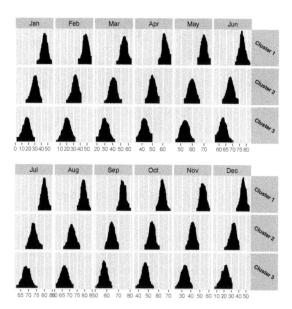

FIGURE 5.3
USA temperature dataset. Prototypes of the three clusters obtained by DCA.

TABLE 5.4
DCA: $Q(P_3) = 0.8269$ decomposed for each variable (month).

Month	$Q(P_3)(j)$	Month	$Q(P_3)(j)$
January	0.1319	July	0.0219
February	0.1292	August	0.0281
March	0.1009	September	0.0444
April	0.0652	October	0.0511
May	0.0443	November	0.0718
June	0.0338	December	0.1040

5.4.2 USA temperature dataset: Hierarchical clustering

On the same dataset, we performed a hierarchical clustering analysis[6]. Using four linkage criteria and the L_2 Wasserstein distance, we obtained the dendrograms shown in Figures 5.4, 5.5, 5.6 and 5.7. As usual, it is possible to partition the 48 states into k clusters by cutting each dendrogram at a suitable level of the value of the aggregation criterion.

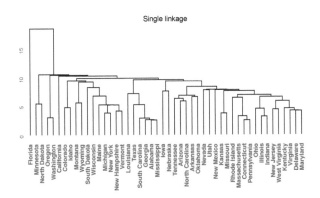

FIGURE 5.4
USA temperature dataset. Single linkage dendrogram.

A generally used rule-of-thumb for extracting clusters from a tree is to cut the dendrogram where the higher difference in the values of the aggregation criterion is observed, but in this case, all the cuts lead to partitions consisting of only two clusters. Other ways of cutting a dendrogram exist. Trying to obtain more than two clusters, we use again the CH index for selecting the best k number of clusters.

---

[6]We used the WH_hclust() function implemented in the HistDAWass package.

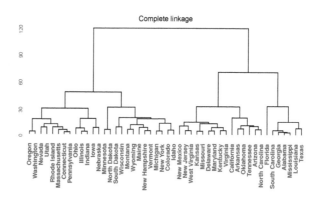

FIGURE 5.5
USA temperature dataset. Complete linkage dendrogram.

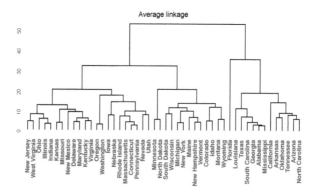

FIGURE 5.6
USA temperature dataset. Average linkage dendrogram.

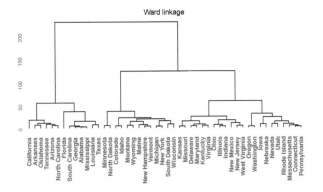

FIGURE 5.7
USA temperature dataset. Ward linkage dendrogram.

TABLE 5.5
Hierarchical clustering: CH indices computed for different
values of k and for each linkage used for building the
dendrograms. The best CH index for each linkage is
highlighted in bold.

Linkage	k = 2	k = 3	k = 4	k = 5	k = 6	k = 7
Single	6.46	6.65	4.82	3.99	14.05	**32.13**
Complete	82.42	76.94	**90.78**	75.91	87.10	88.63
Average	80.44	44.44	79.66	66.53	71.79	**80.85**
Ward	80.44	**100.34**	83.89	87.56	81.95	80.85

Table 5.5 reports the CH index values computed for different values of k
ranging from 2 to 7. Figure 5.8 shows the maps of the US colored accordingly
to the number of clusters suggested by the CH indices in bold in Table 5.5.
It is interesting to observe that the Ward-linkage partition into three clusters
corresponds to the DCA's solution with $k = 3$. It is related to a similar
criterion adopted in the two methods: DCA relocates objects such that a
within-cluster sum of squares criterion is minimized. In contrast, a hierarchical
agglomerative method based on the Ward-linkage criterion joins two clusters
such that a minimum increase of the within-cluster sum of squares is obtained.

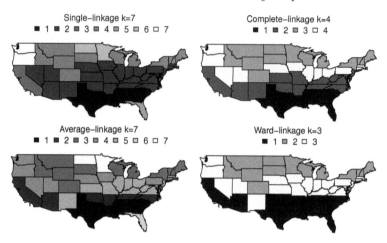

FIGURE 5.8

Maps of clusters of the partitions obtained by cutting the dendrograms according to the highest value of $CH(k)$ index.

5.5 Conclusions

In this chapter, we have presented a partitive and a hierarchical clustering algorithm for data described by distributions. We showed an application of the clustering procedures on data described by histograms, which are particular types of distributional descriptions. Both algorithms rely on the L_2 Wasserstein distance between two distributions. This particular distance allowed for a very straightforward interpretation of the algorithms' output. Further, the interpretation of the clusters' prototypes allows for discovering new insights when clustering a set of distributional data. In the literature, other clustering techniques that exploit the interpretative advantages of the $L2$ Wasserstein distance have been proposed [7, 12, 13, 20] and implemented in the most common open-source tools for data analysis (like R or Python), but for the sake of brevity, we did not present them here.

5.6 Appendix: decomposition of the TSS for grouped distributional data

Let S be the set of n objects described by p distributional variables and clustered in k clusters. We show the decomposition of the *Total Sum of Squared*

(TSS) distances of the objects to the general prototype into the *Within-cluster Sum of Squared* (WSS) distances of the objects of the cluster to the respective prototype and the *Between-cluster Sum of Squared* (BSS) distances of the cluster prototypes to the global prototype. We recall the following notation:

- Ψ_i is a quantile function, namely, the inverse of the F_i *cdf* associated with the f_i *pdf* for the i-th object ($i = 1, \ldots, n$, where n is the number of objects);

- n_h is the number of objects in the h-th cluster ($h = 1, \ldots, k$, where k is the number of clusters partitioning the set of objects);

- C_h denotes the h-th cluster of objects;

- μ_i is the mean of the distribution having f_i as *pdf*;

- $\bar{\Psi}$ is the *average quantile function* of the set of n quantile functions and its mean is denoted with μ;

- Ψ_i^c is the *centered quantile function* with respect μ_i, i.e. $\Psi_i^c = \Psi_i - \mu_i$, for the i-th object. The mean of the distribution, which is associated with the quantile function Ψ_i^c, is 0.

Taking into account that the (squared) L_2 Wasserstein distance is additive with respect to the variables, it is enough to prove the decomposition for a single distributional variable.

$$
\begin{aligned}
TSS &= \sum_{i=1}^n \int_0^1 \left(\Psi_i(t) - \bar{\Psi}(t) \right)^2 dt = \sum_{i=1}^n \int_0^1 \left[(\mu_i + \Psi_i^c(t)) - (\mu + \bar{\Psi}^c(t)) \right]^2 dt \\
&= \sum_{i=1}^n \int_0^1 \left[(\mu_i - \mu) - (\Psi_i^c(t) - \bar{\Psi}^c(t)) \right]^2 dt \\
&= \sum_{i=1}^n \int_0^1 \left[(\mu_i - \mu)^2 + (\Psi_i^c(t) - \bar{\Psi}^c(t))^2 - \underbrace{2 (\mu_i - \mu) (\Psi_i^c(t) - \bar{\Psi}^c(t))}_{=0} \right] dt \\
&= \sum_{i=1}^n \int_0^1 (\mu_i - \mu)^2 dt + \sum_{i=1}^n \int_0^1 (\Psi_i^c(t) - \bar{\Psi}^c(t))^2 dt \\
&= \sum_{i=1}^n (\mu_i - \mu)^2 + \sum_{i=1}^n \int_0^1 (\Psi_i^c(t) - \bar{\Psi}^c(t))^2 dt
\end{aligned}
\tag{5.22}
$$

Consider data partitioned in $k > 1$ groups and the following relations between the local and the global means:

$$
\begin{aligned}
\bar{\Psi}(t) &= \tfrac{1}{n} \sum_{i=1}^n \Psi_i(t) = \tfrac{1}{n} \sum_{i=1}^n [\mu_i + \Psi_i^c(t)] = \\
&= \tfrac{1}{n} \sum_{i=1}^n \mu_i + \tfrac{1}{n} \sum_{i=1}^n \Psi_i^c(t) = \mu + \bar{\Psi}^c(t).
\end{aligned}
\tag{5.23}
$$

If data are partitioned in k groups each one of size $n_h > 1$ such that $\sum_{h=1}^{k} n_h = n$, we have

$$\bar{\Psi}(t) = \frac{1}{n}\sum_{i=1}^{n} \Psi_i(t) = \frac{1}{n}\sum_{i=1}^{n}\sum_{h=1}^{k} I_{ih}\Psi_{ih}(t) \quad I_{ih} = \begin{cases} 1 & if \quad s_i \in C_h \\ 0 & if \quad s_i \notin C_h \end{cases} \quad (5.24)$$

$$\bar{\Psi}_h(t) = \frac{1}{n_h}\sum_{s_i \in C_h} \Psi_i(t) = \frac{1}{n_h}\sum_{s_i \in C_h} [\mu_i + \Psi_i^c(t)]$$

$$= \frac{1}{n_h}\sum_{s_i \in C_h} \mu_i + \frac{1}{n_h}\sum_{s_i \in C_h} \Psi_i^c(t) = \bar{\mu}_h + \bar{\Psi}_h^c(t). \quad (5.25)$$

It follows that

$$\bar{\Psi}(t) = \frac{1}{n}\sum_{i=1}^{n}\sum_{h=1}^{k} I_{ih}\Psi_i(t) = \frac{1}{n}\sum_{i=1}^{n}\sum_{h=1}^{k} I_{ih}[\mu_i + \Psi_i^c(t)]$$

$$= \frac{1}{n}\sum_{i=1}^{n}\sum_{h=1}^{k} I_{ih}\mu_i + \frac{1}{n}\sum_{i=1}^{n}\sum_{h=1}^{k} I_{ih}\Psi_i^c(t)$$

$$= \frac{1}{n}\sum_{h=1}^{k} n_h\bar{\mu}_h + \frac{1}{n}\sum_{h=1}^{k} n_h\bar{\Psi}_h^c(t) = \frac{1}{n}\sum_{h=1}^{k} n_h[\bar{\mu}_h + \bar{\Psi}_h^c(t)] \quad (5.26)$$

$$= \frac{1}{n}\sum_{h=1}^{k} n_h\bar{\Psi}_h(t) = \frac{\sum_{h=1}^{k} n_h\bar{\Psi}_h(t)}{\sum_{h=1}^{k} n_h}$$

$$TSS = \sum_{i=1}^{n}\sum_{h=1}^{k} I_{ih}\int_0^1 \left(\Psi_i(t) - \bar{\Psi}(t)\right)^2 dt$$

$$= \sum_{i=1}^{n}\sum_{h=1}^{k} I_{ih}\int_0^1 \left(\Psi_i(t) - \bar{\Psi}(t) + \bar{\Psi}_h(t) - \bar{\Psi}_h(t)\right)^2 dt$$

$$= \sum_{i=1}^{n}\sum_{h=1}^{k} I_{ik}\int_0^1 \left[(\Psi_i(t) - \bar{\Psi}_h(t)) - (\bar{\Psi}(t) - \bar{\Psi}_h(t))\right]^2 dt$$

$$= \sum_{i=1}^{n}\sum_{h=1}^{k} I_{ih}\int_0^1 \left(\Psi_i(t) - \bar{\Psi}_h(t)\right)^2 dt + \sum_{i=1}^{n}\sum_{h=1}^{k} I_{ih}\int_0^1 \left(\bar{\Psi}(t) - \bar{\Psi}_h(t)\right)^2 dt$$

$$- 2\sum_{i=1}^{n}\sum_{h=1}^{k} I_{ih}\int_0^1 \left(\Psi_i(t) - \bar{\Psi}_h(t)\right)\left(\bar{\Psi}(t) - \bar{\Psi}_h(t)\right) dt$$

$$= \sum_{i=1}^{n} \sum_{h=1}^{k} I_{ih} \underbrace{\left[\left(\mu_i - \bar{\mu}_h\right)^2 + \int_0^1 \left(\Psi_i^c(t) - \bar{\Psi}_h^c(t)\right)^2 dt \right]}_{WSS = Within\ Sum\ of\ Squares}$$

$$+ \sum_{h=1}^{k} n_h \int_0^1 \left(\bar{\Psi}(t) - \bar{\Psi}_h(t)\right)^2 dt$$

$$- 2 \sum_{i=1}^{n} \sum_{h=1}^{k} I_{ih} \int_0^1 \left(\Psi_i(t) - \bar{\Psi}_h(t)\right) \left(\bar{\Psi}(t) - \bar{\Psi}_h(t)\right) dt$$

$$= WSS + \sum_{h=1}^{k} n_h \underbrace{\left[\left(\mu - \bar{\mu}_h\right)^2 + \int_0^1 \left(\bar{\Psi}^c(t) - \bar{\Psi}_h^c(t)\right)^2 dt \right]}_{BSS = Between\ Sum\ of\ Squares}$$

$$- 2 \sum_{i=1}^{n} \sum_{h=1}^{k} I_{ih} \int_0^1 \left(\Psi_i(t) - \bar{\Psi}_h(t)\right) \left(\bar{\Psi}(t) - \bar{\Psi}_h(t)\right) dt$$

$$= WSS + BSS - 2 \sum_{i=1}^{n} \sum_{h=1}^{k} I_{ih} \int_0^1 \left(\Psi_i(t) - \bar{\Psi}_h(t)\right) \left(\bar{\Psi}(t) - \bar{\Psi}_h(t)\right) dt. \tag{5.27}$$

Considering that

$$2 \int_0^1 \sum_{i=1}^{n} \sum_{h=1}^{k} I_{ih} \left(\Psi_i(t) - \bar{\Psi}_h(t)\right) \left(\bar{\Psi}(t) - \bar{\Psi}_h(t)\right) dt$$

$$= 2 \int_0^1 \sum_{h=1}^{k} \sum_{i=1}^{n} I_{ih} \left(\Psi_i(t) - \bar{\Psi}_h(t)\right) \left(\bar{\Psi}(t) - \bar{\Psi}_h(t)\right) dt$$

$$= 2 \int_0^1 \left[\sum_{h=1}^{k} \left(\bar{\Psi}(t) - \bar{\Psi}_h(t)\right) \sum_{i=1}^{n} I_{ih} \left(\Psi_i(t) - \bar{\Psi}_h(t)\right) dt \right]$$

$$= 2 \int_0^1 \left\{ \sum_{h=1}^{k} \left(\bar{\Psi}(t) - \bar{\Psi}_h(t)\right) \left[\sum_{i=1}^{n} I_{ih}\Psi_i(t) - \sum_{i=1}^{n} I_{ih}\bar{\Psi}_h(t) \right] dt \right\}$$

$$= 2 \int_0^1 \left\{ \sum_{h=1}^{k} \left(\bar{\Psi}(t) - \bar{\Psi}_h(t)\right) \left[n\bar{\Psi}(t) - \sum_{s=1}^{k} n_s\bar{\Psi}_s(t) \right] dt \right\} \tag{5.28}$$

$$= 2 \int_0^1 \left\{ \sum_{h=1}^{k} \left(\bar{\Psi}(t) - \bar{\Psi}_h(t)\right) \left[n\bar{\Psi}(t) - \frac{n}{n} \sum_{s=1}^{k} n_s\bar{\Psi}_s(t) \right] dt \right\}$$

$$= 2 \int_0^1 \left\{ \sum_{h=1}^{k} \left(\bar{\Psi}(t) - \bar{\Psi}_h(t)\right) \left[n\bar{\Psi}(t) - n \sum_{s=1}^{k} \frac{n_s}{n}\bar{\Psi}_s(t) \right] dt \right\}$$

$$= 2 \int_0^1 \left\{ \sum_{h=1}^{k} \left(\bar{\Psi}(t) - \bar{\Psi}_h(t)\right) \underbrace{\left[n\bar{\Psi}(t) - n\bar{\Psi}(t) \right]}_{0} dt \right\} = 0$$

Equation (5.27) is

$$TSS = WSS + BSS. \tag{5.29}$$

This result confirms that the Δ criterion adopted in the DCA of distributional data minimizing the WSS guarantees that the BSS is maximized, too, and thus that it searches for a partition of data minimizing the internal heterogeneity of objects in the cluster and, at the same time, maximizing the separation between clusters.

Bibliography

[1] A. Appice, C. d'Amato, F. Esposito, and D. Malerba. Classification of symbolic objects: A lazy learning approach. *Intelligent Data Analysis*, 10(4):301–324, 2006.

[2] P. Brito and F.A.T. De Carvalho. Hierarchical and pyramidal clustering. In E. Diday and M. Noihomme-Fraiture, editors, *Symbolic Data Analysis and the SODAS Software*, pages 157–179. John Wiley & Sons, Ltd, 2007.

[3] T. Caliński and J. Harabasz. A dendrite method for cluster analysis. *Communications in Statistics*, 3(1):1–27, 1974.

[4] M. Chavent, F.A.T. De Carvalho, Y. Lechevallier, and R. Verde. New clustering methods for interval data. *Computational Statistics*, 21(2):211–229, 2006.

[5] A. da Silva, Y. Lechevallier, and F.A.T. De Carvalho. Comparing clustering on symbolic data. In N. Nedjah et al., editors, *Intelligent Text Categorization and Clustering*, volume 164 of *Studies in Computational Intelligence*, pages 81–94. Springer Berlin Heidelberg, 2009.

[6] F.A.T. De Carvalho, Y. Lechevallier, and R. Verde. Clustering methods in symbolic data analysis. In E. Diday and M. Noihomme-Fraiture, editors, *Symbolic Data Analysis and the SODAS Software*, pages 181–203. John Wiley & Sons, Ltd, 2007.

[7] F.D.A.T. De Carvalho, A. Balzanella, A. Irpino, and R. Verde. Co-clustering algorithms for distributional data with automated variable weighting. *Information Sciences*, 549:87–115, 2021.

[8] E. Diday and J.C. Simon. Clustering analysis. In K.S. Fu, editor, *Digital Pattern Recognition*, volume 10 of *Communication and Cybernetics*, pages 47–94. Springer Berlin Heidelberg, 1976.

[9] A.L. Gibbs and F.E. Su. On choosing and bounding probability metrics. *International Statistical Review*, 70(3):419–435, 2002.

[10] C.R. Givens and R.M. Shortt. A class of Wasserstein metrics for probability distributions. *Michigan Math. J.*, 31(2):231–240, 1984.

[11] A. Irpino and R. Verde. Basic statistics for distributional symbolic variables: a new metric-based approach. *Advances in Data Analysis and Classification*, 9(2):143–175, 2015.

[12] A. Irpino, R. Verde, and F.D.A.T. De Carvalho. Dynamic clustering of histogram data based on adaptive squared Wasserstein distances. *Expert Systems with Applications*, 41(7):3351–3366, 2014.

[13] A. Irpino, R. Verde, and F.D.A.T. De Carvalho. Fuzzy clustering of distributional data with automatic weighting of variable components. *Information Sciences*, 406-407:248–268, 2017.

[14] G. McLachlan and D. Peel. *Finite Mixture Models*. John Wiley & Sons, Ltd, 2005.

[15] A. Montanari and D.G. Calò. Model-based clustering of probability density functions. *Advances in Data Analysis and Classification*, 7(3):301–319, 2013.

[16] F. Murtagh and P. Legendre. Ward's hierarchical agglomerative clustering method: Which algorithms implement Ward's criterion? *Journal of Classification*, 31(3):274–295, 2014.

[17] Y. Rozenholc, T. Mildenberger, and U. Gather. Combining regular and irregular histograms by penalized likelihood. *Computational Statistics & Data Analysis*, 54(12):3313 – 3323, 2010.

[18] L. Rüshendorff. Wasserstein metric. In *Encyclopedia of Mathematics*. Springer, 2001.

[19] R. Verde and A. Irpino. Dynamic clustering of histogram data: Using the right metric. In P. Brito et al., editors, *Selected Contributions in Data Analysis and Classification*, Studies in Classification, Data Analysis, and Knowledge Organization, pages 123–134. Springer Berlin Heidelberg, 2007.

[20] R. Verde and A. Irpino. Comparing histogram data using a Mahalanobis-Wasserstein distance. In P. Brito, editor, *COMPSTAT 2008 - Proceedings in Computational Statistics, 18th Symposium*, pages 77–89. Physica-Verlag HD, 2008.

[21] J.H. Ward Jr. Hierarchical grouping to optimize an objective function. *Journal of the American Statistical Association*, 58(301):236–244, 1963.

.

6

Divisive Clustering of Histogram Data

Marie Chavent

Inria Bordeaux Sud-Ouest, ASTRAL team - IMB, UMR CNRS 5251, University of Bordeaux, France

Paula Brito

Faculty of Economics, University of Porto & LIAAD-INESC TEC, Porto, Portugal

CONTENTS

This chapter presents a divisive top-down clustering method designed for histogram-valued data. The method provides a hierarchy on a set of units together with a characterization of each cluster in the form of a conjunction of properties on the descriptive variables, which are necessary and sufficient conditions for cluster belonging (monothetic clustering). At each step, a cluster is chosen to be split, to minimize intra-cluster dispersion, which is measured by the sum of squared Mallows distances between pairs of members of each cluster. The criterion is minimized across the bipartitions induced by a set of binary questions. Since interval-valued variables constitute a special case of histogram-valued variables, the method applies to data described by either kind of variables.

DOI: 10.1201/9781315370545-6

6.1   Introduction

In this chapter, we present a divisive clustering method for histogram-valued data. Whereas agglomerative algorithms proceed bottom-up, starting with single-element clusters and merging at each step the two most similar clusters until one containing all units is formed, divisive algorithms proceed top-down, starting with one single cluster gathering all units, and perform a bipartition of one cluster at each step.

We extend the divisive algorithm proposed in [4] and [5] to data described by histogram-valued variables; interval-valued data may be adressed analogously, since intervals are special case of histograms with just one class (see also [3]). The method successively splits one cluster into two sub-clusters, according to a condition expressed as a binary question on the values of one variable. The cluster to be sub-divided and the variable and condition to be considered at each step are selected so as to minimize intra-cluster dispersion at the next step. Therefore, each formed cluster is directly described by a conjunction of necessary and sufficient conditions for cluster membership (the conditions that lead to its formation by successive splits) and we obtain a monothetic clustering [21] on the given dataset.

Many divisive clustering methods have been proposed in the literature [14]. To obtain a global optimal solution one should, at each step, consider all the possible bipartitions of a cluster C units into two non-empty subsets, but such a complete enumeration procedure is computationally prohibitive. Therefore, divisive clustering methods do not usually consider all possible bipartitions. [16] propose an iterative divisive procedure that uses an average dissimilarity between an unit and a group of units; in [10] a disaggregative clustering method based on the concept of mutual nearest neighborhood is used. Monothetic divisive clustering methods were first developed for binary data [15], [22]. Other approaches may be referred in the context of information-theoretic clustering [8] and spectral clustering [1], [9]. More recently, Hofmeyr and Pavlidis [11] use low dimensional representations of the data which maximize the quality of a binary partition, and use this bi-partitioning recursively to generate clustering models. The authors in [20] propose a divisive hierarchical maximum likelihood clustering DRAGON, which has been applied on mutation and microarray data, and proved to be computationally efficient. Chen and Billard [6] present a study of divisive clustering using Hausdorff distances for interval data. Roux [19] provides a comparative study of divisive and agglomerative hierarchical clustering algorithms.

Divisive methods have been widely developed in the context supervised classification and are usually known as tree structured classifiers, well known examples are CART [2] and ID3 [17]. Ciampi [7] considers that trees offer a natural approach for both class formation (clustering) and development of classification rules.

In the next section, we present the divisive clustering algorithm for histogram-valued data, detailing the criterion, proposed distances and the rules for binary splitting and assignement. Section 6.3 presents two applications, finally Section 6.4 concludes.

6.2 Divisive clustering

Divisive clustering algorithms proceed top-down, starting with the set to be clustered S, and performing a bipartition of one cluster of the current partition at each step. At step t, a partition of S in t clusters is present, one of which will be further divided in two sub-clusters. The cluster to be divided and the splitting rule are chosen so that the next partition in $t+1$ clusters minimizes intra-cluster dispersion.

6.2.1 The criterion

The "quality" of a given partition $P_t = \left\{ C_1^{(t)}, C_2^{(t)}, \ldots, C_t^{(t)} \right\}$ is measured by a criterion $Q(t)$, the sum of intra-cluster dispersion for each cluster :

$$Q(t) = \sum_{h=1}^{t} I(C_h^{(t)}) = \sum_{h=1}^{t} \sum_{s_i, s_{i'} \in C_h^{(t)}} D^2(s_i, s_{i'}) \tag{6.1}$$

with $D^2(s_i, s_{i'}) = \sum_{j=1}^{p} d^2(Y_j(s_i), Y_j(s_{i'}))$, where d is a distance between distributions that allows for the Huyghens decomposition in within and between sum of squares. That is, for each cluster, intra-cluster dispersion is defined as the sum of all pairwise squared-distances D^2 between the cluster elements. We consider distances D^2 additive on the descriptive variables.

At each step of the algorithm, one cluster is chosen to be split in two sub-clusters, so that $Q(t+1)$ is minimized, or, equivalently, $Q(t) - Q(t+1)$ maximized, as Q always decreases at each step.

Distances

Several distances may be considered to evaluate the dissimilarity between distributions. Let, as usual, each observation be written as
$Y_j(s_i) = H_{Y_j(s_i)} = ([\underline{I}_{ij1}, \overline{I}_{ij1}[, p_{ij1}; \ldots; [\underline{I}_{ijm_j}, \overline{I}_{ijm_j}], p_{ijm_j})$. The two following distances may be used for this purpose, as they both allow for the Huyghens decomposition (see Chapter 5 as concerns the Mallows distance):

1. Mallows (or L_2 Wasserstein) distance

$$d_M^2(Y_j(s_i), Y_j(s_{i'})) = \int_0^1 (\Psi_{ij}(t) - \Psi_{i'j}(t))^2 dt \tag{6.2}$$

where $\Psi_{ij}, \Psi_{i'j}$ are the quantile functions corresponding to the distributions $Y_j(s_i), Y_j(s_{i'})$, respectively. If both quantile functions are written with m pieces, and the same set of cumulative weights, and the Uniform distribution within subintervals is assumed, the square of the Mallows distance between these distributions is given by

$$d_M^2(Y_j(s_i), Y_j(s_{i'})) = \sum_{\ell=1}^{m} p_\ell \left[(c_{ij\ell} - c_{i'j\ell})^2 + \frac{1}{3}(r_{ij\ell} - r_{i'j\ell})^2 \right]$$

(6.3)

where, $c_{ij\ell}, c_{i'j\ell}$ and $r_{ij\ell}, r_{i'j\ell}$ with $\ell \in \{1, \ldots, m\}$ are the centers and half ranges of the subinterval ℓ of the distributions $Y_j(s_i)$ and $Y_j(s_{i'})$, respectively (see Chapter 1).

2. Euclidean distance

$$d_E^2(Y_j(s_i), Y_j(s_{i'})) = \sum_{\ell=1}^{m_j} (p_{ij\ell} - p_{i'j\ell})^2 \qquad (6.4)$$

The Mallows distance has been used in agglomerative hierarchical clustering in [13], as well as in Chapter 5 of this book.

We notice that the former requires that the distributions to be compared are written with the same quantile partition, while the latter applies to histograms with the same sub-intervals.

6.2.2   Binary Questions and Assignment

The bipartition to be performed at each step of the algorithm is defined by one single variable. The binary questions, as in the classical case, are defined by conditions of the type

$R_j := Y_j \leq v_j, j = 1, \ldots, p$.

Each condition R_j leads to a bipartition of a cluster, sub-cluster 1 gathers the elements who verify the condition, sub-cluster 2 those who do not.

An element $s_i \in S$ verifies the condition $R_j = Y_j \leq v_j$ if and only if the probability/frequency associated with values of $Y_j(s_i)$ below v_j is larger or equal than 0.5.

Therefore, to decide whether s_i meets the condition R_j it is enough to check the median of the distribution $Y_j(s_i)$, $Me_j(s_i)$: if $Me_j(s_i) \leq v_j$ then condition R_j is satisfied. The median of each distribution $Y_j(s_i)$ is determined from the uniformity assumption within each sub-interval of the observed histograms.

As a consequence, for each variable $Y_j, j = 1, \ldots, p$, the values v_j that need to be considered, and that lead to different assignments, will be the midpoints (or any other intermediate values) between any two medians of the distributions $Y_j(s_i), i = 1, \ldots, n$.

The sequence of conditions met by the elements of each cluster constitutes a necessary and sufficient condition for cluster membership. The obtained clustering is therefore monothetic, i.e. each cluster is represented by a conjunction of properties in the descriptive variables.

Example

Consider the distribution of the arrival delays (in minutes) for three different airlines, A, B and C, represented in Table 6.1.

TABLE 6.1
Distribution of arrival delays for three airlines.

Airline	Delay (min)
A	$\{[0,10[,0.33;[10,30[,0.33;[30,60],0.33\}$
B	$\{[0,10[,0.2;[10,30[,0.6;[30,60],0.2\}$
C	$\{[0,10[,0;[10,30[,0.5;[30,60[,0.45;[60,90],0.05\}$

Assuming a uniform distribution within each sub-interval, the medians are, $Me(A) = 20.3, Me(B) = 20, Me(C) = 30$. Therefore, we may define two binary questions:

$R1$: Delay ≤ 20.15 and $R2$: Delay ≤ 25

Rule $R1$ separates the airlines in the following two clusters $C_1 = \{B\}, C_2 = \{A, C\}$, whereas rule $R2$ separates the airlines in the clusters $C_1 = \{A, B\}, C_2 = \{C\}$.

6.2.3 Algorithm

The proposed divisive clustering algorithm may be summarized as follows. Let $P_t = \{C_1^{(t)}, \ldots, C_t^{(t)}\}$ be the current partition in t clusters at step t.
Initialization : $P_1 = \{C_1^{(1)} \equiv S\}$.

At step t: Determine the cluster $C_{h^*}^{(t)}$ and the binary question $R_j := Y_j \leq v_j$, such that the new resulting partition $P_{t+1} = \{C_1^{(t+1)}, \ldots, C_{t+1}^{(t+1)}\}$, in $t+1$ clusters, minimizes intra-cluster dispersion, given by

$$Q(t+1) = \sum_{h=1}^{t+1} \sum_{s_i, s_{i'} \in C_h^{(t+1)}} \sum_{j=1}^{p} d^2(Y_j(s_i), Y_j(s_{i'})), \text{ among partitions in } t+1$$

clusters obtained by splitting a cluster of P_t in two clusters.

Notice that minimizing $Q(t + 1)$ is equivalent to maximizing $\Delta Q = I(C_{h^*}^{(t)}) - (I(C_a^{(t+1)}) + I(C_b^{(t+1)}))$, where C_a and C_b are the two sub-clusters obtained by the split of C_{h^*}.

At each step t the maximum number N of alternative "cuts" to be assessed is given by $N = \sum_{h=1}^{t}(n_h - 1)$, if n_h is the cardinal of cluster $C_h^{(t)}$.

When the desired, pre-fixed, number of clusters is attained, or the partition P has n clusters each with a single unit $s_i, i = 1, \ldots, n$ (step n), the algorithm stops.

6.3 Application: Crime dataset

The "Crime" dataset concerns official data on violent crime occurrence in the USA, together with some social indicators. The original dataset "Communities and Crime Data Set" is available at the UCI ML Repository [1]. The data has been recorded per county, and then aggregated at state level in the form of histogram-valued data, with four intervals per histogram defined by the quartiles of the observed data. Here we analyse a set of 30 states (AL, AR, AZ, CA, CO, CT, FL, GA, IA, IN, KY, LA, MA, MO, NC, NH, NJ, NY, OH, OK, OR, PA, RI, SC, TN, TX, UT, VA, WA, WI) for which it has been possible to obtain enough information to describe the variables' distribution (at least 20 records per state). Four variables are considered: violent crimes (total number of violent crimes per 100 000 habitants) (violentPerPop), percentage of people aged 25 years and above with less than 9th grade education (pctLowEdu); percentage of people aged 16 years and above who are employed (pctEmploy); percentage of population who are divorced (pctAllDivorc) and percentage of immigrants who immigrated within the last 10 years (pctFgnImmig.10).

We have applied the method described above to this data, whose values have been standardized according to the procedure available in the R package HistDAWass [12], using the definitions of mean and standard deviation presented in Chapter 3, and using the Mallows distance (see (6.3)).

Figure 6.1 depicts the obtained tree. We notice that the first split is defined by variable "percentage of population who are divorced". The next division is made on the basis of "total number of violent crimes per 100 000 habitants". The first four binary questions, leading to the successive splits are as follows:

Q1: $pctAllDivorc < 10.74$ or $pctAllDivorc \geq 10.74$;
Q2: $violentPerPop < 485.94$ or $violentPerPop \geq 485.94$;
Q3: $pctLowEdu < 12.52$ or $pctLowEdu \geq 12.52$;
Q4: $violentPerPop < 697.82$ or $violentPerPop \geq 697.82$.

To decide on the appropriate number of clusters, we rely on the Silhouette Index [18]. Figure 6.2 depicts the corresponding values for $k = 2$ to $k = 10$. The maximum value is attained for $k = 2$, then we observe a local maximum for $k = 5$. The partition in two clusters provides an explained inertia (ratio of between to total sum of squares) of just 32.9%, whereas the partition in five clusters explains 58.8% of the total inertia. We therefore retain the latter. Figure 6.3 depicts the corresponding Silhouette values.

The partition in five clusters is presented in Table 6.2 and Figure 6.4. Their description, directly obtained from the divisive tree is as follows:

$C1 : pctAllDivorc < 10.74$;
$C2 : pctAllDivorc \geq 10.74 \wedge violentPerPop < 485.94$;
$C3 : pctAllDivorc \geq 10.74 \wedge violentPerPop \geq 485.94 \wedge pctLowEdu < 12.52$;

---

[1]https://archive.ics.uci.edu/ml/datasets

$C4 : pctAllDivorc \geq 10.74 \wedge violentPerPop \geq 485.94 \wedge pctLowEdu \geq 12.52 \wedge$
$violentPerPop < 697.82;$
$C5 : pctAllDivorc \geq 10.74 \wedge pctLowEdu \geq 12.52 \wedge violentPerPop \geq 697.82.$

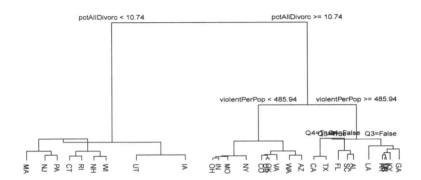

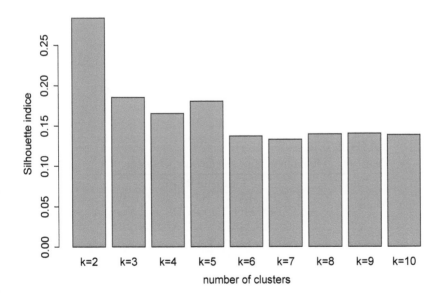

FIGURE 6.1
Tree produced by the divisive clustering algorithm on the Crimes dataset.

FIGURE 6.2
Values of the Silhouette Index for partitions in $k = 2$ to 10 clusters.

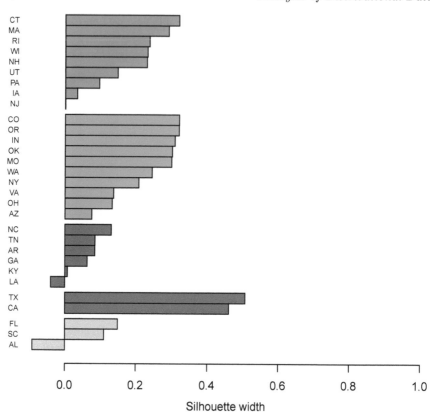

FIGURE 6.3
Values of the Silhouette Index for the partition in $k = 5$ clusters.

TABLE 6.2
Partition in five clusters of the Crimes dataset.

Cluster	Composition
C1	CT IA MA NH NJ PA RI UT WI
C2	AZ CO IN MO NY OH OK OR VA WA
C3	AR GA KY LA NC TN
C4	CA TX
C5	AL FL SC

We notice that cluster C1 gathers states with low values of divorce rate, cluster C2 states where divorce is more frequent, but violent crime is not high; in cluster C3 both divorce and violent crimes are more frequent, but low education is not very frequent; cluster C4, formed just by California and Texas, is

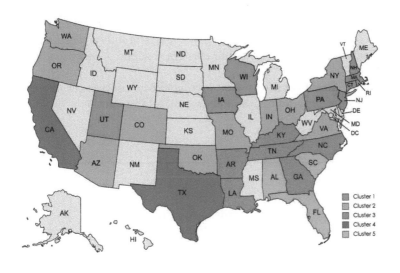

Created with mapchart.net

FIGURE 6.4
Partition of the analyzed states in five clusters.

characterized by high rates of divorce and low education, but somehow intermediate values of violent crimes, whereas cluster C5, comprehending Florida, South Carolina and Alabama, features high values of all variables.

The R code allowing to apply the method presented here, together with an explicative vignette, may be found at `https://github.com/chavent/divdiss`.

6.4 Conclusion

In this chapter we have presented a divisive clustering method for histogram-valued data; since intervals are special cases of histograms with just one class, it allows treating interval-valued data within the same framework. The method uses within-cluster sum of squares of distances as quality criterion, to decide on the split at each step of the algorithm. In that sense it bears some similarity with Ward hierarchical clustering, which uses the same type of quality measure. However, divisive clustering produces clusters which are directly interpretable, since they are by construction described by conjunctions of properties on the descriptive variables (i.e., they are monothetic clusters). Therefore, and as compares with Ward clustering, it gains in interpretability, even if it may sometimes lose a bit in within clusters cohesion / between-clusters separation.

Further developments of the method presented here concern the extension to take into account both numerical and categorical distributional variables.

Bibliography

[1] D.L. Boley. Principal direction divisive partitioning. *Data Mining and Knowledge Discovery*, 2(4):325–344, 1998.

[2] L. Breiman, J.H. Friedman, R.A. Olshen, and C.J. Stone. *Classification and Regression Trees*. Wadsworth, Belmont, CA, 1984.

[3] P. Brito and M. Chavent. Divisive monothetic clustering for interval and histogram-valued data. In *Proc. ICPRAM 2012-1st International Conference on Pattern Recognition Applications and Methods*, pages 229–234, 2012.

[4] M. Chavent. A monothetic clustering method. *Pattern Recognition Letters*, 19(11):989–996, 1998.

[5] M. Chavent, Y. Lechevallier, and O. Briant. DIVCLUS-T: A monothetic divisive hierarchical clustering method. *Computational Statistics and Data Analysis*, 52(2):687–701, 2007.

[6] Y. Chen and L. Billard. A study of divisive clustering with Hausdorff distances for interval data. *Pattern Recognition*, 96:106969, 2019.

[7] A. Ciampi. Classification and discrimination: the RECPAM approach. In R. Dutter and W. Grossmann, editors, *Proc. COMPSTAT'94*, pages 129–147, Heidelberg, 1994. Physica Verlag.

[8] I.S. Dhillon, S. Mallela, and R. Kumar. A divisive information-theoretic feature clustering algorithm for text classification. *Journal of Machine Learning Research*, (3):1265–1287, 2003.

[9] H. Fang and Y. Saad. Farthest centroids divisive clustering. In *Proc. ICMLA*, pages 232–238, 2008.

[10] K.C. Gowda and G. Krishna. Disaggregative clustering using the concept of mutual nearest neighborhood. *IEEE Trans. SMC*, 8:888–895, 1978.

[11] D. Hofmeyr and N. Pavlidis. Maximum clusterability divisive clustering. In *2015 IEEE Symposium Series on Computational Intelligence*, pages 780–786. IEEE, 2015.

[12] A. Irpino. *HistDAWass: Histogram-Valued Data Analysis, R package, version 1.0.6*, 2021. https://cran.r-project.org/web/packages/HistDAWass/index.html.

[13] A. Irpino and R. Verde. A new Wasserstein based distance for the hierarchical clustering of histogram symbolic data. In V. Batagelj et al., editors, *Proc. IFCS 2006*, pages 185–192, Heidelberg, 2006. Springer.

[14] L. Kaufman and P.J. Rousseeuw. *Finding Groups in Data*. Wiley, New York, 1990.

[15] G.N. Lance and W.T. Williams. Note on a new information statistic classification program. *The Computer Journal*, 11:195–197, 1968.

[16] P. MacNaughton-Smith. Dissimilarity analysis: A new technique of hierarchical subdivision. *Nature*, 202:1034–1035, 1964.

[17] J.R. Quinlan. Induction of decision trees. *Machine Learning*, 1:81–106, 1986.

[18] P.J. Rousseeuw. Silhouettes: a graphical aid to the interpretation and validation of cluster analysis. *Journal of Computational and Applied Mathematics*, 20:53–65, 1987.

[19] M. Roux. A comparative study of divisive and agglomerative hierarchical clustering algorithms. *Journal of Classification*, 35(2):345–366, 2018.

[20] A. Sharma, Y. López, and T. Tsunoda. Divisive hierarchical maximum likelihood clustering. *BMC Bioinformatics*, 18(16):139–147, 2017.

[21] P.H. Sneath and R.R. Sokal. *Numerical Taxonomy*. Freeman, San Francisco, 1973.

[22] W.T. Williams and J.M. Lambert. Multivariate methods in plant ecology. *J. Ecology*, 47:83–101, 1959.

7

Clustering of Modal Valued Data

Vladimir Batagelj

Institute of Mathematics, Physics and Mechanics, Ljubljana, Slovenia
University of Primorska, Andrej Marušič Institute, Koper, Slovenia
National Research University Higher School of Economics, Moscow, Russia

Simona Korenjak-Černe

University of Ljubljana, School of Economics and Business, Ljubljana, Slovenia
Institute of Mathematics, Physics and Mechanics, Ljubljana, Slovenia

Nataša Kejžar

University of Ljubljana, Faculty of Medicine, Institute for Biostatistics and Medical Informatics, Ljubljana, Slovenia

CONTENTS

In dealing with big data we often reduce their complexity by aggregating observations over subsets of units. In the traditional (classical) approach each variable is represented by its mean value over the subset. In a symbolic data

DOI: 10.1201/9781315370545-7

analysis approach more information about the values of each variable over the subset is preserved by summarizing the values using a selected symbolic description such as multi-valued, interval-valued, modal multi-valued, and modal interval-valued [4,10]. In this chapter, we limit our discussion to the representation of values by the corresponding distribution.

The partition of the set of units into subsets can be given – determined by some standard classification such as place, time intervals, type of unit, etc. Otherwise, we can determine the partition using some clustering method on the original set of units.

The clustering approach presented in this chapter can be used for symbolic data that are represented as symbolic modal values with empirical probabilities or frequencies (probability/frequency distribution) over the corresponding set of values. Modal valued symbolic variables can be further categorized into *multi-valued*, which values are defined over a finite set of categories, and *histogram-valued*, which values are defined over a finite set of intervals [4,10]. Values of traditional nominal variables having a single value on each unit can also be represented in this way – with a distribution assigning probability 1 to the subset containing a given value. In general, we can transform a variable measured in any traditional measurement scale type (nominal, ordinal, numerical) into a nominal variable by partitioning its range (set of possible values) into a smaller number of subsets – categories. In this case, some information is lost (i.e., ordinality). But on the other hand, more variability is preserved and such a description enables us to consider together variables of all types. If we partition numerical variables into histogram-valued set of intervals, they can be aggregated into histograms and thus preserve ordinality, however in the following clustering approach we treat all variable partitionings as non-ordered categories.

An additional advantage of our approach is that we can preserve in the description of modal symbolic values also the size of the subset of units from which the symbolic value was aggregated, e.g. when countries are symbolic units and their population is observed, the size of the population contributes weights to the final result.

For clustering modal symbolic objects we adapted the leaders' method [1, 9,13] and the agglomerative hierarchical clustering method [31]. The chaining of both methods enables fine-tuned clustering of very large data sets.

Using an appropriate setting of weights the clustering criterion function used in the clustering process can be adapted in a way that each cluster representative is also composed of distributions of variables' values in the whole cluster.

Both methods are implemented as an R package called **clamix** [3]. The package contains also data used in the section Application.

7.1   Distributional symbolic data

Modal valued symbolic data are getting more and more common. Computerization of most people's activities provides us with large (huge) data sets. To get this kind of data into a more manageable form an aggregation of the raw data is usually done. If we want to preserve information about the variability and/or uncertainty inherited from the raw data, variables have to be represented with more than a single value. Such aggregation is very common in data provided by official institutions where due to privacy issues the access and use of raw data is usually not at researchers' disposal.

In this chapter, a unit $S_i \in \mathbf{S}$ (a case, a description of a symbolic object s_i) in a data set of n units $\mathbf{S}$ is a list of values of modal valued variables Y_j, $j = 1, \ldots, p$. A symbolic variable Y_j has m_j categories (subsets). Each variable is therefore described by a list of values for corresponding categories:

$$Y_j = \left[L_{j1} : y_{j1}, \ldots, L_{jm_j} : y_{jm_j} \right].$$

where $L_{j\ell}$ is a category label and $y_{j\ell}$ is its value. These values can be counts/frequencies, probabilities, amounts of money, areas, or other. A description of symbolic object s_i is a list of modal values $S_i = [\mathbf{y}_{i1}, \ldots, \mathbf{y}_{ip}]$ where $\mathbf{y}_{ij} = \left[L_{j1} : y_{ij1}, \ldots, L_{jm_j} : y_{ijm_j} \right]$ is the modal value corresponding to symbolic variable Y_j for symbolic object s_i. Since for a symbolic variable Y_j the categories are fixed we usually fix their order and omit them from the modal value list – the category is determined by the value's position $\mathbf{y}_{ij} = \left[y_{ij1}, \ldots, y_{ijm_j} \right]$.

When the original data include a variable with a single value for each observed unit but with a large number of possible values we can transform it into a nominal (categorical) variable by partitioning its range into a small number of subsets. If the partition is not predefined, the range of values can be divided by using one of the standard statistical approaches – for example, for an ordinal variable, based on the quantiles of the distribution of a variable over all units. An introduction to discretization methods is given in [32] and for a survey on this topic see [11].

When the raw data for a variable contains missing values the corresponding symbolic variable can be extended with an additional category NA – not available. The use of (multiple) imputation methods is an alternative option.

7.1.1   Example: Tourism

If we were interested in studying data about tourism in Slovenian regions, we can find for each Slovenian commune the data on accommodation facilities, tourist arrivals, and overnight stays by groups of tourist accommodations at the web page of the Slovenian Statistical Office [28]. They have as values the following categories which are used as subsets in symbolic variables' descriptions:

Symbolic variable	Categories
Y_{rooms} *number of rooms by accommodation type* :	hotels and similar accommodation camping sites other accommodation facilities
Y_{arrivals} *tourist arrivals* :	domestic foreign
$Y_{\text{overnight stays}}$ *tourist overnight stays* :	domestic foreign

In April 2015 there were 3,222 rooms in hotels and similar accommodation facilities, 1,000 camping sites, and 2,019 other accommodation facilities in Gorenjska region. They had 7,465 arrivals of domestic tourists and 34,345 of foreign tourists, and there were 15,281 overnight stays of domestic and 62,897 of foreign tourists, therefore this region can be described with the symbolic description:

$$S_{\text{Gorenjska}} = \underbrace{[[3222, 1000, 2019]}_{Y_{\text{rooms}}}, \underbrace{[7465, 34345]}_{Y_{\text{arrivals}}}, \underbrace{[15281, 62897]]}_{Y_{\text{overnight stays}}}$$

The extended form of $\mathbf{y}_{\text{Gorenjska,rooms}}$ is

$$[\text{hotels} : 3222, \text{camping} : 1000, \text{other} : 2019]$$

7.1.2   Example: Structure of population

Population data are usually collected in structures, such as is the age-sex distribution of the population of a particular country or region at a particular point in time. Symbolic data descriptions allow us to preserve such structures and enable us to use these richer descriptions also in further analysis. For instance, if we were also interested in the age of the tourists in the previous example, we would use age-groups instead of one of the central values (i.e., average, median, or mode) for this variable which would preserve more variability of the original data. Similarly, as in the previous example, we could represent regions' population with age-sex structures with two symbolic variables:

Symbolic variable	Categories
Y_{men} *number of men by economic age-groups* :	0–14 years 15–64 years 65 + *years*
Y_{women} *number of women by economic age-groups* :	0–14 years 15–64 years 65 + *years*

and describe the age-sex distribution of the population of the Gorenjska region on 1 January 2011 as:

$$S_{\text{Gorenjska}} = \underbrace{[[16052, 70751, 13678],}_{Y_{\text{men}}} \underbrace{[14981, 67856, 20109]]}_{Y_{\text{women}}}$$

Of course, there are many examples in other fields where such representations better describe the collected data than the traditional data descriptions. For example self-monitoring through Quantified Self movement where, instead of analyzing all the saved heart rate values that exhibit quite a lot of random variation, a 5-minute heart rate values/intervals in the jogging course is a better choice.

7.1.3   Example: Ego-centered networks and TIMSS

Symbolic data descriptions enable us to describe also more complex data. In socio-economic studies so-called ego-centered (or personal) networks are often used. They consist of main units – *egos* (E) and additional units – *alters* (A), that are linked with the corresponding ego and sometimes also among them. The relation between an ego and its alters is illustrated in Figure 7.1.

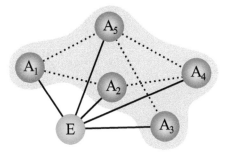

FIGURE 7.1
Ego and its alters.

Such networks combine two related classical data sets: egos data set and alters data set with the same or with some or all different observed variables. For example, if we want to study teaching approaches based on questionnaires for teachers and also for their students, teachers can be treated as egos and their students as alters. This approach was used for clustering TIMSS (Trends in International Mathematics and Science Study) data in [26]. In this approach, each teacher is described by frequency distributions over teachers' and students' variables. For example, the teacher with identification code 18, who is 45 years old, male, and who discussed concepts with his teaching colleagues daily or almost daily, can be presented with the following classical description – a list of values (18, 45 years, male, (almost) daily). His students' answers to three questions about (a) gender, (b) whether they enjoy learning math, and (c) whether they use calculators in math classes, can be represented as frequency counts:

Symbolic variable	Categories	Frequencies
A_{sex}	female	15
sex :	male	10
A_{enjoy}	agree a lot	2
enjoy learning math :	agree	7
	disagree	12
	disagree a lot	4
A_{usage}	every or almost every lesson	8
use calculators :	about half lessons	6
	some lessons	11
	never	0

If we use the following partition/categories for the ranges of teachers' variables

Symbolic variable	Categories
E_{age}	under 25
age :	25–29
	30–39
	40–49
	50–59
	60 or older
E_{sex}	female
sex :	male
E_{interact}	(almost) never
interactions teachers/discuss concepts :	2–3 times per month
	1–3 times per week
	(almost) daily

the teacher with id 18 can be represented as a modal-valued symbolic description with

$$S_{18} = \underbrace{[[0,0,0,1,0,0]}_{E_{\text{age}}}, \underbrace{[0,1]}_{E_{\text{sex}}}, \underbrace{[0,0,0,1]}_{E_{\text{interact}}}, \underbrace{[15,10]}_{A_{\text{sex}}}, \underbrace{[2,7,12,4]}_{A_{\text{enjoy}}}, \underbrace{[8,6,11,0]]}_{A_{\text{usage}}}$$

The modal values can be visually presented as a histogram (on intervals) or as a bar plot (on categories). For example, the teacher 18 could be represented with one histogram (for variable E_{age}, based on the numerical variable *age*) and five bar plots, where two are based on the originally nominal variable *sex* and three on the originally ordinal variables *interact*, *enjoy*, and *usage*. See Figure 7.2 for two variables. Note the different scales on ordinate axes.

Such a representation is particularly useful when variables are of different types (numeric, ordinal, nominal) as in the case with TIMSS data sets, because it enables the use of variables of all types simultaneously.

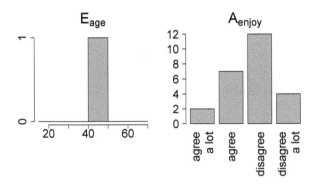

FIGURE 7.2
Histogram and bar plot for the variables E_{age} and A_{enjoy}.

7.2 Clustering modal-valued symbolic data

Cluster analysis as an exploratory data analysis tool can be very useful in detecting possible patterns in very large data sets, especially if it uses richer data descriptions that preserve more information from the original data.

We present clustering algorithms using the general leaders' method [1,9,13] and agglomerative hierarchical clustering method [31] adapted for the previously presented data descriptions. The criterion functions for both clustering procedures are the same. Therefore they can be used in combination, for example,

- to use leaders' method to reduce the size of the set of units (get a (still large) number of leader representatives);

- and further, use the agglomerative clustering method on the leader representatives.

Special attention is given to the clustering criterion function where weights are added to enable the preservation of the frequency distribution of variables' values in each cluster. The weights corresponding to each variable separately account for the cases where different variables are based on a different number of original units (as is for example the number of men and women in an observed region when observing the age-sex distribution of a population). The inclusion of the basic dissimilarity δ in the definition of dissimilarity measure (see (7.2)) allows a tunning of the clustering procedure to obtain more homogenous final clusters (e.g., the squared Euclidean distance for probability distributions favors components with the largest values – the cluster representatives are biased [20]). Some examples of basic dissimilarities δ can be found in Table 7.1, second column. For details on basic dissimilarities see the paper [19]. In our application in this chapter we will use the basic dissimilarity δ_1.

7.2.1 Overview of the current literature on clustering modal-valued symbolic data

Many clustering approaches have been developed for clustering sets of symbolic objects. Those that are dealing with modal-valued symbolic data can be found in the general SDA literature [4, 6, 10] and some earlier articles [7, 12, 14, 17, 24, 25, 30]. The methodology for clustering symbolic data is collected in the recently published book from Billard and Diday [5]. Here, only some of them are shortly reviewed. The paper of Irpino et al. [18] and Chapter 5 of this book describe the dynamic clustering approach for interval-valued modal variables based on the L_2 Wasserstein distance. The weights of how relevant is a variable for the clustering can be automatically computed. Carvalho et al. [8] propose dynamic clustering for any type of symbolic data. Kim & Billard [21] propose an extension of the Ichino-Yaguchi dissimilarity measure to interval-valued modal data and in the paper one year later [22] the Gowda-Diday and Ichino-Yaguchi extensions to all modal valued data. They use a divisive clustering algorithm and propose two cluster validity indices to improve the decision about the optimal number of final clusters. Some generalizations of dissimilarity measures are proposed in the paper of Kim & Billard (2013) [23].

7.2.2 Clustering problem and algorithms

A non-empty subset C of units, $\emptyset \subset C \subseteq \mathbf{S}$, is called a *cluster*. A set of clusters $P = \{C_1, C_2, \ldots, C_k\}$ is called a *clustering*. A clustering P is a *partition* iff each unit of $\mathbf{S}$ belongs to exactly one of its clusters. We denote by Π_k the set of all partitions of the set of units $\mathbf{S}$ into k clusters.

For a modal value $\mathbf{y}_j$ we define its *size* or *total* value as

$$t_j = \sum_{\ell=1}^{m_j} y_{j\ell}.$$

If $y_{j\ell}$ are frequencies we get the probability distribution with $\mathbf{p}_j = \mathbf{y}_j/t_j$. In the following exposition (and in the R package **clamix**) the modal value $\mathbf{y}_j$ is represented by a couple $(t_j, \mathbf{p}_j)$.

This representation of modal values allows us to compute the couples $(t_{(uv)j}, \mathbf{P}_{(uv)j})$ for a cluster $C_{uv} = C_u \cup C_v$ from couples for two disjoint clusters C_u and C_v. It holds

$$t_{(uv)j} = t_{uj} + t_{vj}$$

$$\mathbf{y}_{(uv)j} = \mathbf{y}_{uj} + \mathbf{y}_{vj} = t_{uj}\mathbf{p}_{uj} + t_{vj}\mathbf{p}_{vj}$$

$$\mathbf{p}_{(uv)j} = \frac{\mathbf{y}_{(uv)j}}{t_{(uv)j}}$$

where $\mathbf{p}_{(uv)j}$ denotes the relative distribution of the variable Y_j in the joint cluster, $\mathbf{y}_{(uv)j}$ the frequency distribution of variable Y_j in the joint cluster and $t_{(uv)j}$ the weight (count of values) for that variable in the joint cluster.

We approach the *clustering problem* as an optimization (minimization) problem $(\varPhi, \mathrm{Err}, \mathrm{Min})$ over the *set of feasible clusterings* $\varPhi$ for the *criterion function* Err. We assume that the criterion function has the following form

$$\mathrm{Err}(P) = \sum_{C \in P} \mathrm{err}(C).$$

The *total error* $\mathrm{Err}(P)$ of the clustering P is the sum of *cluster errors* $\mathrm{err}(C)$ of its clusters $C \in P$. The set of feasible clusterings is the set of all partitions of the set of units into k clusters, $\varPhi = \varPi_k$, where k is a selected number.

Cluster error might be expressed in many different ways. Here we assume that $\mathrm{err}(C)$ is a sum of dissimilarities of its units from the *cluster representative* R. For a given representative R and a cluster C we therefore define:

$$\mathrm{err}(C, R) = \sum_{S \in C} d(S, R),$$

where d is a selected dissimilarity measure. The best representative R_C is called a *leader*

$$R_C = \arg \min_R \mathrm{err}(C, R). \tag{7.1}$$

We assume that such a representative exists and is unique.

We can finally define

$$\mathrm{err}(C) = \mathrm{err}(C, R_C) = \min_R \sum_{S \in C} d(S, R).$$

To construct a dissimilarity measure d we shall also assume that the representatives R have the same form of description as the symbolic objects – i.e., they are represented with the list of non-negative vectors of size m_j for each variable Y_j

$$\mathbf{r}_j = \left[r_{j1}, \dots, r_{jm_j} \right].$$

We denote the set of all possible representatives by $\mathcal{R}$.

We assume the additive model also for the dissimilarity measure between a symbolic object S and a representative R

$$d(S, R) = \sum_{j=1}^{p} \alpha_j d_j(S, R), \quad \alpha_j \geq 0, \quad \sum_{j=1}^{p} \alpha_j = 1,$$

and for the variable Y_j contribution d_j to the dissimilarity d

$$d_j(S, R) = \sum_{\ell=1}^{m_j} w_{Sj\ell} \cdot \delta(p_{Sj\ell}, r_{j\ell}), \quad w_{Sj\ell} \geq 0 \tag{7.2}$$

where α_j are weights for variables. They can be used to specify the importance of variables. Their "default" values are $\alpha_j = 1/p$. $w_{Sj\ell}$ are weights for each variable's component. Weights can differ for each unit, each variable Y_j, and each of its components. Dissimilarity δ is called a *basic dissimilarity*. Some examples of basic dissimilarities are presented in the second column of Table 7.1. For the basic dissimilarity δ_1 we get a kind of generalized squared Euclidean dissimilarity. Note that the basic dissimilarities are intended to work on probabilities **p**. The information about the size of the symbolic object can be considered in the weights w.

For example, in our illustration in Section 7.1, the weights are not equal for all variables: for variables constructed from teacher data they are $w = 1$, and for those constructed from students' data $w =$ the number of teacher's students that answered the questionnaire.

As another example, in clustering data from the world's countries' population pyramids [27] two symbolic variables (based on population counts for each gender) for each country s_i were considered. In this case, the weight w_{i1} is the number of all men and w_{i2} is the number of all women in the country.

Using weights the information about the variable's value distribution and their size can be retained throughout the clustering process.

A distribution $\mathbf{y}_j$ does not necessarily count the raw data elements. For example, if a shop collects all information about the purchases of customers (via their loyalty cards) the shop analytics could be done about the money spent for different types of items bought (variable categories). For a customer s_i, w_i would be the money that he/she spent in a given time period, and $\mathbf{y}$ the distribution of the money spent on bought items.

For solving the clustering optimization problem for modal symbolic objects we adapted two standard clustering methods: the agglomerative hierarchical method and the leaders' method.

7.2.3 Leaders' method

The leaders' method is a generalization of the k-means algorithm [1, 9, 13]. It iteratively calculates the best clustering into a given number, k, of clusters. In each step, the best cluster representatives (leaders) are determined for each cluster in the clustering from the previous step. Then the new clustering is determined by assigning each unit to the nearest leader. The process stops when the result does not change anymore.

Algorithm 1 Leaders' method

---

determine an initial clustering P_0; $P := P_0$;
repeat
    for each $C \in P$: determine its leader R_C;
    assign each unit to the nearest new leader – producing a new clustering P
    until the leaders stabilize.

TABLE 7.1
The basic dissimilarities and the corresponding cluster leader, the leader of the merged clusters, and the dissimilarity between merged clusters. Indices j and ℓ are omitted.

$\delta(\mathbf{ps}, \mathbf{r})$	$\mathbf{r}^*_{\mathbf{C}}$	$\mathbf{z}$	$\mathbf{D(C_u, C_v)}$
δ_1 $(ps - r)^2$	$\dfrac{P_C}{w_C}$	$\dfrac{w_u u + w_v v}{w_u + w_v}$	$\dfrac{w_u \cdot w_v}{w_u + w_v}(u - v)^2$
δ_2 $\left(\dfrac{ps-r}{r}\right)^2$	$\dfrac{Q_C}{P_C}$	$\dfrac{uP_u + vP_v}{P_u + P_v}$	$\dfrac{P_u}{u}\left(\dfrac{u-z}{z}\right)^2 + \dfrac{P_v}{v}\left(\dfrac{v-z}{z}\right)^2$
δ_3 $\dfrac{(ps-r)^2}{r}$	$\sqrt{\dfrac{Q_C}{w_C}}$	$\sqrt{\dfrac{u^2 w_u + v^2 w_v}{w_u + w_v}}$	$w_u \dfrac{(u-z)^2}{z} + w_v \dfrac{(v-z)^2}{z}$
δ_4 $\left(\dfrac{ps-r}{ps}\right)^2$	$\dfrac{H_C}{G_C}$	$\dfrac{H_u + H_v}{\dfrac{H_u}{u} + \dfrac{H_v}{v}}$	$G_u(u - z)^2 + G_v(v - z)^2$
δ_5 $\dfrac{(ps-r)^2}{ps}$	$\dfrac{w_C}{H_C}$	$\dfrac{w_u + w_v}{H_u + H_v}$	$w_u \dfrac{(u-z)^2}{u} + w_v \dfrac{(v-z)^2}{v}$
δ_6 $\dfrac{(ps-r)^2}{ps\,r}$	$\sqrt{\dfrac{P_C}{H_C}}$	$\sqrt{\dfrac{P_u + P_v}{\dfrac{P_u}{u^2} + \dfrac{P_v}{v^2}}}$	$\dfrac{P_u}{u}\dfrac{(u-z)^2}{uz} + \dfrac{P_v}{v}\dfrac{(v-z)^2}{vz}$

$$w_C = \sum_{S \in C} w_S$$

$$P_C = \sum_{S \in C} w_S ps$$

$$Q_C = \sum_{S \in C} w_S ps^2$$

$$H_C = \sum_{S \in C} \frac{w_S}{ps}$$

$$G_C = \sum_{S \in C} \frac{w_S}{ps^2}$$

The basic scheme of the leaders' method is simple – see Algorithm 1. In the elaboration of the algorithm we have to answer to the following questions:

- how to determine the leader R_C of a cluster C;

- how to determine the new clustering given the new leaders.

For each basic dissimilarity δ, the corresponding optimal leader is given in Table 7.1, column three. The derivations of these results, with some options for solving the problem in the case of zero denominators, can be found in [19].

7.2.4 Agglomerative hierarchical method

The agglomerative hierarchical method is based on a step-by-step merging of the two closest clusters. The clustering procedure has $n - 1$ steps. At the beginning, every unit forms its own cluster. In each of the following steps, the two closest clusters are joint.

Algorithm 2 Agglomerative hierarchical clustering procedure

---

each unit forms a cluster: $P_n = \{\{S\} : S \in \mathbf{S}\}$;
they are at level 0: $h(\{S\}) = 0, S \in \mathbf{S}$;
for $k = n - 1$ **to** 1 **do**
    determine the closest pair of clusters
        $(u, v) = \arg\min_{p,q:\ p \neq q}\{D(C_p, C_q): C_p, C_q \in P_{k+1}\}$;
    join the closest pair of clusters $C_{(uv)} = C_u \cup C_v$
        $P_k = (P_{k+1} \setminus \{C_u, C_v\}) \cup \{C_{(uv)}\}$;
        $h(C_{(uv)}) = D(C_u, C_v)$
        determine the dissimilarities $D(C_{(uv)}, C_s), C_s \in P_k$
endfor

The standard agglomerative hierarchical clustering procedure can be described with Algorithm 2. P_k is a partition of the finite set of units $\mathbf{S}$ into k clusters. The function $h(C)$ is a *level* function of cluster C.

In the elaboration of this procedure for modal symbolic objects, we have to specify the details of

- the computation of dissimilarities between the new (joint) cluster and the remaining clusters.

Dissimilarity between clusters

We express the dissimilarity between clusters in terms of their leaders and the selected basic dissimilarity. To obtain the compatibility with the adapted leaders' method, we define the dissimilarity between disjoint clusters C_u and C_v, $C_u \cap C_v = \emptyset$, as [2]

$$D(C_u, C_v) = \text{err}(C_u \cup C_v) - \text{err}(C_u) - \text{err}(C_v).$$

The derivations can be found in [19] and the results for selected basic dissimilarities are given in the fifth column of Table 7.1. The fourth column gives the expressions for $\mathbf{z}$ (the leader of joint cluster $C_u \cup C_v$).

Note that for δ_1 we get

$$D(C_u, C_v) = \sum_j \alpha_j \sum_\ell \frac{w_{uj\ell} \cdot w_{vj\ell}}{w_{uj\ell} + w_{vj\ell}}(u_{j\ell} - v_{j\ell})^2$$

a *generalized Ward's relation*. This relation holds also for singletons $C_u = \{S\}$ or $C_v = \{T\}$, $S, T \in \mathbf{S}$.

For the basic dissimilarity δ_1 (squared Euclidean distance) and assuming that the weights are constant inside a modal value, $w_{ij\ell} = w_{ij}$, $\ell = 1, \ldots, m_j$ we can prove that [19]:

- the leaders' components $\mathbf{r}_j^*$ are also probability distributions;
- setting the weights $w_{ij\ell} = t_{ij}$ we get $\mathbf{r}_j^* = \mathbf{p}_j$ – the leader of each cluster is composed by the distributions of variables' values over the cluster.

7.2.5 Huygens Theorem for δ_1

Huygens theorem has a very important role in many fields. In physics, it is about the moment of inertia. In statistics, it is related to the decomposition of the sum of squares, on which the analysis of variance is based. In clustering, it is used for "connecting" clustering criteria. It has the form

$$TI = WI + BI, \tag{7.3}$$

where, using physics terms, TI is the *total inertia*, WI is the *inertia within clusters* and BI is the *inertia between clusters*.

Let $R_\mathbf{S}$ denote the leader of the cluster consisting of all units $\mathbf{S}$. Then we define [2]

$$
\begin{aligned}
TI &= \sum_{S \in \mathbf{S}} d(S, R_\mathbf{S}) \\
WI &= \mathrm{Err}(P) = \sum_{C \in P} \sum_{S \in C} d(S, R_C) \\
BI &= \sum_{C \in P} d(R_C, R_\mathbf{S})
\end{aligned}
$$

For a selected dissimilarity d and a given set of units $\mathbf{S}$ the value of total inertia TI is fixed. Therefore, if Huygens theorem holds, the minimization of the within inertia $WI = \mathrm{Err}(P)$ is equivalent to the maximization of the between inertia BI. For basic dissimilarity δ_1, the equality (7.3) is easily verifiable. For the others, however, the ratio WI/TI can be used as a measure of the normalized "tension" of partitioning the initial set $\mathbf{S}$ into more than a single cluster.

7.3 Interpretation of results

The interpretation of the obtained results is one of the important issues in data analysis in general and in clustering in particular. As mentioned earlier, an important advantage of our adapted clustering methods is that when using the basic dissimilarity δ_1 and setting the weights $w_{ij\ell} = t_{ij}$ the obtained clusters' leaders are also described with distributions. This option was used in the clustering of the world's countries in [27] and is a recommended option for clustering of modal valued data with known sizes of original raw data subsets. We still have no good answer to the question: which symbolic variables and corresponding components are characteristic for a given cluster?

For detecting a cluster's characteristics we mainly use the following steps:

- identification of the maximal and/or minimal value(s) of selected symbolic variable(s);

- comparisons of the clusters' descriptions across clusters;

- comparisons of the clusters' descriptions of the obtained clusters with the description of the whole data set.

For the basic dissimilarity δ_1, a possible approach to the comparison of descriptions of the obtained clusters with the description of the whole data set is to define an index, which we call *specificity*, for each symbolic variable Y_j and for each cluster C as

$$\mathrm{spec}(Y_j, C) = \frac{1}{2} \sum_{\ell=1}^{m_j} |r_{\mathbf{S}j\ell} - r_{Cj\ell}|,$$

where $r_{\mathbf{S}j\ell}$ is the ℓ-th component of the symbolic variable Y_j for the leader of the whole set of units $\mathbf{S}$, and $r_{Cj\ell}$ is the ℓ-th component of the symbolic variable Y_j for the leader of cluster C. The leader of the whole set of units is defined as the best representative of the initial data set, see Equation (7.1). Geometrically $\mathrm{spec}(Y_j, C)$ is the half area of the symmetric difference of the areas below the distribution of values of Y_j on the set of units $\mathbf{S}$ and the distribution of values of Y_j on the cluster C.

The specificity index $\mathrm{spec}(Y_j, C)$ has the following properties:

- $0 \leq \mathrm{spec}(Y_j, C) \leq 1$; this can be seen from
$$\mathrm{spec}(Y_j, C) = \frac{1}{2} \sum_{\ell=1}^{m_j} |r_{\mathbf{S}j\ell} - r_{Cj\ell}| \leq \frac{1}{2} \sum_{\ell=1}^{m_j} (r_{\mathbf{S}j\ell} + r_{Cj\ell}) = 1$$

- $\mathrm{spec}(Y_j, C) = 0$ iff $\mathbf{r}_{Cj} = \mathbf{r}_{\mathbf{S}j}$;

- $\mathrm{spec}(Y_j, C) = 1$ iff $\mathbf{r}_{Cj} \cdot \mathbf{r}_{\mathbf{S}j} = 0$.

Note that $\mathbf{r}_{Cj} \cdot \mathbf{r}_{\mathbf{S}j} = 0$ is not possible since $r_{\mathbf{S}j\ell} = 0 \Rightarrow r_{Cj\ell} = 0$. It may happen, however, for the dissimilarity Δ between clusters C_1 and C_2 for symbolic variable Y_j, defined as

$$\Delta(C_1, C_2; Y_j) = \frac{1}{2} \sum_{\ell=1}^{m_j} |r_{C_1 j\ell} - r_{C_2 j\ell}|.$$

Note that $\mathrm{spec}(Y_j, C) = \Delta(\mathbf{S}, C; Y_j)$. Δ can be used for comparing clusters.

To identify the most characteristic components ℓ of the symbolic variable Y_j on the cluster C we compute the *contrast* indices

$$c_{j\ell} = \begin{cases} \frac{r_{Cj\ell}}{r_{Sj\ell}} & \text{if } r_{Cj\ell} \geq r_{Sj\ell} \\ -\frac{r_{Sj\ell}}{r_{Cj\ell}} & \text{otherwise} \end{cases}, \quad \ell = 1, \ldots, m_j$$

and select among them those with the highest absolute values.

By definition $|c_{j\ell}| \geq 1$. A value of $|c_{j\ell}|$ close to 1 means that this component has on cluster C almost the same probability as on the whole set $\mathbf{S}$. There are two special cases:

- $c_{j\ell} = -\infty$, this corresponds to $r_{Cj\ell} = 0$ and $r_{Sj\ell} > 0$. The ℓ-th category of variable Y_j does not appear as a value in data in cluster C.

- $c_{j\ell} = \text{NaN}$ (undefined), this corresponds to $r_{Cj\ell} = r_{Sj\ell} = 0$. The ℓ-th category of variable Y_j does not appear as a value in the data set.

Another approach would be to apply the ideas from the generalized ANOVA procedure available in the R-package TraMineR [29].

7.4 Application

A data set used as an example in the package **clamix** [3] is a subset from a data set about teaching and learning mathematics (based on a few selected teachers' and students' answers) from the TIMSS Advanced 2008 study [16] (data from [15]) described in Subsection 7.1.3. The study itself explores student achievement in advanced mathematics (and physics) in the final year of secondary school across countries. There were 10 countries included in the study, from teacher answers 147 variables were produced and from student answers, additional 77 variables were obtained.

In this example, we concentrate on Slovenian teachers (egos) and their students (alters). There were 94 Slovenian teachers included in the data set and 2,156 students. Because of missing values on all questions, one teacher (with 16 students) and 20 additional students were excluded from the data set. Therefore 93 teachers with 2,120 students were finally considered.

For the purpose of our example analysis a research question that we would like to answer is: do the seniority of a teacher, the amount of his/her discussion with other colleagues, the gender of students and the work on students' homework assignments characterize the learning of mathematics? Therefore the answers to the following questions were of our primary interest:

E_1 : **seniority**: years teaching (Q3A) – numerical variable MT2GTAUT;

E_2 : **professional interaction with coworkers**: discussion of how to teach a particular concept (Q9A) – ordinal variable MT2GOTDC;

A_1 : **sex**: student's sex (Q2) – nominal variable MS2GSEX;

A_2 : **math lessons**: reviewing homework (Q15E) – ordinal variable MS2MACRH;

A_3 : **homework**: how many minutes per week (Q18A) – numerical variable MS2MHTIM.

  The statistical units are (as described in the Introduction) teachers. Five variables are used in the clustering process: two obtained from teachers' answers and three from students (taught by that teacher). Both numerical variables were categorized according to the five equidistant quantiles leading to symbolic variables E_1 and A_3. In A_3 also the category NA is considered. For example, the teacher with ID 601 whose seniority (E_1 = years of teaching) is in the interval $(21, 26.6]$, with professional interaction with coworkers 2-3 times per month (E_2), with 28 student answers, where 57 % of them are girls and 43 % are boys (A_1), where 10 % of them answered that they review homework every or almost every lesson, 36 % answered that they do this in about half of lessons, 36 % answered to do this in some lessons and 18 % claimed that they never review homework in math lessons (A_2), and where 46 % of students answered that they didn't spent for homework more than 30 minutes per week, 21 % of them spent for homework more than 30 minutes and up to 60 minutes per week, 11 % of them needed more than one hour and up to two hours, none of them needed more than two hours and up to three hours per week, 4 % of them needed more than three hours and up to ten hours, and 18 % of them didn't provide any answer (A_3), is represented as a modal-valued symbolic description with pairs:

$$
S_{601} = \quad \underbrace{[(1, [0,0,0,1,0]),}_{E_1} \quad \underbrace{(1, [0,1,0,0]),}_{E_2}
$$

$$
\underbrace{(28, [.57, .43]),}_{A_1} \quad \underbrace{(28, [.1, .36, .36, .18]),}_{A_2} \quad \underbrace{(28, [.46, .21, .11, 0, .04, 0.18])]}_{A_3}
$$

Clustering results

Because there is only a (relatively) small number of statistical units included in the data set, only hierarchical clustering was used. The dissimilarity measure used here is the squared Euclidean distance (δ_1). All variables were weighted equally ($\alpha_j = 1/5$). The obtained dendrogram is presented in Figure 7.3. The labels of units in the dendrogram are the teacher's IDs from the TIMSS data set. Three main clusters were identified. Their representatives and the

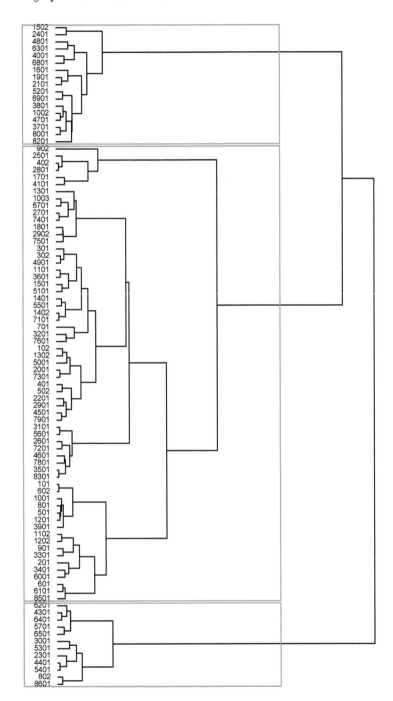

FIGURE 7.3
The dendrogram on 93 units.

representative of the whole set of units **S** are given in Table 7.2. They are visually presented as bar plots in Figures 7.4 and 7.5. Category labels' true names denoted as c_i; $i \in \{1, \ldots, m_j\}$ in Table 7.2 can be seen on the x axes in the figures.

Note the scales of ordinate axes in the bar plots. They are different according to the number of teachers/students in the cluster; but for each symbolic variable the total areas of the columns are the same for all four clusters.

TABLE 7.2

The descriptions of the leaders for the three obtained clusters C_1, C_2 and C_3, and for the whole data set **S**. Rows represent symbolic variables and columns their values for the appropriate number of categories.

	$Y \backslash \ell$	c_1	c_2	c_3	c_4	c_5	c_6
	E_1	21	23	14	16	19	
	E_2	13	50	24	6		
S	A_1	1203	917				
	A_2	319	364	940	483	14	
	A_3	442	348	379	190	327	434
	E_1	7	2	2	3	3	
	E_2	3	10	3	1		
C_1	A_1	238	132				
	A_2	178	94	68	25	5	
	A_3	47	52	69	54	100	48
	E_1	13	16	11	9	15	
	E_2	7	32	20	5		
C_2	A_1	931	573				
	A_2	126	228	770	375	5	
	A_3	330	256	264	125	212	317
	E_1	1	5	1	4	1	
	E_2	3	8	1	0		
C_3	A_1	34	212				
	A_2	15	42	102	83	4	
	A_3	65	40	46	11	15	69

To answer the research question from the beginning of the analysis we can visually observe the following: in the final clustering the first cluster includes mostly teachers with less teaching experience, however the pattern of concepts discussion among coworkers is very similar for the three clusters (there is slightly more frequent communication in the second cluster). According to students' answers there is a noticeable difference in gender distribution. Cluster 3 represents mainly male students while the other two are more gender balanced, with female students slightly prevailing. In homework handling, it can be observed that students of junior teachers (cluster 1) spend the largest amount of time doing their homework and these assignments are also most

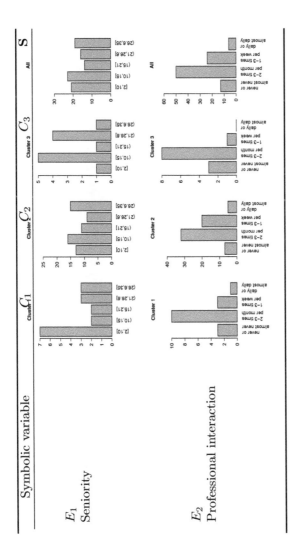

FIGURE 7.4
Barplots of the teachers symbolic variables for leaders of the three obtained clusters and on all units for TIMSS data.

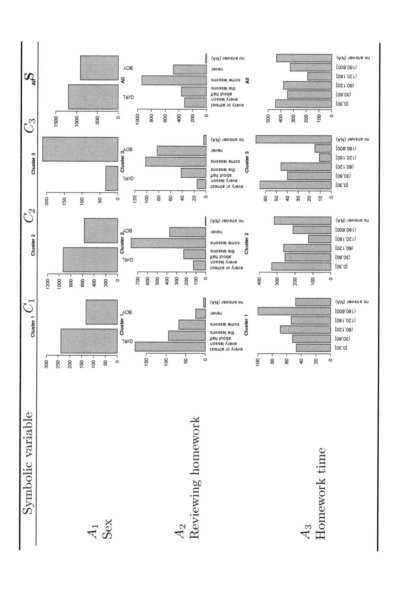

FIGURE 7.5
Barplots of the students symbolic variables for leaders of the three obtained clusters and on all units for TIMSS data.

frequently reviewed at math lessons. There are non-negligible numbers of missing answers for the last variable A_3 (homework time), especially in clusters 2 and 3.

For a more detailed analysis of the obtained clusters, we apply the approach proposed in Section 7.3. From Table 7.2, we compute the specificities of symbolic variables for all three clusters and for each category also the corresponding contrast index. These results are arranged in Table 7.3.

TABLE 7.3

Specificities and contrasts of clusters.

	Y	spec	c_1	c_2	c_3	c_4	c_5	c_6
	E_1	0.19	1.83	-2.10	-1.28	1.03	-1.16	
	E_2	0.09	1.26	1.09	-1.46	-1.10		
C_1	A_1	0.08	1.13	-1.21				
	A_2	0.42	3.20	1.48	-2.41	-3.37	2.05	
	A_3	0.18	-1.64	-1.17	1.04	1.63	1.75	-1.58
	E_1	0.05	-1.11	1.01	1.14	-1.22	1.15	
	E_2	0.07	-1.28	-1.08	1.21	1.21		
C_2	A_1	0.05	1.09	-1.14				
	A_2	0.09	-1.80	-1.13	1.16	1.09	-1.99	
	A_3	0.02	1.05	1.04	-1.02	-1.08	-1.09	1.03
	E_1	0.33	-2.71	1.69	-1.81	1.94	-2.45	
	E_2	0.24	1.79	1.24	-3.10	$-\infty$		
C_3	A_1	0.43	-4.11	1.99				
	A_2	0.12	-2.47	-1.01	-1.07	1.48	2.46	
	A_3	0.14	1.27	-1.01	1.05	-2.00	-2.53	1.37

From this table we get the following characterizations of clusters:

Cluster 2:

It is the largest of the three clusters – it contains 64 (69%) teachers and 1,504 (71%) students. The specificities of all symbolic variables in this cluster are small. This means that the cluster C_2 is similar to the whole set of units **S** – the teachers rarely discuss the concepts with their colleagues, they are reviewing math homework rarely, there are more girls than boys in the cluster.

Cluster 1:

It contains 17 (18%) teachers and 370 (17%) students. In C_1 the variable with the largest specificity (0.42) is A_2 (reviewing math homework) with category 1 (every or almost every lesson) more than 3 times more frequent than the average (i.e., in the set of all units **S**), and the category 4 (never) more than 3 times less frequent than the average. Also other A_2 values differ considerably from the average. The next largest specificity (0.19) concerns the variable E_1 (teaching years) with category 1 (2-10) almost twice more frequent than the average, and category 2 (10-15) twice less frequent than the average. Therefore we can describe the cluster C_1 as containing mostly teachers at the beginning of their career and reviewing math homework on almost every lesson.

Cluster 3:

It contains 12 (13%) teachers and 246 (12%) students. The variable with the largest specificity (0.43) is A_1 (gender of students). In C_3 the girls are more than 4 times less frequent than the average and the boys are almost twice more frequent than the average. The next variable is E_1 (teaching years) with specificity 0.33 – in C_3 there are 2.5 times fewer teachers than on average at the beginning (2-10) or at the end (26.6-35) of their careers. No one among them is discussing concepts with colleagues daily, and also they are doing this 1-3 times per week 3 times less frequently than the average. From A_3 we see that students are spending more than 120 min on homework more than twice less frequently than on average. We can describe the cluster C_3 as classes with mostly male students, with middle-aged teachers that are not discussing concepts with their colleagues frequently and are not requiring very much time for homework.

7.5   Conclusions

In this chapter, we presented the adaptations of the leaders' method and the agglomerative hierarchical clustering method for symbolic data described by modal valued symbolic variables. Both adapted methods are based on the same criterion function, therefore they can be used in combination which enables fine-tuned clustering of very large data sets. The definition of the dissimilarity includes weights to allow preserving frequency distributions of variables' values also in clusters.

The most important advantages of the presented clustering methods are that they provide meaningful descriptions of units and clusters that preserve more variability than classical data descriptions with only one value for each variable, and meaningful cluster leaders that enable easier interpretations. To discern the differences among the obtained clusters, we proposed to compute a cluster specificity index for each variable, comparing the values of the cluster representative to the whole-data-set representative.

Acknowledgments

The authors are very grateful to Barbara Japelj Pavešić, Slovenian national research coordinator of TIMSS in The Educational Research Institute, for helpful information and links to the TIMSS data.

This work was supported in part by the Slovenian Research Agency (research programs P1-0294 and P3-0154 and research projects J1-9187, J1-1691, and J5-2557), project COSTNET (COST Action CA15109), and prepared within the framework of the HSE University Basic Research Program.

Bibliography

[1] M.R. Anderberg. *Cluster Analysis for Applications*. Academic Press, New York, 1973.

[2] V. Batagelj. Generalized Ward and related clustering problems. In *Classification and Related Methods of Data Analysis*, pages 67–74. North-Holland, Amsterdam, 1988.

[3] V. Batagelj and N. Kejžar. clamix—Clustering symbolic objects R package. https://r-forge.r-project.org/projects/clamix/, 2010.

[4] L. Billard and E. Diday. *Symbolic Data Analysis: Conceptual Statistics and Data Mining*. John Wiley, Chichester, 2006.

[5] L. Billard and E. Diday. *Clustering Methodology for Symbolic Data*. Wiley Series in Computational Statistics, 2020.

[6] H.H. Bock and E. Diday. *Analysis of Symbolic Data: Exploratory Methods for Extracting Statistical Information from Complex Data*. Springer Science & Business Media, 2000.

[7] P. Brito. Clustering of symbolic data. In C. Hennig, M. Meila, F. Murtagh, and R. Rocci, editors, *Handbook of Cluster Analysis*. Chapman and Hall/CRC, 2015.

[8] F.A.T. de Carvalho, P. Brito, and H.H. Bock. Dynamic clustering for interval data based on L2 distance. *Computational Statistics*, 21(2):231–250, 2006.

[9] E. Diday. *Optimisation en Classification Automatique*. Institut national de recherche en informatique et en automatique, Rocquencourt, 1979.

[10] E. Diday and M. Noirhomme-Fraiture. *Symbolic Data Analysis and the SODAS software*. Wiley Online Library, 2008.

[11] S. García, J. Luengo, J.A. Sáez, V. López, and F. Herrera. A survey of discretization techniques: Taxonomy and empirical analysis in supervised learning. *IEEE Transactions on Knowledge and Data Engineering*, 25(4):734–750, 2013.

[12] K.C. Gowda and E. Diday. Symbolic clustering using a new dissimilarity measure. *Pattern Recognition*, 24(6):567–578, 1991.

[13] J.A. Hartigan. *Clustering Algorithms*. Wiley-Interscience, New York, 1975.

[14] M. Ichino and H. Yaguchi. Generalized Minkowski metrics for mixed feature-type data analysis. *IEEE Transactions on Systems, Man and Cybernetics*, 24(4):698–708, 1994.

[15] IEA. Data repository for TIMSS Advanced 2008, 2009. Accessed July 22, 2015, http://www.iea.nl/data.html.

[16] IEA. TIMSS Advanced 2008, 2009. Accessed July 22, 2015 http://timssandpirls.bc.edu/timss_advanced/index.html.

[17] A. Irpino and R. Verde. A new Wasserstein based distance for the hierarchical clustering of histogram symbolic data. In *Data Science and Classification*, pages 185–192. Springer, 2006.

[18] A. Irpino, R. Verde, and F.A.T. De Carvalho. Dynamic clustering of histogram data based on adaptive squared Wasserstein distances. *Expert Systems with Applications*, 41(7):3351–3366, 2014.

[19] N. Kejžar, S. Korenjak-Černe, and V. Batagelj. Clustering of modal-valued symbolic data. *Advances in Data Analysis and Classification*, 15:513–541, 2021.

[20] N. Kejžar, S. Korenjak-Černe, and V. Batagelj. Clustering of distributions: A case of patent citations. *Journal of Classification*, 28(2):156–183, 2011.

[21] J. Kim and L. Billard. A polythetic clustering process and cluster validity indexes for histogram-valued objects. *Computational Statistics & Data Analysis*, 55(7):2250–2262, 2011.

[22] J. Kim and L. Billard. Dissimilarity measures and divisive clustering for symbolic multimodal-valued data. *Computational Statistics & Data Analysis*, 56(9):2795–2808, 2012.

[23] J. Kim and L. Billard. Dissimilarity measures for histogram-valued observations. *Communications in Statistics: Theory and Methods*, 42(2):283–303, 2013.

[24] S. Korenjak-Černe and V. Batagelj. Clustering large datasets of mixed units. In A. Rizzi, M. Vichi, and H.H. Bock, editors, *Advances in Data Science and Classification. Proceedings of IFCS*, volume 98, pages 21–24, 1998.

[25] S. Korenjak-Černe and V. Batagelj. Symbolic data analysis approach to clustering large datasets. In K. Jajuga, A. Sokolowski, and H.H. Bock, editors, *Proc. 8th Conference of the International Federation of Classification Societies, July 16-19, 2002, Cracow, Poland. Classification, Clustering and Data Analysis.* Springer, 2002.

[26] S. Korenjak-Černe, V. Batagelj, and B. Japelj Pavešić. Clustering large data sets described with discrete distributions and its application on TIMSS data set. *Statistical Analysis and Data Mining*, 4(2):199–215, 2011.

[27] S. Korenjak-Černe, N. Kejžar, and V. Batagelj. A weighted clustering of population pyramids for the world's countries, 1996, 2001, 2006. *Population Studies*, 69(1):105–120, 2015.

[28] Republic of Slovenia Statistical Office. SI-STAT data portal, 2015. Accessed July 22, 2015
http://pxweb.stat.si/pxweb/Database/Regions/Regions.asp.

[29] M. Studer, G. Ritschard, A. Gabadinho, and N.S. Müller. Discrepancy analysis of complex objects using dissimilarities. In F. Guillet, G. Ritschard, D.A. Zighed, and H. Briand, editors, *Advances in Knowledge Discovery and Management*, volume 292 of *Studies in Computational Intelligence*, pages 3–19. Springer, Berlin, 2010.

[30] R. Verde and A. Irpino. Ordinary least squares for histogram data based on Wasserstein distance. In *Proceedings of COMPSTAT'2010*, pages 581–588. Springer, 2010.

[31] J.H. Ward. Hierarchical grouping to optimize an objective function. *Journal of the American Statistical Association*, 58(301):236–244, 1963.

[32] Y. Yang, G.I. Webb, and X. Wu. Discretization methods. In O. Maimon and L. Rokach, editors, *Data Mining and Knowledge Discovery Handbook*, chapter 6, pages 113–130. Springer, 2005.

8

Mixture Models for Distributional Data

Richard Emilion

Denis Poisson Institute, University of Orléans, France

CONTENTS

From a theoretical point of view, distributional data are considered here as realizations or 'outcomes' of a *random distribution,* that is a random variable taking values in a space of probability measures, any probability measure being considered, as usual, as the distribution of a random variable. From a practical point of view, distributional data are not directly observed and measured, they are statistical summaries or statistical estimations derived from collected data. They are definitely of interest to the user in many applications involving high speed sensors or complex systems. For example in [27], the size

DOI: 10.1201/9781315370545-8

distribution of aerosol particles is estimated by histograms at each time point. The aim of the present chapter is to propose some probabilistic models, specially mixture models, that can be fit to distributional data, considered as describing some statistical units. The problem is not new: in a paper introducing the famous Poisson-Dirichlet random distribution, J.F.C. Kingman [13] mentions in 1975 several earlier works where 'people are interested in describing the probability distribution of objects which are themselves probability distributions'.

A similar problem raised in "Symbolic Data Analysis", a paradigm introduced in 1987 by E. Diday [8], who called 'symbol'of a statistical unit any mathematical object describing the variability internal to that unit. For example, a symbol of a class of real data, the latter considered as a statistical unit, can be an interval (the class range), a function (e.g. the class empirical cdf or an histogram built from that class), or a probability distribution (e.g. a theoretical distribution estimated from that class data). In this latter case it is then seen that a probability distribution on random symbols turns out to be the distribution of a random distribution.

Probabilistic models for distributional data and symbols can be of interest in various real world problems: it was recently shown that the use of symbols substantially reduces the computation time when dealing with big datasets [4], [25], [26], [29].

The chapter is structured as follows. In Section 8.1, we propose a theoretical setting in which distributional data can be defined and handled. Section 8.2 concerns the mixture problem. EM algorithm applied to the estimation of a Dirichlet distribution mixture is detailed in Section 8.3. In Section 8.4, we propose the use of Kernel Mixtures and npEM algorithm. Section 8.5 concerns an example of non-finite mixture, the Dirichlet Process Mixture, (DPM). Three mixture models for dependent variables are presented in Section 8.6. Section 8.7 concerns a mixture model for intervals and their associated frequencies. The chapter ends with a conclusion section.

8.1 Distributional Data

Let $(\Omega, \mathcal{F}, \mathbb{P})$ be a probability space.

8.1.1 Random Distributions

Let $\mathbb{V}$ be a separable space and $\mathcal{V}$ a σ-algebra on $\mathbb{V}$. Let us denote $\mathcal{M}_1(\mathbb{V})$ the space of all probability measures on $(\mathbb{V}, \mathcal{V})$ endowed, as usual, with the smallest σ-algebra making measurable for all $A \in \mathcal{V}$, the mappings

$$\phi_A : \mathcal{M}_1(\mathbb{V}) \longrightarrow [0,1]$$
$$P \longrightarrow P(A)$$

Definition 1
A random distribution P on $\mathbb{V}$ is a measurable mapping from Ω to $\mathcal{M}_1(\mathbb{V})$.

$$P : \Omega \longrightarrow \mathcal{M}_1(\mathbb{V}). \tag{8.1}$$

First note that the distribution $\mathbb{P}_P$ of a random distribution P is a probability distribution on $\mathcal{M}_1(\mathbb{V})$.
Next, observe that the σ-algebra chosen on $\mathcal{M}_1(\mathbb{V})$ is such that a mapping $P : \Omega \longrightarrow \mathcal{M}_1(\mathbb{V})$ is a random distribution if and only if

$$P(A) = \phi_A \circ P : \Omega \longrightarrow [0,1]$$
$$\omega \longrightarrow P(\omega)(A) \tag{8.2}$$

is measurable for all $A \in \mathcal{V}$. This condition is very useful as seen in the following examples.

1. Let $X : \Omega \longrightarrow \mathbb{V}$ be a random variable with distribution $\mathbb{P}_X$, let $X^{(i)} \overset{i.i.d.}{\sim} \mathbb{P}_X$, $i = 1, \ldots, n$, and let $\delta_{X^{(i)}}$ denote the standard Dirac probability measure at the random point $X^{(i)}$. Then, (8.2) shows that $\delta_{X^{(i)}}$ is a random distribution so that the standard empirical measure

$$\varepsilon_n = \frac{\sum_{i=1}^{n} \delta_{X^{(i)}}}{n}$$

is a random distribution too.

2. Let $\kappa : \Omega \times \mathcal{V} \longrightarrow [0,1]$ be a Markov kernel such that $\omega \longrightarrow \kappa(\omega, A)$ is measurable for all $A \in \mathcal{V}$, and $A \longrightarrow \kappa(\omega, A)$ is a probability measure for all $\omega \in \Omega$, then, (8.2) shows that $P(\omega)(A) = \kappa(\omega, A)$ defines a random distribution P.

3. Let $X : \Omega \longrightarrow \mathbb{V}$ and $C : \Omega \longrightarrow \mathbb{C}$ be two random variables, $(\mathbb{C}, \mathcal{C})$ being a measurable space. Assume the existence of a regular conditional probability which ensures that the mapping

$$(c, A) \longrightarrow \mathbb{P}_{X|C=c}(A) \tag{8.3}$$

is measurable w.r.t. the σ-algebra $\mathcal{C} \otimes \mathcal{V}$ and the Borel σ-algebra of [0,1]. Since the mapping $c \longrightarrow \mathbb{P}_{X|C=c}(A)$ is measurable, (8.2) shows that the mapping

$$T : \mathbb{C} \longrightarrow \mathcal{M}_1(\mathbb{V})$$
$$c \longrightarrow \mathbb{P}_{X|C=c} \tag{8.4}$$

is a random distribution.

4. Let m be a positive integer. If $\mathbb{V}$ is a finite set, say $\mathbb{V} = \{1, \ldots, m\}$, then $\mathcal{M}_1(\mathbb{V})$ can be identified to $\mathbb{T}_m$, the m-simplex of probability vectors:

$$\mathbb{T}_m = \{y = (y_1, \ldots, y_m) \in \mathbb{R}_+^m : \sum_{l=1}^{m} y_l = 1\}, \qquad (8.5)$$

Any random variable taking values in $\mathbb{T}_m$ is then a random distribution on $\mathbb{V} = \{1, \ldots, m\}$. Probability distributions on $\mathbb{T}_m$ include Dirichlet distributions, Aitchison distributions, Normal Distributions on the simplex, see a nice presentation e.g. in [18] pp. 112-127.

5. Let α be a positive finite measure on V, Ferguson [12] has proved the existence of a random distribution $P : \Omega \longrightarrow \mathcal{M}_1(\mathbb{V})$, the celebrated Dirichlet Process with parameter α, such that for any positiver integer and any measurable finite partition $A_1, \ldots, A_m$ of $\mathbb{V}$, $(P(A_1), \ldots, P(A_m)) \sim Dd(\alpha(A_1), \ldots, \alpha(A_m))$, where Dd stands for the Dirichlet distribution on $\mathbb{T}_m$. Dirichlet Process is a very rich, powerful, and popular tool in non-parametric Bayesian statistics.

6. Stick-Breaking process [21], which approximates the Dirichlet Process, and Poisson-Dirichlet distribution [13] are two celebrated discrete random distributions.

8.1.2 Sample of Distributional Data

Definition 2

Let n be a positive integer. A n-sample $p_1, \ldots, p_n$ of distributional data is a sample from a random distribution P on $\mathbb{V}$, that is

$$p_i = P^{(i)}(\omega) \in \mathcal{M}_1(\mathbb{V}), \quad P^{(i)} \overset{i.i.d.}{\sim} \mathbb{P}_P, \; i = 1, \ldots, n$$

for some $\omega \in \Omega$ expressing the randomness of the sample.

We now extend the definition of distributional data by considering not only distributions but also functions of distributions. As shown in the next section examples, this extension will cover the case of interval data, histogram data, functional data and the like. Another advantage is that we will deal with a space $\mathbb{S}$ which is more simple than the rather complex space $\mathcal{M}_1(\mathcal{V})$.

Definition 3

Let n be a positive integer. A n-sample $s_1, \ldots, s_n$ of distributional data is a sample from a measurable function $f(P)$ of a random distribution P on $\mathbb{V}$, that is

$$s_i = f(P^{(i)})(\omega) \in \mathbb{S}, \quad P^{(i)} \overset{i.i.d.}{\sim} \mathbb{P}_P, \; i = 1, \ldots, n \qquad (8.6)$$

where

$$f : \mathcal{M}_1(\mathcal{V}) \longrightarrow \mathbb{S} \qquad (8.7)$$

is a measurable function taking values in some measurable space $(\mathbb{S}, \mathcal{S})$.

8.1.3 Examples of Distributional Data

1. *Intervals.* Assume that $\mathbb{V} = \mathbb{R}$ and that the support of $p = P(\omega)$, for each $\omega \in \Omega$, is a bounded interval $[l_p, u_p]$ with $l_p < u_p$, then

$$f(p) = [l_p, u_p] \tag{8.8}$$

defines an interval distributional data while

$$f(p) = (\frac{l_p + u_p}{2}, \frac{u_p - l_p}{2}) \tag{8.9}$$

defines a $\mathbb{R} \times \mathbb{R}_+$ distributional data used e.g. in [5], [15], [20] and

$$f(p) = (c_p, r_p) \tag{8.10}$$

where $c_p = \frac{l_p + u_p}{2}$, $r_p = u_p - l_p > 0$, defines a $\mathbb{R} \times \mathbb{R}_+$ distributional data also used e.g. in [5], [15].

2. *Probability vectors.* Assume that $\mathbb{V} = \mathbb{R}$ and that there exists a bounded interval containing the support of $p = P(\omega)$ for all $\omega \in \Omega$. Let $A_1, \ldots, A_m$ be a fixed partition of that interval into m adjacent intervals of finite length. Then

$$f(p) = (p(A_1), \ldots, p(A_m)) \tag{8.11}$$

defines a probability vector data or an histogram distributional data used e.g. in [6], [22], [23].

3. *Histograms.* Let m be fixed but the partition in the preceding example depending on p, then the following equality defines distributional data as proposed in [7] and [15].

$$f(p) = (l_{p,1}, u_{p,1}, q_{p,1}, \ldots, l_{p,m}, u_{p,m}, q_{p,m}) \tag{8.12}$$

where $[l_{p,\ell}, u_{p,\ell}]$ is the ℓ-th interval of the partition, $\ell = 1, \ldots, m$, and $q_{p,\ell}$ its corresponding frequency, with $q_{p,\ell} \geq 0$ and $q_{p,1} + \ldots + q_{p,m} = 1$.

4. *Functional data.* Assume that $\mathbb{V} = \mathbb{R}$ and that $p = P(\omega)$ has density d_p, c.d.f. F_p, and quantile function $\Psi_p = F_p^{-1}$, respectively. Then

$$f(p) = d_p, f(p) = F_p, f(p) = \Psi_p \tag{8.13}$$

respectively, define three function-valued distributional data as used in e.g. [24].

8.2   The Mixture Problem

Finite mixtures, defined below in (8.14), of well-known parametric probability measures, often provide appropriate models for the distribution of real world datasets collected from a statistical population with *subpopulations*.

Assume here that $f : \mathcal{M}_1(\mathbb{V}) \longrightarrow \mathbb{S} = \mathbb{R}^{m_1}$ and that $S = f(P) : \Omega \longrightarrow \mathbb{R}^{m_1}$ has density d_S w.r.t. the Lebesgue measure λ_{m_1} in $\mathbb{R}^{m_1}$.

Given a n-sample $s_1, \ldots, s_n \in \mathbb{S}$ as defined in (8.6), our problem consists in proposing some appropriate probabilistic mixture models for d_S and to estimate their parameters.

8.2.1   Finite Mixtures

Formally, let $k \geq 2$ be an integer and let $D_h, h = 1, \ldots, k$ be k probability measures on a measurable space $\mathbb{V}$. A *finite mixture* of the D_h's is just a convex combination

$$\sum_{h=1}^{k} q_h D_h, \text{ where } q_h \geq 0 \text{ and } \sum_{h=1}^{k} q_h = 1. \tag{8.14}$$

Note that a finite mixture is a probability measure. Countable (resp. uncountable) mixtures are also well-defined.

For a mixture as described in (8.14), the D_h's are commonly called the *components* of the mixture, k the number of components, while q_h is the *weight* of component h.

8.2.2   Identifiability

A finite mixture is said to be *identifiable* if the above representation (8.14) is unique up to a permutation, that is, if

$$\sum_{h=1}^{k} q_h D_h = \sum_{h=1}^{k'} q_h' D_h'$$

implies that

$$k = k' \text{ and } \exists \sigma \in \boldsymbol{s}_k : (q_h, D_h) = (q_{\sigma(h)}', D_{\sigma(h)}') \tag{8.15}$$

where $\boldsymbol{s}_k$ denotes the group of permutations of $\{1, \ldots, k\}$.

8.2.3   Mixture and Classification

An essential aspect consists in introducing an unobserved r.v. C

$$C : \Omega \longrightarrow \{1, 2, \ldots, k\} \text{ with } \mathbb{P}(C = h) = q_h \tag{8.16}$$

such that

$$\mathbb{P}_{S|C=h} = D_h. \tag{8.17}$$

Then, applying the total probability formula, it is seen that (8.14) is satisfied so that the mixture problem can be seen as an unsupervised classification problem. Indeed, $C(\omega) \in \{1 \dots k\}$, is the unobserved class label of ω and $q_h = \mathbb{P}(C = h)$ is the weight of class h. Further, it is assumed that the distribution of S in the unobserved class h is D_h, which is expressed mathematically by the h-th conditional distribution in (8.17).

However, while the outcomes s_i of S are given, the sub-populations of origin of the s_i's, that is the outcomes of C, are missing so that the estimation problem is not obvious. The unobserved r.v. C is said *latent* and any sample of S appears as a sample of *incomplete data*, while any sample of the pair (S, C) as a sample of *complete data*. The mixture problem is therefore a part of the general problem of statistical inference for incomplete data.

The distribution of C is usually called the *mixing measure* of the mixture.

8.2.4 Examples

The most popular example of mixture is the *Gaussian mixture*, widely used for modeling real world data distributions:

$$\mathbb{V} = \mathbb{R}^d, d = 1, 2, \dots, \ D_h = \mathcal{N}(\mu_h, \Sigma_h) \tag{8.18}$$

where $\mathcal{N}(\mu_k, \Sigma_k)$ denotes a Gaussian distribution with mean $\mu_k \in \mathbb{R}^d$, and variance Σ_k, a $d \times d$ symmetric non-negative definite matrix.

More generally, mixtures of distributions belonging to the *exponential family* are frequently used. This family is the set of distributions having a density w.r.t. a reference measure that can be expressed in the form

$$f(x|\Theta) = h(x)g(\Theta)exp(\eta(\Theta).T(x)), \ x \in \mathbb{V} \tag{8.19}$$

where $\Theta = (\theta_1, \dots, \theta_r)$ is a vector parameter and $\eta(\Theta) = (\eta_1(\Theta), \dots, \eta_s(\Theta))$ and $T(x) = (T_1(x), \dots, T_s(x))$ are vector-valued functions, the dot product of vectors being denoted by '.'.

The exponential family includes most of the standard discrete and continuous distributions including the distribution of categorical variables, Bernoulli, Poisson, exponential, Gaussian, log-normal, inverse Gaussian, gamma, chi-square, Pareto, beta, and Dirichlet distributions. However it does not include the Student t, the Fisher F, the Cauchy, the hypergeometric distributions, the logistic distributions, neither finite mixtures and compound distributions (infinite mixtures).

It is well-known that a mixture of distributions belonging to a same exponential family is identifiable.

8.3 Mixtures of Dirichlet Distributions

It is assumed here that the we have a sample of histogram data as defined in
(8.11). We consider finite mixtures of the popular Dirichlet Distribution, but
finite mixtures of any other distribution on the simplex, as those mentioned
in [18] pp. 112-127, can also be considered.

Let $\underline{\alpha} = (\alpha_1, \ldots, \alpha_\ell, \ldots, \alpha_{m_1})$ with $\alpha_\ell \geq 0$. A T_{m_1} - valued r.v.
$Y = (Y_1, \ldots, Y_{m_1})$ follows a Dirichlet distribution $D(\underline{\alpha})$ whenever $Y = (Y_1, \ldots, Y_{m_1-1})$ has for density the popular Dirichlet density

$$Dd(\underline{\alpha})(y) = \frac{\Gamma(\alpha_1 + \ldots + \alpha_{m_1})}{\Gamma(\alpha_1) \ldots \Gamma(\alpha_{m_1})} y_1^{\alpha_1 - 1} \ldots y_{m_1-1}^{\alpha_{m_1-1} - 1} (1 - \sum_{\ell=1}^{m_1-1} y_\ell)^{\alpha_{m_1} - 1} I_{\mathbb{U}_{m_1-1}}(y)$$

(8.20)

where

$$\mathbb{U}_{m_1-1} = \{y = (y_1, \ldots, y_{m_1-1}) \in \mathbb{R}_+^{m_1-1} : \sum_{\ell=1}^{m_1-1} y_\ell \leq 1\}.$$

(8.21)

Estimation of Dirichlet distribution parameters has been done in [16], [17],
[19].

Since the Dirichlet density belongs to the exponential family, a mixture

$$d_S(.|(\underline{\alpha}_h, q_h)_h) = \sum_{h=1}^{k} q_h Dd(\underline{\alpha}_h), \text{ where } \underline{\alpha}_h = (\alpha_{h,1}, \ldots, \alpha_{h,m_1})$$

(8.22)

of such distributions, shortly a Dd mixture, is identifiable. Its parameters
$(\underline{\alpha}_h, q_h)_h$ can be estimated for example by the popular EM algorithm and its
variants CEM, SEM, SAEM, MCEM and the like.

In our notations below, for any observed probability vector $\underline{s} = (s_1, \ldots, s_l, \ldots, s_{m_1})$ we will agree that

$$Dd(\underline{\alpha})(\underline{s}) = Dd(\underline{\alpha})(s_1, \ldots, s_{m_1-1})$$

(8.23)

where the right-hand side expression is defined in (8.20).

Finite mixtures of Dirichlet Distributions were used in various applied
fields, see e.g. [6], [22], [23].

8.3.1 EM algorithm for a Mixture of Dirichlet Distributions

We consider here that we have a n-sample $s_1, \ldots, s_n$ of distributional data as
defined in (8.11) such that

$$\underline{s}_i = (s_{i,1}, \ldots, s_{i,m_1}) \in \mathbb{T}_{m_1} \ i = 1, \ldots, n.$$

(8.24)

The estimation method introduces a latent class variable C such that

$$
\begin{aligned}
C : \Omega &\longrightarrow \{1, \ldots, k\} \\
\mathbb{P}(C = h) &= q_h, \quad h = 1, \ldots, k \\
\mathbb{P}_{S|C=h} &= Dd(\underline{\alpha}_h).
\end{aligned}
$$

The complete variable (S, C) likelihood for an observed 1-sample $(\underline{s}, c)$ is

$$
\begin{aligned}
L(\underline{s}, c) &= L_C(c) L_{S|C=c}(\underline{s}) \\
&= q_c Dd(\underline{\alpha}_c)(\underline{s}) = \prod_{h=1}^{k} (q_h Dd(\underline{\alpha}_h)(\underline{s}))^{1_{h=c}}
\end{aligned}
$$

and the likelihood for a n-sample $(\underline{s}_i, c_i)_{i=1,\ldots,n}$ is

$$
L((\underline{s}_i, c_i)_i) = \prod_{i=1}^{n} \prod_{h=1}^{k} (q_k Dd(\underline{\alpha}_h)(\underline{s}_i))^{1_{c_i=h}}.
$$

The Log Likelihood is then

$$
LL((\underline{s}_i, c_i)_i, \alpha, q) = \sum_{i=1}^{n} \sum_{h=1}^{k} 1_{c_i=h} (\log(DD(\underline{\alpha}_h)(\underline{s}_i)) + \log(q_h))
$$

The 'E' part in the EM algorithm consists in the following Expectation computation w.r.t. the conditional distribution of C given S and given a value α_{old}, q_{old} of the parameters α, q that have to be estimated:

$$
\begin{aligned}
&E_{\mathbb{P}_{C|S,\alpha_{old},q_{old}}} (LL((\underline{s}_i, c_i)_i, \alpha, q)) \\
&= \sum_{i=1}^{n} \sum_{h=1}^{k} E_{\mathbb{P}_{C|S,\alpha_{old},q_{old}}} (1_{c_i=h})(\log(DD(\underline{\alpha}_h)(\underline{s}_i)) + \log(q_h)) \\
&= \sum_{i=1}^{n} \sum_{h=1}^{k} t_{i,h} (\log(DD(\underline{\alpha}_h)(\underline{s}_i)) + \log(q_h)) \quad (8.25)
\end{aligned}
$$

with

$$
t_{i,h} = \frac{q_{h,old} DD(\underline{\alpha}_{old,h})(\underline{s}_i)}{\sum_{h=1}^{k} q_{h,old} DD(\underline{\alpha}_{old,h})(\underline{s}_i)} \quad (8.26)
$$

and

$$
\begin{aligned}
&\log(DD(\underline{\alpha}_h)(\underline{s}_i)) \\
&= \log \Gamma(\sum_{\ell=1}^{m_1} \alpha_{h,\ell}) - \sum_{\ell=1}^{m_1} \log \Gamma(\alpha_{h,\ell}) + \sum_{\ell=1}^{m_1} (\alpha_{h,\ell} - 1) \log(s_{i,\ell}). \quad (8.27)
\end{aligned}
$$

The 'M' part of EM consists in finding the new parameters that Maximize the above expectation. Derivations yield m_1 equations for component h:

$$\sum_{i=1}^{n} t_{i,\ell,\alpha_{old},q_{old}} \left(F\left(\sum_{\ell=1}^{m_1} \alpha_{h,\ell}\right) - F(\alpha_{h,\ell}) + \log(s_{i,\ell}) \right) = 0, \quad \ell = 1, \ldots, m_1 \quad (8.28)$$

where $F = \frac{\Gamma'}{\Gamma}$ denotes the standard logarithmic derivative of the Γ function, known as the digamma function.

The solution requires some numerical methods. An implementation, using R software package BB and the *dfsane* function, was done in [28].

8.3.2   Clustering quality and Consistency

Likelihood is a mixture quality criterion but, another criterion, more based on the clustering aspect, is defined in [28]. Indeed EM-like algorithms provide fuzzy classes since $t_{i,h}$, which is the posterior probability of $C = h$ given $S = \underline{s}_i$, can be seen as the membership degree of $\underline{s}_i$ in the fuzzy class h. Hence, defining the fuzzy centroid of class h as

$$\frac{\sum_{i=1}^{n} t_{i,h} \underline{s}_i}{\sum_{i=1}^{n} t_{i,h}} \quad (8.29)$$

and choosing a distance on the set of probability vectors, we can proceed in computing the within variance of each class and the variance between classes, getting a clustering quality of the mixture. Such a criterion can be used to compare various algorithms, see [28].

On the other hand, the probability vectors s_i in (8.24), used in the above EM-like algorithms, were computed w.r.t. a partition of $\mathbb{R}$ into m_1 adjacent intervals. Assuming that the support of X is included in a bounded interval, the partition intervals can be taken of finite length and a natural question concerns the behavior of the mixture when this length goes to 0, the number m_1 of adjacent intervals increasing to infinity. The answer is a consistency result proved in [9] for a model derived from a mixture of Dirichlet processes, using the martingale theorem and a nice property [14] of Dirichlet processes introduced by Ferguson [12].

8.4   Kernel Finite Mixtures

Instead of proposing parametric models, a semi-parametric approach is possible.

When the random variable S takes values in the simplex $\mathbb{T}_{m_1}$, a *Dirichlet Kernel* estimator was proposed in [10].

Definition 4

Given an observed sample of probability vectors $\underline{s}_i, i = 1, \ldots, n$ of size m_1, the multivariate Dirichlet kernel is the function

$$K_H(x) = |H|^{-\frac{1}{2}} \sum_{i=1}^{n} Dd(H^{-\frac{1}{2}}(x - \underline{s}_i)), x \in T_{m_1-1} \qquad (8.30)$$

where H is a $(m_1 - 1) \times (m_1 - 1)$ symmetric positive definitive bandwith matrix to be estimated, $|H|$ denoting the determinant of H.

Then, we can proceed to the estimation of a Dirichlet kernel mixture using the npEM algorithm, an extension of the EM algorithm to kernels proposed in [2] and implemented in the R package mixtools [3]. We omit the details. As noticed in [10], we can use other kernels replacing in (8.30) the Dirichlet density Dd by any other distribution on the simplex as those mentioned in [18] pp. 112-127.

8.5 Dirichlet Process Mixture (DPM)

In the finite mixture of k Dirichlet distributions or of k Kernels, the mixing class variable, denoted by C, takes values in $\{1, \ldots, k\}$. A more flexible approach proposed in [11] is the hierarchical model called Dirichlet Process Mixture (DPM) which is an infinite mixture of Dd's, the mixing distribution being an outcome of a Dirichlet Process:

$$
\begin{aligned}
s_i | \alpha_i &\sim & Dd(\underline{\alpha}_i) \\
\underline{\alpha}_i | P &\sim & P \\
P | c, P_0 &\sim & D(cP_0).
\end{aligned}
\qquad (8.31)
$$

Here too, the Dd's can be replaced by any distribution on the simplex or any Kernel such as the one in (8.30) and in the remark thereafter.

Estimation in such a hierarchical model can be done using Gibbs sampling algorithm and the fact that the posterior distribution of a DPM is a MDP (Mixture of Dirichlet Processes) [1] : If the s_i's and P are as in (8.31) then the posterior is such that

$$P | s_1, \ldots, s_n \sim \int DP(cP_0 + \sum_{i=1}^{n} \delta_{\theta_i}) d\mathbb{P}_{\theta_1, \ldots, \theta_n | s_1, \ldots, s_n} \qquad (8.32)$$

8.6   Dependent random distributions

We are now considering the case of $\mathbf{p}$ dependent random distributions $P_1, \ldots, P_{\mathbf{p}}$ with $P_j \in \mathcal{M}(\mathbb{V}_j), j = 1, \ldots, \mathbf{p}$ and $\mathbf{p}$ an integer such that $\mathbf{p} \geq 2$. Thus we are dealing with a distributional data table of n rows and $m = m_1 + \ldots + m_{\mathbf{p}}$ columns, entries in the first m_1 columns being probability vectors of size $m_1, \ldots$, and so on, entries in the last $m_{\mathbf{p}}$ columns being probability vectors of size $m_{\mathbf{p}}$:

$$s_i = (f_1(P_1^{(i)}), \ldots, f_{\mathbf{p}}(P_{\mathbf{p}}^{(i)}))(\omega) \in \mathbb{S}, \;\; P_j^{(i)} \overset{i.i.d.}{\sim} \mathbb{P}_{P_j}, \; j = 1, \ldots, \mathbf{p}, \; i = 1, \ldots, n$$
$$(8.33)$$

where $f_1, \ldots, f_p$ are as in (8.11), that is

$$f_j(p) = (p(A_{1,j}), \ldots, p(A_{m_j,j})) \in \mathbb{T}_j \tag{8.34}$$

$A_{1,j}, \ldots, A_{m_j,j}$ being a measurable partition of $\mathbb{V}_j$. The following constructions of $\mathbf{p}$ dependent Dirichlet distributions, proposed in [10], take in account the dependency between the $\mathbf{p}$ columns of distributional data.

8.6.1   Dependent Dirichlet Distributions, Model 1

This first model is just a finite mixture with independence conditionally to a class variable C taking values $h = 1, \ldots, k$ with probability q_h, resp.

$$\sum_{h=1}^{k} q_h Dd_{m_1,h} \otimes Dd_{m_2,h} \otimes \ldots \otimes Dd_{m_{\mathbf{p}},h}$$

where $Dd_{m_j,c}$ denotes a Dirichlet distribution in dimension m_j with parameters depending on c. Estimation using EM algorithm is straightforward. Note that independence conditionally to C does not require independence of the $\mathbf{p}$ histogram variables.

8.6.2   Dependent Dirichlet Distributions, Model 2

Let $X \sim \mathcal{N}(0,1)$ with cdf $\Phi : \Phi(x) = \frac{1}{\sqrt{2\pi}} \int_{-\infty}^{x} e^{-\frac{t^2}{2}} dt$. Let $\alpha > 0$, since $\Phi(X) \sim Uniform(0,1)$, if F_α denotes the cdf of $\Gamma(\alpha, 1)$, and $g_\alpha = F_\alpha^{-1} \circ \Phi$, then $g_\alpha(X) \sim \Gamma(\alpha, 1)$

Let $m = m_1 + \ldots + m_{\mathbf{p}}$ and Σ be a $m \times m$ symmetric positive definitive matrix having as diagonal blocks $\mathbf{p}$ Identity matrices of size m_j, respectively.

For example if $\mathbf{p} = 2, m_1 = 2, m_2 = 3$, then $m = 5$ and

$$\Sigma = \begin{pmatrix} 1 & 0 & a & b & c \\ 0 & 1 & d & e & f \\ a & d & 1 & 0 & 0 \\ b & e & 0 & 1 & 0 \\ c & f & 0 & 0 & 1 \end{pmatrix}$$

Our construction starts with Gaussians but starting from Dependent Gammas is another possible option.

Let $G = (G_1, \ldots, G_m) \sim Gauss(0, \Sigma)$ be a Gaussian vector in dimension m. Due to the pattern of Σ, $G_1, \ldots, G_{m_1}$ are i.i.d. $\mathcal{N}(0, 1)$, and the same holds for $G_{m_1+1}, \ldots, G_{m_1+m_2}$ and so on, up to $G_{m_1+\ldots+m_{p-1}+1}, \ldots, G_m$. However these **p** groups of vectors are dependent from each other if the correlation parameters are $\neq 0$.

Let $\alpha_{1,1}, \ldots, \alpha_{1,m_1}$ be some positive parameters and let

$$X_1 = g_{\alpha_{1,1}}(G_1), \ldots, X_{m_1} = g_{\alpha_{1,m_1}}(G_{m_1})$$

$X_1, \ldots, X_{m_1}$ are independent Gammas with parameters $(\alpha_1, 1), \ldots (\alpha_{m_1}, 1)$, respectively. Define similarly independent Gammas $X_{m_1+1}, \ldots, X_{m_1+m_2}$ and so on, up to $X_{m_1+\ldots+m_{p-1}+1}, \ldots, X_m$. Observe that these groups of Gamma vectors are however dependent from each other.

The joint density of the X_j's can be derived from the Gaussian density:

$$\frac{1}{\sqrt{(2\pi)^m |\Sigma|}} \exp^{-\frac{1}{2} v^t \Sigma^{-1} v} \prod_{\ell=1}^{m} (g_{\alpha_\ell}^{-1})'(x_\ell)$$

where $v = (g_{\alpha_1}^{-1}(x_1), \ldots, g_{\alpha_m}^{-1}(x_m))^t$ and

$$(g_{\alpha_\ell}^{-1})'(x_\ell) = \frac{\sqrt{2\pi}}{\Gamma(\alpha_\ell)} \exp^{\frac{[\Phi^{-1}(F_{\alpha_\ell}(x_\ell))]^2}{2} - x_\ell} x_\ell^{\alpha_\ell - 1}$$

Then, dividing each of these Gammas by the sum of their respective group, we get **p** Dirichlet vectors which are dependent.

Consider the following bijections

$$Q_{1:m_1} : \mathbb{R}_+^{m_1} \longrightarrow \mathbb{T}_{m_1-1} \times \mathbb{R}_+$$

used when **p** $= 1$

$$Q_{1:m_1}(x_1, \ldots, x_{m_1}) = (y_1, \ldots, y_{m_1-1}, z_1)$$

with

$$y_1 = \frac{x_1}{x_1 + \ldots + x_{m_1}}, \ldots, y_{m_1-1} = \frac{x_{m_1-1}}{x_1 + \ldots + x_{m_1}}, z_1 = x_1 + \ldots + x_{m_1}$$

The Jacobian of $Q_{1:m_1}^{-1}$ is $z_1^{m_1-1}$.

Similarly, let

$$Q_{m_1+1:m_1+m_2} : \mathbb{R}_+^{m_2} \longrightarrow \mathbb{T}_{m_2-1} \times \mathbb{R}_+$$

$$Q_{m_1+1:m_1+m_2}(x_{m_1+1}, \ldots, x_{m_1+m_2}) = (y_{m_1+1}, \ldots, y_{m_1+m_2-1}, z_2)$$

with

$$y_{m_1+1} = \frac{x_{m_1+1}}{x_{m_1+1} + \ldots + x_{m_1+m_2}}, \ldots, y_{m_1+m_2-1} = \frac{x_{m_1+m_2-1}}{x_{m_1+1} + \ldots + x_{m_1+m_2}}$$

Let $z_2 = x_{m_1+1} + \ldots + x_{m_1+m_2}$. The Jacobian of $Q^{-1}_{m_1+1:m_1+m_2}$ is $z_2^{m_2-1}$.

And so on, define similarly $Q_{m_1+\ldots+m_{p-1}+1:m}$.

Applying these bijections to the X_j's, we get the density of the random vector

$$(Y_1, \ldots, Y_{k_1-1}, Z_1, Y_{k_1+1}, \ldots, Y_{k_1+k_2-1}, Z_2, \ldots, Y_{k_1+\ldots+k_{p-1}+1}, \ldots, Y_{m-1}, Z_p) :$$

$$\frac{1}{\sqrt{(2\pi)^m|\Sigma|}} \exp^{-\frac{1}{2}w^t\Sigma^{-1}w} \prod_{\ell=1}^{m}(g_{\alpha_\ell}^{-1})'(u_\ell)z_1^{m_1-1}z_2^{m_2-1}\ldots z_p^{m_p-1}$$

where $w = (g_{\alpha_1}^{-1}(u_1), \ldots, g_{\alpha_j}^{-1}(u_\ell), \ldots, g_{\alpha_\ell}^{-1}(u_m))^t$ and
$u_j = y_\ell z_1$ if $1 \le \ell < m_1$, $u_{k_1} = z_1(1 - y_1 - \ldots - y_{m_1-1})$
$u_\ell = y_\ell z_2$ if $k_1 + 1 \le \ell < m_1 + m_2$,
$u_{m_1+m_2} = z_2(1 - y_{m_1+1} - \ldots - y_{m_1+m_2-1}) \ldots$
$u_j = y_\ell z_p$ if $m_1 + \ldots + m_{p-1} + 1 \le \ell < m = m_1 + \ldots + m_p$,
$u_m = z_p(1 - y_{m_1+\ldots+m_{p-1}+1} - \ldots - y_{m-1})$.

This yields the joint density of the Dependent Dirichlet

$$(Y_1, \ldots, Y_{k_1-1}, Y_{k_1+1}, \ldots, Y_{k_1+k_2-1}, \ldots, Y_{k_1+\ldots+k_{p-1}+1}, \ldots, Y_{m-1})$$

by integrating out the above density w.r.t. $z_1, z_2, \ldots, z_p$.

This integral can be estimated by simulation since (Y, Z) can be easily simulated. Estimation of the parameters Σ and α_ℓ's can be done by the EM algorithm.

8.6.3   Dependent Dirichlet Distributions, Model 3

Let d be large enough $(d > \max(m_1, m_2, \ldots, m_p))$, let $G = (G_1, \ldots, G_d)^t$:
$G_j \overset{ind}{\sim} \Gamma(\alpha_j, 1)$.

Choose m_1 integers in $\{1, \ldots, d\}$, divide each of the m_1 corresponding G_j's by their sum getting a Dirichlet vector in dimension m_1. Proceed similarly with $m_2, \ldots, m_p$

If some integers in these choices are common we get **p** Dependent Dirichlet vectors.

Here is an example for $d = 3, \mathbf{p} = 2 = m_1 = m_2$. Choosing 1, 2 and then 2, 3, we get 2 dependent Dirichlet vectors $(Y_1 = \frac{X_1}{X_1+X_2}, Y_2 = \frac{X_2}{X_1+X_2})$ and $(Y_3 = \frac{X_2}{X_2+X_3}, Y_4 = \frac{X_3}{X_2+X_3})$.

The joint density of (Y_1, Y_3) is a continuous mixture:

$$f_{(Y_1,Y_3)}(y_1, y_3) = \int_{\mathbb{R}_+} f_{X_1}\left(\frac{y_1 x_2}{1 - y_1}\right) f_{X_3}\left(\frac{y_2 x_2}{1 - y_2}\right) f_{X_2}(x_2) dx_2$$

Finally, note that multi-histograms and finite mixture of kernels built from such density functions provide non-parametric models for $\mathbf{p} \ge 2$. Infinite mixtures of Kernels can be derived using a DPM just as in the case $\mathbf{p} = 1$.

8.7 Mixture models for general histograms

So far, the models we have proposed only work for vectors of probability, that is for histograms built on a same fixed partition as in Section 8.1.3, example 2. In this last section, we propose some models which work when the partition depends on the histogram, as in Section 8.1.3, example 3.

For technical reasons, instead of considering the ℓ-th interval range $r_{i,\ell}$ of the i-th histogram data, rather consider its log-range $\log(r_{i,\ell})$, and its midpoint $c_{i,\ell}$, so that our data are

$$s_i = (c_{i,1}, \log(r_{i,1}), q_{i,1}, \ldots, c_{i,m}, \log(r_{i,m}), q_{i,m}), \quad i = 1, \ldots, n$$

If intervals and their associated frequencies are assumed independent, then finite mixtures or DPM based on

$$G_0 = \mathcal{N}(\mu, \Sigma) \otimes Dirichet(\alpha_1, \ldots, \alpha_m)$$

are of course appropriate models.

Here is an example of model where the Gaussian sub-vector and the Dirichlet sub-vector are correlated.

Let

$$LR = (c_{.,1}, \log(r_{.,1}), \ldots, c_{.,m}, \log(r_{.,m})) \sim \mathcal{N}_{2m}(\mu, \Sigma)$$

be a Gaussian vector in dimension $2m$ with mean vector μ and invertible variance-covariance matrix Σ so that the vector

$$SLR = \Sigma^{-\frac{1}{2}}(LR - \mu)\Sigma^{\frac{1}{2}} \sim \mathcal{N}_{2m}(0, I)$$

is a standard Gaussian vector $(SLR_1, SLR_2, \ldots, SLR_{2m-1}, SLR_{2m})$ in dimension $2m$ and

$$(\frac{SLR_1 + SLR_2}{2}, \ldots, \frac{SLR_{2m-1} + SLR_{2m}}{2}) \sim \mathcal{N}_m(0, I)$$

is a standard Gaussian vector in dimension m, which is of course correlated to the vector LR.

Then transform the latter into independent Gammas using the cdf of the standard Gaussian in dimension 1 and the inverse cdf of each Gamma distribution $\Gamma(1, \alpha_\ell), \ell = 1, \ldots, m$. Finally dividing each Gamma by the sum of the Gammas we get a Dirichlet vector $D = T(LR)$ in dimension m with parameter $(\alpha_1, \ldots, \alpha_m)$, which is of course correlated to LR as a transform $T(LR)$ of LR.

A finite mixture of k such vectors (LR, D) densities can therefore be fit to distributional data

$$(c_{.,1}, \log(r_{.,1}), \ldots, c_{.,m}, \log(r_{.,m}), q_{.,1}, \ldots, q_{.,m}).$$

Parameter estimation can be done by applying EM-like procedures on the margins LR and D since they are mixtures of K Gaussian distributions (resp. of k Dirichlet distributions).

8.8 Conclusion

We have proposed a mathematical setting in which distributional data can be rigorously defined and handled, and then some mixture models which can be used for a model-based classification of such data. Dirichlet distribution mixtures have been tested on various real datasets such as internet flow data, renewal energy data, movies data, but testing the other models presented here, and proposing some new models, could be an interesting topic for future works.

Bibliography

[1] C.E. Antoniak. Mixtures of Dirichlet processes with applications to bayesian nonparametric problems. *The Annals of Statistics*, 2(6):1152–1174, 1974.

[2] T. Benaglia, D. Chauveau, and D.R Hunter. An em-like algorithm for semi- and non-parametric estimation in multivariate mixtures. *J. Comput. Graph. Statist.*, 18(2):505–526, 2009. http://dx.doi.org/10.1198/jcgs.2009.07175.

[3] T. Benaglia, D. Chauveau, D.R. Hunter, and D.S. Young. Mixtools : An R package for analyzing mixture models. *Journal of Statistical Software*, 32(4):1–29, 2009. http://www.jstatsoft.org/v32/i06/.

[4] B. Beranger, H. Lin, and S.A. Sisson. New models for symbolic data analysis. *arXiv e-prints*, 2020. https://arxiv.org/abs/1809.03659.

[5] P. Brito and A.P. Duarte Silva. Modeling interval data with normal and skew-normal distributions. *Journal of Applied Statistics*, 39(1):3–20, 2012.

[6] R. Calif, R. Emilion, and T. Soubdhan. Classification of wind speed distributions. *Renewable Energy*, 36(11):3091–3097, 2011.

[7] S. Dias and P. Brito. Linear regression model with histogram-valued variables. *Statistical Analysis and Data Mining: The ASA Data Science Journal*, 8(2):75–113, 2015.

[8] E. Diday. The symbolic approach in clustering and related methods of data analysis. In H-H. Bock, editor, *Proceedings of IFCS 1987, Classification and Related Methods of Data Analysis*, pages 673–384. North-Holland, 1988.

[9] R. Emilion. Unsupervised classification and analysis of objects described by nonparametric probability distributions. *Statistical Analysis and Data Mining (SAM)*, 5(5):388–398, 2012.

[10] R. Emilion. Models in sda: Dependent Dirichlet distributions and kernels. In *Proceedings of the International Workshop on Advances in Data Science, Beijing*, 2016.

[11] R. Emilion. Random intervals as random distributions. In *Proceedings of the Third International Symposium on Interval Data Modeling, Beijing*, 2017.

[12] T.S. Ferguson. A Bayesian analysis of some nonparametric problems. *The Annals of Statistics*, 1(2):209–230, 1973.

[13] J.F.C. Kingman. Random discrete distributions. *Journal of the Royal Statistical Society. Series B*, 37(1):1–22, 1975.

[14] R.M. Korwar and M. Hollander. Contributions to the theory of Dirichlet processes. *The Annals of Probability*, 4(4):705–711, 1973.

[15] J. Le-Rademacher and L. Billard. Likelihood functions and some maximum likelihood estimators for symbolic data. *Journal of Statistical Planning and Inference*, 141(4):1593–1602, 2011.

[16] T.P. Minka. Estimating a Dirichlet distribution, 2000. https://www.microsoft.com/en-us/research/publication/estimating-{D}irichlet-distribution/.

[17] A. Naryanan. Algorithm AS 266: Maximum likelihood estimation of the Dirichlet distribution. *Applied Statistics*, 40:365–374, 1991.

[18] V. Pawlowsky-Glahn, J.J. Egozcue, and R. Tolosana-Delgado. *Modeling and Analysis of Compositional Data*. Wiley, Chichester, United Kingdom, 2015.

[19] G. Ronning. Maximum-likelihood estimation of Dirichlet distribution. *Journal of Statistical Computation and Simulation*, 32:215–221, 1989.

[20] S. Sankararaman and S. Mahadevan. Likelihood-based representation of epistemic uncertainty due to sparse point data and/or interval data. *Reliability Engineering and System Safety*, 96(5):814–824, 2011.

[21] J. Sethuraman. A constructive definition of Dirichlet priors. *Statist. Sinica*, 4:639–650, 2011.

[22] T. Soubdhan, R. Emilion, and R. Calif. Classification of daily solar radiation distribution using a mixture of Dirichlet distributions. *Solar Energy*, 83:1056–1063, 2009.

[23] A. Soule, K. Salamatian, N. Taft, R. Emilion, and K. Papagiannaki. Flow classification by histograms or how to go on safari in the internet. In *Proceedings of ACM Sigmetrics'04, New York*, 2004.

[24] R. Verde and A. Irpino. Dynamic clustering of histogram data: using the right metric. In *Selected Contributions in Data Analysis and Classification*, pages 123–134. Springer, 2007.

[25] T. Whitaker, B. Beranger, and S.A. Sisson. Composite likelihood methods for histogram-valued random variables. *arXiv e-prints*, 2019. `https://arxiv.org/abs/1908.11548`.

[26] T. Whitaker, B. Beranger, and S.A. Sisson. Logistic regression models for aggregated data. *arXiv e-prints*, 2019. `https://arxiv.org/abs/1912.03805`.

[27] D. Wraith, K. Mengersen, C. Alston, J. Rousseau, and T. Hussein. Using informative priors in the estimation of mixtures over time with application to aerosol particle size distributions. *The Annals of Applied Statistics*, 8(1):232–258, 2014.

[28] B. Xia, H. Wang, R. Emilion, and E. Diday. Em algorithm for Dirichlet samples and its application to movie data. In *Proceedings of the Symposium on The Service Innovation Under The Background of Big Data & IEEE Workshop on Analytics and Risk*, 2017.

[29] X. Zhang, B. Beranger, and S.A. Sisson. Constructing likelihood functions for interval-valued random variables. *Scandinavian Journal of Statistics*, 47(1):1–35, 2020.

9

Classification of Continuous Distributional Data Using the Logratio Approach

Ivana Pavlů

Palacký University, Department of Mathematical Analysis and Applications of Mathematics, Czech Republic

Peter Filzmoser

TU Wien, Institute of Statistics and Mathematical Methods in Economy, Austria

Alessandra Menafoglio

Politecnico di Milano, MOX, Department of Mathematics, Italy

Karel Hron

Palacký University, Department of Mathematical Analysis and Applications of Mathematics, Czech Republic

CONTENTS

The classification of observations into two or more predefined groups belongs to one of the basic tasks in statistical modeling [20]. These observations can be standard multivariate observations, but also functions [32]. Motivated by

DOI: 10.1201/9781315370545-9

the generation of massive data, an increasing interest is devoted to analyzing distributional data. Currently, there are two main approaches to distributional data modeling. SYMBOLIC DATA ANALYSIS (SDA) as introduced in [2,29] provides a unified approach to analyze distributional data, resulting from capturing intrinsic variability of groups of individuals as input observations. In parallel to the SDA approach, a concise methodology has been developed since the early 1980s to deal with *compositional data* – i.e., data carrying only relative information [1,15,30] – through the logratios of their parts. There exists a unified methodology in COMPOSITIONAL DATA ANALYSIS which is based on the Bayes space approach [12,37], and this methodology aims to treat multivariate observations which can be identified with probability functions of discrete distributions, and it also captures the specific features of continuous distributions (densities). The aim of this chapter is to describe briefly a general setting that includes both the discrete and the continuous setting, and then to focus on the latter one in the context of classification of probability density functions. The theoretical developments are illustrated with real-world sedimentological data.

9.1   Introduction

There are several types of variables in symbolic data analysis that naturally induce a probability distribution. For the discrete case, the most representative case are categorical modal variables. A categorical modal variable Y with a finite domain $O = \{m_1, \ldots, m_D\}$ is a multi-state variable such that for each element of Y, a category set is given, and for each category m_l, a weight (e.g. a frequency or probability) is provided. If the weight is a frequency, it represents the proportion of individuals of the underlying microdata set characterized by this category. Consequently, from the probabilistic point of view, we would get a probability function over a set of categories. Although the usual representation is taken with weights summing up to one, this constraint is rather a convention than a real need. For example, the weights could also contain concentrations of chemical elements, or household expenditures in local currency. Alternatively, probability functions can also be obtained if one considers histogram-valued variables with either absolute or relative frequencies [24]. In this case, the "observations" (histograms) are again of the same data type, and each class of the histograms forms a part (category).

In all the above cases, the main point is that the weights (frequencies) express relative contributions on a whole. The concrete representation of the weights (probabilities, concentrations, ppm, etc.) can be chosen arbitrarily without any loss of information. This idea could also be adapted to the continuous case, where the domain of symbolic variables is characterized by a subset of the real line, typically a bounded or unbounded interval. Then the probability function is replaced by a density, a non-negative Borel measurable function with unit integral constraint. An example are age/income distributions in a certain region or particle size distributions of a sediment sample, i.e., the finite domain is replaced by an infinite one. And again, even if we used a representation of densities that would lead to another integral value, the main feature that the density conveys relative contributions of Borel sets (subsets of the domain) to the overall probability (weight, frequency) remains. In other words, both compositional data and density functions as distributional variables share the property of *scale invariance* which needs to be taken into account for their statistical processing. A key point is a proper sample space of relative data which is reflected by the Bayes space methodology [37], that leads to the Aitchison geometry on the simplex [13] when considering the special case of discrete distributions (expressed through compositional data).

In this chapter we focus on the continuous case of distributional variables, on probability density functions, and demonstrate the use of the Bayes space methodology in the development of a classification procedure generalizing the classical method of *linear discriminant analysis*. Therefore, the next section is devoted to describe Bayes spaces, and this forms the milestone to introduce the statistical analysis of density functions through the logratio approach. Concrete aspects of their classification, together with computational aspects related to their spline approximation are discussed in Section 9.3. A data set of particle size distributions, sampled in different localities, is employed in Section 9.4 to illustrate the methodological developments. Finally, Section 9.5 concludes.

9.2 Bayes spaces

As it was stated in Section 9.1, a proper choice of the sample space for probability density functions (PDFs) is essential. In particular, it was demonstrated e.g. in [17, 35] that analyzing PDFs within the usual L^2 space, for which most methods of Functional Data Analysis (FDA) [32] have been designed, may lead to meaningless results. Instead, the relative character of densities can be captured through Bayes spaces that rely upon an appropriate Hilbert space structure to deal with the data constraints.

Here we recall basics of the Bayes space methodology. To this end, we consider two positive functions f and g with the same support to be equivalent, if $f = c \cdot g$, for a positive constant c. Recalling the scale invariance of PDFs, this implies that densities (not necessarily unit-integral densities, i.e., PDFs) within an equivalence class provide the same relative information, or, equivalently, that ratios between contributions of Borel sets to the whole mass measure do not change. The Bayes space $\mathcal{B}^2(I)$ consists of (equivalence classes of) densities f on a domain I for which the logarithm is square-integrable. The

Bayes space methodology [37] is general and allows dealing with unbounded supports I as well as to choose a reference measure that is different from the Lebesgue one. For the unbounded support it is a must [37], however, it is also possible for a bounded support because it enables to weight the domain according to the aim of the analysis [34]. For the purpose of this chapter, the focus is on the case of a compact support $I = [a, b] \subset \mathbf{R}$ with the Lebesgue measure, which has been demonstrated to be of broad applicability by several authors [7, 17, 25–27, 35].

In $\mathcal{B}^2(I)$, the counterparts of sum and multiplication by a scalar are called *perturbation* and *powering*, which are defined, for $f, g \in \mathcal{B}^2(I)$ and $c \in \mathbf{R}$, as

$$(f \oplus g)(t) = \frac{f(t)g(t)}{\int_a^b f(s)g(s)ds} = \mathcal{C}(fg)(t); \qquad (c \odot f)(t) = \frac{f^c(t)}{\int_a^b f^c(s)ds} = \mathcal{C}(f^c)(t),$$

where $t \in I = [a, b]$ and $\mathcal{C}(f)$ stands for the unit-integral representation within the equivalence class of proportional densities, also named *closure*. Note that $e(t) = \frac{1}{b-a}$ (uniform density on $[a, b]$) is the neutral element of perturbation. The Bayes inner product is finally defined as

$$\langle f, g \rangle_{\mathcal{B}} = \frac{1}{2\eta} \int_a^b \int_a^b \ln \frac{f(t)}{f(s)} \cdot \ln \frac{g(t)}{g(s)} dt \, ds,$$

where η stands for the length of interval the I, i.e., $\eta = b-a$. The corresponding norm and distance are

$$\|f\|_{\mathcal{B}} = \sqrt{\langle f, f \rangle_{\mathcal{B}}}; \quad d_{\mathcal{B}}(f, g) = \|f \ominus g\|_{\mathcal{B}},$$

respectively, where $\ominus$ stands for *perturbation-subtraction* of f by g, $(f \ominus g)(t) = [f \oplus (-1) \odot g](t)$, for t in I.

In [12] and [37] it was shown that the Bayes space $(\mathcal{B}^2(I), \oplus, \odot, \langle \cdot, \cdot \rangle_{\mathcal{B}})$ forms a separable Hilbert space. Therefore, there exists an isometric isomorphism between the Bayes space $\mathcal{B}^2(I)$ and the space $L^2(I)$ of square integrable real functions on a compact support I. An instance of such isometric isomorphism is called *centered log-ratio (clr) transformation*, defined for a PDF $f \in \mathcal{B}^2(I)$ as

$$f_c(t) = \operatorname{clr}[f](t) = \ln f(t) - \frac{1}{\eta} \int_I \ln f(s) \, ds, \quad t \in I. \tag{9.1}$$

The clr representation of a PDF is featured by a zero-integral constraint on I, i.e., $\int_I f_c(t)dt = 0$, and accordingly we refer to the space $L_0^2(I) \subset L^2(I)$. When analyzing clr transforms of densities, the zero integral constraint may give rise to computational issues and thus needs to be properly accounted for. Given a clr transform $f_c \in L_0^2(I)$, the corresponding density $f \in \mathcal{B}^2(I)$ can be obtained through the inverse transformation, $f(t) = \operatorname{clr}^{-1}[f_c](t) = \mathcal{C}(\exp[f_c])(t), t \in I$, where $\mathcal{C}$ denotes the closure. Finally, we point out that the following important properties of the isometric isomorphism (9.1) hold,

$$\operatorname{clr}(f \oplus g)(t) = f_c(t) + g_c(t), \quad \operatorname{clr}(c \odot f)(t) = c \cdot f_c(t), \quad \langle f, g \rangle_{\mathcal{B}} = \langle f_c, g_c \rangle_2,$$

where $\langle \cdot, \cdot \rangle_2$ denotes the inner product in $L_0^2(I)$. Intuitively, the clr transformation translates operations and metrics of the Bayes space into the usual operations and metrics of the L^2 space. Then, by taking into account the zero integral constraint of clr-transformed densities, any method of FDA including those aimed at classification of functional data can be applied.

Note that also in symbolic data analysis it is popular to search for a proper representation of distributional data which would enable their further statistical processing. One particular case is the quantile representation of distributional data [18]. The principle is to express the observed variable values by some predefined quantiles of the underlying distribution. For example, for categorical multi-valued variables, quantiles may be determined from the ranking of the categories based on their frequencies, or other designed methods. In the simplest case, when quartiles are chosen, the representation for each variable is defined by the (Min, Q_1, Q_2, Q_3, Max), which form a kind of coordinates of the variable. The same representation can also be applied to continuous distributions, and it is possible to use the whole quantile function for the purpose of FDA [9,10,16,38,39]. Although one can honor scale invariance by following this approach, it lacks a deeper geometric reasoning.

A further peculiarity, characterizing the FDA approach, is the necessity of smoothing the input raw data, if these represent a discretized set of observations of the underlying continuous functions (densities in our case). For this purpose, a spline representation [6,11] is typically used. Specifically, when a B-spline representation of clr-transformed densities $f_1, \ldots, f_n$ is concerned, we denote by $\{\beta_l, l = 1, \ldots, L\}$ a given basis system and express $\mathrm{clr}(f_i)(t)$, $i = 1, \ldots, n$ on such basis through spline coefficients η_{il} as

$$\mathrm{clr}(f_i(t)) = \sum_{l=1}^{L} \eta_{il} \beta_l(t) \tag{9.2}$$

or, in matrix notation, $\mathrm{clr}(f_i(t)) = \boldsymbol{\beta}(t)^T \boldsymbol{\eta}_i$. For clr-transformed densities it is in addition necessary to respect the zero-integral constraint of functions in the $L_0^2(I)$ space. This problem was addressed in [22, 23] and the resulting *compositional splines* will form the input data also for the classification procedure developed in the next section.

9.3 Functional linear discriminant analysis for density functions

Data classification is considered as one of the main tasks not only in conventional multivariate analysis but also in FDA. There is a number of methods available in the literature which rely on Fisher's linear discriminant analysis [5], k-nearest neighbor classification [3], mixture modeling [4], or data depth [28]. Here, we adapt one of the pioneering methods for classification of functional data, based on linear discriminant analysis (LDA) [19], to PDFs. In fact, it is possible to develop any of the latter classification procedures directly on clr-transformed PDFs. The key issue is a proper representation of PDFs using splines that honor the zero-integral constraint.

9.3.1   The FLDA model

For development of functional linear discriminant analysis (FLDA) we assume that $s_{ih}(t)$, $t \in I$, are the observed densities for the i-th individual from the h-th class. Then, $s_{ih}(t)$ can be expressed as

$$s_{ih}(t) = \mu_h(t) \oplus \epsilon_{ih}(t), \quad i = 1, \ldots, n_h, \quad h = 1, \ldots, k, \qquad (9.3)$$

where $\mu_h(t)$ is a common effect (density) of the h-th class and $\epsilon_{ih}(t)$ is a functional measurement error. However, as previously stated, observing functional data (observations in the form of functions) in real world is often impossible. Therefore, in the following, we shall refer our developments to discretely observed data, more precisely, to the discrete observation of the clr-transformed density functions.

The FLDA model is built in several steps [19]. First, consider I to be a series of points where the raw data are measured. Then, let $s_{ih}(t)$ be the observed value at $t \in I$ for the i-th individual from the h-th class. This way, each vector s_{ih} represents one (discretized) functional observation (i.e., the clr transformed density). For simplicity of notation, we assume that the measuring took place at the same points for all densities, say corresponding to representatives of histogram classes [23]. Otherwise, an extended grid would be necessary to include all measuring points. An alteration of the presented method can then be used even for irregularly measured data [19].

With the original assumption, s_{ih} can be further decomposed as

$$s_{ih} = g_{ih} + \epsilon_{ih}, \quad i = 1, \ldots, n_h, \quad h = 1, \ldots, k,$$

where g_{ih} is the vector of true values and ϵ_{ih} is the vector of measurement errors. Originally, the natural cubic spline functions with an equidistant sequence of knots were chosen to represent the functional data in the general case [19] as they can be easily computed using statistical software. Through the spline representation, g_{ih} can be written as

$$g_{ih} = B\eta_{ih}, \qquad (9.4)$$

where

$$B = (\beta(t_1), \ldots, \beta(t_n))^T$$

represents the basis matrix at values $t_1, \ldots, t_n \in I$ and η_{ih} is an L-dimensional vector of spline coefficients, where L stands for the number of basis functions. The natural cubic splines, used in [19], will be replaced by compositional splines here (see Section 9.3.3 for further details), however, the formal representation remains the same. Note that, for a given order of the spline functions and sequence of knots [6], the same spline basis stands for all observations as long as they have been measured on the same grid of representative points. Consequently, a common spline matrix B for all observations will be used further in the following.

For the next step, assume that coefficients η_{ih} can be modeled using a Gaussian distribution, that is,

$$\eta_{ih} = \mu_h + \gamma_{ih}, \quad \gamma_{ih} \sim N(\mathbf{0}, \boldsymbol{\Gamma}),$$

where $\boldsymbol{\Gamma}$ is the covariance matrix. In the general functional case, the assumption of normally distributed independent measurement errors ϵ_{ih} would lead to

$$s_{ih} = B(\mu_h + \gamma_{ih}) + \epsilon_{ih}, \quad i = 1, \ldots, n_h, \quad h = 1, \ldots, k, \quad (9.5)$$

$$\epsilon_{ih} \sim N(\mathbf{0}, \sigma^2 \boldsymbol{I}), \quad \gamma_{ih} \sim N(\mathbf{0}, \boldsymbol{\Gamma})$$

Due to the zero integral constraint, however, this assumption cannot be made for the clr densities, as the relation between the errors should reflect the zero integral constraint. Therefore, we expect the measurement errors to have a singular normal distribution [21], that is,

$$\epsilon_{ih} \sim N(\mathbf{0}, \sigma^2 \boldsymbol{V}\boldsymbol{V}^T),$$

with $\boldsymbol{V}\boldsymbol{V}^T = \boldsymbol{I}_{n_h} - \mathbf{1}_{n_h \times n_h}/n_h$ being an idempotent matrix. Here the matrix $\boldsymbol{V}$ stands for a transformation matrix between clr coefficients of (multivariate) compositional data and their respective isometric logratio (orthonormal) coordinates [15]. This assumption thus indicates that if the constrained clr data would be expressed as coordinates with respect to *any* orthonormal basis, then the measurement errors would be normally distributed with independent components.

According to [19], the following constraint can be placed onto the class means in the model to provide a form of normalization for the linear discriminants,

$$\mu_h = \lambda_0 + \Lambda\alpha_h, \quad \Lambda^T C^{-1}\Lambda = I, \quad \sum_{h=1}^{k} \alpha_h = \mathbf{0}. \quad (9.6)$$

Here, the q-dimensional parameters α_h enable the original observations to be represented in a reduced q-dimensional discriminant space. The results obtained using this constraint are identical to reducing the rank of the FLDA procedure (similarly as for rank-reduced LDA). This way, the FLDA model can be written in its final form as

$$s_{ih} = B(\lambda_0 + \Lambda\alpha_h + \gamma_{ih}) + \epsilon_{ih}, \quad i = 1, \ldots, n_h, \quad h = 1, \ldots, k, \quad (9.7)$$

$$\gamma_{ih} \sim N(\mathbf{0}, \boldsymbol{\Gamma}), \quad \epsilon_{ih} \sim N(\mathbf{0}, \sigma^2 \boldsymbol{V}\boldsymbol{V}^T),$$

$$\sum_{h=1}^{k} \alpha_h = 0, \quad \Lambda^T B^T C^{-1} B \Lambda = I,$$

as the last constraint ensures that λ_0, Λ and α_h are not confounded. Note the importance of α_h – these coefficients represent the projected linear discriminants and they will be used for the classification itself.

9.3.2   Classification using the FLDA model

Referring to Equation (9.7), each observation s_{ih} has normal distribution with mean $B\mu_h = B(\lambda_0 + \Lambda\alpha_h)$ and covariance matrix $\Sigma = \sigma^2 VV^T + B\Gamma B^T$. Under the assumption of independence, the joint distribution of observed clr-transformed densities could be formally written as

$$\prod_{h=1}^{k} \prod_{i=1}^{n_h} \frac{1}{(2\pi)^{\frac{n_h}{2}} |\Sigma|^{\frac{1}{2}}} \exp\left\{ -\frac{1}{2} (s_{ih} - B[\lambda_0 + \Lambda\alpha_h])^T \Sigma^{-1} (s_{ih} - B[\lambda_0 + \Lambda\alpha_h]) \right\}$$
(9.8)

with unknown parameters $\lambda_0, \Lambda, \alpha_h, \Gamma$ and σ^2. Nevertheless, because the matrix Σ is singular, (9.8) needs to be reformulated by using rank-reduced data as discussed further in the text.

To find the maximum value of (9.8), it is convenient to treat γ_{ih} as latent variables. In that case, the *EM* algorithm [8] can be used to maximize the respective joint likelihood function of s_{ih} and γ_{ih} [19]

$$\prod_{h=1}^{k} \prod_{i=1}^{n_h} \frac{1}{(2\pi)^{\frac{n_h+q}{2}} \sigma^{n_h} |\Gamma|^{\frac{1}{2}}}$$
$$\exp\left[-\frac{1}{2\sigma^2} (s_{ih} - B[\lambda_0 + \Lambda\alpha_h + \gamma_{ih}])^T (s_{ih} - B[\lambda_0 + \Lambda\alpha_h + \gamma_{ih}]) - \frac{1}{2}\gamma_{ih}^T \Gamma^{-1} \gamma_{ih} \right].$$
(9.9)

In the *EM* algorithm, the *E*-step consists of computing

$$E(\gamma_{ih}|s_{ih}, \lambda_0, \Lambda, \alpha_h, \Gamma, \sigma^2) = \left(\sigma^2 \Gamma^{-1} + B^T B \right)^{-1} B^T (s_{ih} - B\lambda_0 - B\Lambda\alpha_h),$$
(9.10)

while the *M*-step maximizes the log-likelihood function

$$-\frac{1}{2} \sum_{h=1}^{k} \sum_{i=1}^{n_h} E\left\{ (s_{ih} - B[\lambda_0 - \Lambda\alpha_h - \gamma_{ih}])^T \right.$$
$$\left. (s_{ih} - B[\lambda_0 - \Lambda\alpha_h - \gamma_{ih}]) / \sigma^2 + n_h \ln(\sigma^2) + \gamma_{ih}^T \Gamma^{-1} \gamma_{ih} + \log|\Gamma| \right\} \quad (9.11)$$

for fixed values γ_{ih}. By using an appropriate stopping criterion, the *EM* algorithm provides the optimal values of parameters $\lambda_0, \Lambda, \alpha_h, \Gamma, \sigma^2$ for which the log-likelihood function reaches its local or global maximum. These parameters are then used for classification. Note that reaching a local rather than the global maximum of the log-likelihood function (9.9) is further discussed in [19]. This problem may appear for very sparsely measured data and can be partially solved by placing a rank constraint on $\boldsymbol{\Gamma}$.

To classify each observation (original or new) into one of the given classes, the Bayes' formula is modified for the functional case. The clr transformed (discretized) density $\boldsymbol{s}$ is assigned to class C_h if h minimizes

$$\arg\min{}_h \left(\boldsymbol{s} - \boldsymbol{B}\boldsymbol{\lambda}_0 - \boldsymbol{B}\boldsymbol{\Lambda}\boldsymbol{\alpha}_h\right)^T \boldsymbol{\Sigma}^{-1} \left(\boldsymbol{s} - \boldsymbol{B}\boldsymbol{\lambda}_0 - \boldsymbol{B}\boldsymbol{\Lambda}\boldsymbol{\alpha}_h\right) - 2\ln\pi_h, \quad (9.12)$$

where π_h stands for the prior probability of $\boldsymbol{s}$ being from class C_h.

Since for this case $\boldsymbol{B}$ (and therefore $\boldsymbol{\Sigma}$) is the same for all functions, the classification rule for $\boldsymbol{s}$ can be further simplified to

$$\arg\min{}_h \left(\|\hat{\boldsymbol{\alpha}}_{\boldsymbol{s}} - \boldsymbol{\alpha}_h\|^2 - 2\ln\pi_h\right), \quad (9.13)$$

where

$$\hat{\boldsymbol{\alpha}}_{\boldsymbol{s}} = \left(\boldsymbol{\Lambda}^T \boldsymbol{B}^T \boldsymbol{\Sigma}^{-1} \boldsymbol{B}\boldsymbol{\Lambda}\right)^{-1} \boldsymbol{\Lambda}^T \boldsymbol{B}^T \boldsymbol{\Sigma}^{-1}(\boldsymbol{s} - \boldsymbol{B}\boldsymbol{\lambda}_0)$$

can be interpreted as the linear discriminant of $\boldsymbol{s}$. Therefore, both $\hat{\boldsymbol{\alpha}}_{\boldsymbol{s}}$ and $\boldsymbol{\alpha}_h$ represent points from $\mathbb{R}^q$. Up to the constant $-2\ln\pi_h$, the distance between the class representatives $\boldsymbol{\alpha}_h$ and $\hat{\boldsymbol{\alpha}}_{\boldsymbol{s}}$ is minimized in (9.13).

To reduce the potentially highly variable fit to the data, a p-rank constraint is placed on $\boldsymbol{\Gamma}$ by expressing $\boldsymbol{\Gamma} = \boldsymbol{\Theta}\boldsymbol{D}\boldsymbol{\Theta}^T$. This decomposition leads to

$$\mathrm{Cov}(\boldsymbol{s}) = \sigma^2 \boldsymbol{I} + \boldsymbol{B}_{\boldsymbol{s}}\boldsymbol{\Theta}\boldsymbol{D}\boldsymbol{\Theta}^T \boldsymbol{B}_{\boldsymbol{s}}^T. \quad (9.14)$$

The FLDA procedure can thus be summarized as follows:

- calculate the appropriate spline matrix $\boldsymbol{B}$;

- compute the initial estimates of the required parameters $\boldsymbol{\lambda}_0, \boldsymbol{\Lambda}, \boldsymbol{\alpha}, \boldsymbol{\Theta}, \sigma^2$ and $\boldsymbol{D}$;

- perform both steps of the *EM* algorithm and calculate the value of the log-likelihood function in each iteration;

- determine the optimal values of the parameters of interest under model (9.7).

To start the iterative process leading to the classification, initial values of $\boldsymbol{\lambda}_0, \boldsymbol{\Lambda}, \boldsymbol{\alpha}, \boldsymbol{\Theta}, \sigma^2$ and $\boldsymbol{D}$ are required. If the correct classification for a significant part of the data set is known, the previous relations (9.6), (9.14) can be used for the parameter estimates. Obtaining the B-spline basis $\boldsymbol{B}$ and the respective coefficients $\boldsymbol{\eta}_{ih}$ is described in detail in Section 9.3.3.

9.3.3 Spline representation of PDFs

The usual starting point in FDA are discretized data, resulting here from the multivariate clr transformation [1] of proportions corresponding to histogram classes [23]. An adequate method is required to fill in the sparsely measured data and to obtain a smooth curve for each observation. A common way of practice for functions from the L^2 space is to use a B-spline representation [32]. The point is that the standard B-spline basis does not honor the zero-integral constraint of clr-transformed densities and the singularity constraint it then propagated into the spline coefficients [35]. In order to avoid this inconvenience, we briefly recall how a B-spline basis is constructed directly in $L_0^2(I)$ [23].

Let $\lambda_0 = a < \lambda_1 < \ldots \lambda_g < b = \lambda_{g+1}$ be a sequence of knots on $I = [a, b]$. Then B-spline of degree $k, k \in \mathbb{N}$, is defined by

$$B_i^{k+1}(x) = \frac{x - \lambda_i}{\lambda_{i+k} - \lambda_i} B_i^k(x) + \frac{\lambda_{i+k+1} - x}{\lambda_{i+k+1} - \lambda_{i+1}} B_{i+1}^k(x),$$

while for $k = 0$

$$B_i^1(x) = \begin{cases} 1, & x \in [\lambda_i, \lambda_{i+1}), \\ 0, & \text{otherwise} \end{cases}$$

with $k + 1$ expressing the order of the B-spline.

Let $Z_i^{k+1}(x)$ be functions defined as

$$Z_i^{k+1}(x) = \frac{\mathrm{d}}{\mathrm{d}x} B_i^{k+2}(x)$$

for $k \in \mathbb{N}_0$. It can be shown that $Z_i^{k+1}(x)$ hold similar properties as $B_i^{k+1}(x)$ of being piecewise polynomials of degree k and having continuous derivatives up to degree $k - 1$. The vector space $\mathcal{S}_k^{\Delta\lambda}[a, b]$ of polynomial splines of degree $k \in \mathbb{N}_0$ with given sequence of knots on I has the dimension $g + k + 1$.

Additional knots need to be added to involve all functions $B_i^{k+1}(x)$ forming the basis – in this case

$$\lambda_{-k} = \cdots = \lambda_{-1} = a, \quad \lambda_{g+2} = \cdots = \lambda_{g+k+1}.$$

This way, every spline s_k from $\mathcal{S}_k^{\Delta\lambda}[a, b]$ can be uniquely represented by

$$s_k(x) \in \sum_{i=-k}^{g} b_i B_i^{k+1}(x).$$

Now consider the spline functions $Z_i^{k+1}(x)$ with the added constraint of zero integral on the relevant vector space $\mathcal{Z}_k^{\Delta\lambda}[a, b]$, i.e.

$$\mathcal{Z}_k^{\Delta\lambda}[a, b] := \left\{ s_k(x) \in \mathcal{S}_k^{\Delta\lambda}[a, b] : \int_I s_k(x)\mathrm{d}x = 0 \right\}.$$

It can be proven [23] that this vector space has dimension $g + k$ and that the corresponding functions $Z_{-k}^{k+1}(x), \ldots, Z_{g-1}^{k+1}(x)$ form its basis.

Furthermore, $s_k(x)$ can be written in a matrix notation [23] as

$$s_k(x) = \boldsymbol{Z}_{k+1}(x)\boldsymbol{z} = \boldsymbol{B}_{k+1}(x)\boldsymbol{D}\boldsymbol{K}\boldsymbol{z}, \tag{9.15}$$

where

$$\boldsymbol{Z}_{k+1}(x) = \left(Z_{-k}^{k+1}(x), \ldots, Z_{g-1}^{k+1}(x)\right), \quad \boldsymbol{B}_{k+1}(x) = \left(B_{-k}^{k+1}(x), \ldots, B_g^{k+1}(x)\right),$$

$$\boldsymbol{D} = (k+1) \operatorname{diag} \left(\frac{1}{\lambda_1 - \lambda_{-k}}, \ldots, \frac{1}{\lambda_{g+k+1} - \lambda_g}\right),$$

$$\boldsymbol{K} = \begin{pmatrix} 1 & 0 & 0 & \cdots & 0 & 0 \\ -1 & 1 & 0 & \cdots & 0 & 0 \\ 0 & -1 & 1 & \cdots & 0 & 0 \\ \vdots & \vdots & \ddots & \ddots & \vdots & \vdots \\ 0 & 0 & 0 & \cdots & -1 & 1 \\ 0 & 0 & 0 & \cdots & 0 & -1 \end{pmatrix} \in \mathbb{R}^{g+k+1, g+k}$$

and $\boldsymbol{z} = (z_{-k}, \ldots, z_{g-1})^T$ being a vector of spline coefficients. Note that $\boldsymbol{Z}_{k+1}(x)$ is characterized fully by the degree k and a given sequence of knots. Thus, every function $s_k(x)$ can be obtained through a common B-spline basis characterized by $\boldsymbol{Z}_{k+1}(x) = \boldsymbol{B}_{k+1}(x)\boldsymbol{DK}$ and a unique vector $\boldsymbol{z}$ while including the zero integral constraint. As a result, $\boldsymbol{Z}_{k+1}(x)$ can be used as a basis for the FLDA classification model, that is,

$$\boldsymbol{S} = \boldsymbol{Z}_{k+1}(x) = \boldsymbol{B}_{k+1}(x)\boldsymbol{DK},$$

with $\boldsymbol{z}$ corresponding to spline coefficients η_{ih} from previous section.

9.4 Classification of particle size distributions

In this section, the introduced methodology is demonstrated on a real example. The data for this application contain particle size distributions (PSDs) measured at four different locations in the Czech Republic (Dobšice, Brodek u Prostějova, Rozvadovice, Ivaň; further denoted as classes 1-4) with locations playing the role of classes [33]. The data set consists of 250 vectors (each corresponding to a unique PSD), and the different classes are represented unevenly, with sample sizes 96, 39, 66, and 49. Note that the site of origin (and therefore the correct classification) of the measured samples is known – thus it is possible to estimate the required parameters and to examine the quality of the classification model. Although not shared publicly, the dataset is available from the authors upon request.

9.4.1 Preprocessing of the raw data

Each observation of the data set of interest is a 102-dimensional vector and describes the distribution of particles in different soil samples. The values are sorted by increasing grain size varying on a scale from 0,088 μm to 2000 μm, and they describe the relative proportion of the given grain size in the sample. Therefore, the elements of each vector sum up to 1 which indicates the compositional character of the data.

To reduce the effect of the units, a logarithmic scale was chosen to represent the particle sizes. Due to the occurrence of zero values, their replacement by non-zero values has to be addressed, as a clr transformation cannot be applied to data containing zeros. A common approach is to substitute zeros by a multiple of the smallest positive value in the data set [30]. In this case, a detection limit 0.0015 was chosen to eliminate the influence of extremely small values.

The empirical probability densities can be obtained from the raw data in the following way. Values of the considered scale (here 102 log-transformed values) can be seen as the class representatives of the histogram. Therefore, each probability stands for the relative proportion of the given particle size corresponding to its histogram class. If the original grid is not equidistant, the representing values (that is, centers of the histogram classes) need to be slightly altered to ensure that the lengths of the resulting histogram classes are equal. The obtained empirical densities can be finally transformed using the clr transformation for discrete compositions [15], which is given for a D-dimensional vector by

$$\operatorname{clr}(\boldsymbol{x}) = \left(\ln \left(\frac{x_1}{g(\boldsymbol{x})} \right), \ldots, \ln \left(\frac{x_D}{g(\boldsymbol{x})} \right) \right), \quad \boldsymbol{x} \in \mathbb{R}^D,$$

where $g(\boldsymbol{x})$ is the geometric mean of the components of $\boldsymbol{x}$. The resulting densities with their clr transformations are displayed in Figure 9.1.

9.4.2   Spline representation

As described in Section 9.3.3, an appropriate spline representation is a key element for performing the FLDA procedure. Cubic spline functions were chosen together with a non-equidistant sequence of knots. While choosing the number and placement of the knots, the length and refinement of the grid should be taken into account as well as the shape of considered curves. It is common to place knots at the endpoints of the grid; for the inner knots, they should be placed densely where the behavior of the curves changes significantly. In this way, the resulting splines better respect the original curves and characterize them more precisely [23]. The sequence of 16 knots chosen for this case is shown in Figure 9.1 by dashed gray vertical lines. Consequently, the resulting spline matrix $\boldsymbol{B}$ is a $102 \times (16 + 1)$ matrix. In Figure 9.2, the Z-spline basis is illustrated as well as its corresponding B-spline basis without the zero integral constraint.

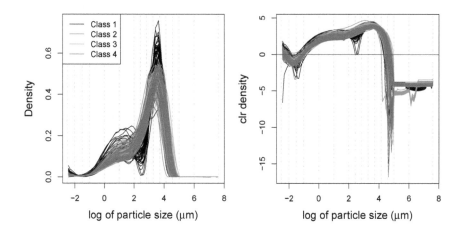

FIGURE 9.1
The empirical densities (left) and the corresponding clr transformations
(right), where the color represents the class membership. The vertical dashed
lines represent the knot positions.

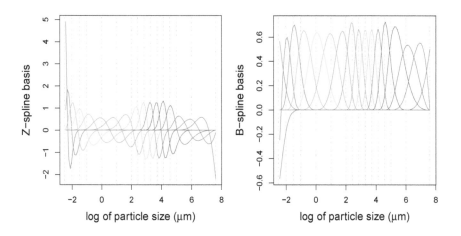

FIGURE 9.2
Z-spline basis (left) and B-spline basis (right) of cubic functions for the given
data.

9.4.3 Running the FLDA procedure

The FLDA model for L^2 data was implemented in the software environment R [31] in [19]. To apply the described FLDA procedure to PDF data, some adaptations were done to the initial code, which is available at http://faculty.marshall.usc.edu/gareth-james/Research/Research.html. The R-package *robCompositions* [36] was used to perform the clr transformation during the preprocessing of the original data. Besides the initial data set, its grid of points and the given classes, the required arguments for the procedure include the order and sequence of the knots for computing the adequate spline basis.

For running the FLDA procedure, the steps described at the end of Section 9.3.2 were followed. The potential risk of the *EM* algorithm failing to find a global maximum was already addressed in Section 9.3.2. Due to the numerical computation of the log-likelihood function, the convergence criterion for the *EM* algorithm should also be discussed. Here we look at the change in the computed log-likelihoods in consecutive iterations and stop the algorithm when the relative improvement is smaller than a chosen tolerance. Another option is to simply define the maximum number of iterations that the algorithm should run – at the risk that convergence may not be achieved if the maximum number of iterations is too low.

The estimated parameters may be further used to determine the expected curves (centroids) μ_h from Equation (9.6). Their display, along with the original observations, can be seen in Figure 9.3. Furthermore, the distances between the observations and the centroids can be examined. It can be shown [19] that $\hat{\alpha}_s$ and α_h from Equation (9.13) stand for the linear discriminants of s and μ_h, respectively, and therefore they can be used for a lower-dimensional representation of the classification problem. The absolute differences between the linear discriminants can be interpreted as the projected distances between observations and centroids representing particular classes, as shown in Figure 9.4 for two-dimensional $\hat{\alpha}_s$ and α_h. Finally, it is possible to make class predictions: the i-th observation is expected to be from class C_h if the distance between the curve and the h-th centroid is minimal among all classes.

9.4.4 Quality of the classification

Since the true origin of the observations is known for the considered data set, prediction error rates can be determined. To evaluate the quality of the class predictions of the established model, repeated K-fold cross-validation (CV) was performed. In each of the $N = 100$ replications, the observations from the original data set were randomly divided into $K = 5$ groups with respect to the uneven proportions of the represented classes. In turn, $K - 1$ groups were used for training the model, and the remaining K-th fold was used as a testing data set. The resulting misclassification rate is around 0.2,

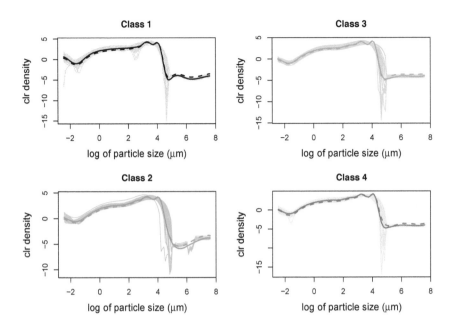

FIGURE 9.3
Observations from the particular classes (light) and their corresponding centroids calculated with the initial (bold dashed) and final (bold solid) parameter estimates.

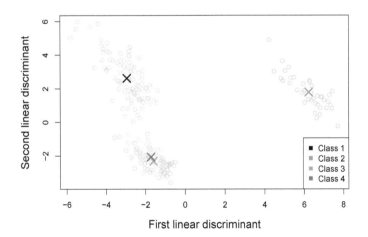

FIGURE 9.4
Two-dimensional projection of observations $\hat{\boldsymbol{\alpha}}_s$ (shown by symbols ○) and the class centroids $\boldsymbol{\alpha}_h$ (shown by symbols ×).

meaning that approximately 80% of the observations are correctly classified (within-class results are summarized in Table 9.1).

TABLE 9.1

Percentages of correctly classified observations from all classes and the respective weighted mean for the described cross-validation procedure.

C_1	C_2	C_3	C_4	overall weighted mean
0.9797	1.0000	0.6069	0.5051	0.7914

Already from Figure 9.1 it could be seen that there are some differences in the considered groups. In the interval $[5, 8]$, the shape of the functions for class C_2 (red) differs clearly, and in the interval $[2, 3]$ the class C_1 functions (black) show some deviation. Also Figure 9.4 clearly reveals a clear separation of C_1 and C_2 from the rest, where C_3 and C_4 seem to be hard to distinguish. Also the average squared distances of the observations to the class centroids show clear class differences, as computed in Table 9.2.

TABLE 9.2

Average squared distances between observations from class C_i (rows) and centroids from class C_j (columns).

	C_1	C_2	C_3	C_4
C_1	1.276	84.421	28.323	23.784
C_2	86.119	2.181	78.510	79.009
C_3	25.865	80.760	1.117	1.140
C_4	25.146	78.966	0.721	0.733

Another approach to consider is to see how well the model distinguishes between observations from each class and the remaining classes joined together. To follow this idea, only one original class is considered, and the remaining observations form the second class. The FLDA model was then used as above with repeated 5-fold CV on the proportionally sampled data to calculate the average accuracy of classification as summarized in Table 9.3. These results show that distinguishing C_1 and especially C_2 from the rest is very successful, but also classes C_3 and C_4 can be well distinguished from the remaining observations.

TABLE 9.3

Average accuracy of classification with $N = 100$ iterations of 5-fold CV, performed for each considered class against the rest.

	Proportion of correctly classified observations	
Original class kept	Original class	Remaining observations
1	0.988	0.966
2	1.000	1.000
3	0.946	0.742
4	0.948	0.696

9.5   Conclusions

Probability density functions represent an increasingly popular type of distributional data. This is because frequently it is necessary to aggregate massive data to a sufficient detail (which is a usual task in symbolic data analysis), or simply because of the distributional character of the observations, such as particle size distributions as analyzed here. The classification of density functions is a challenging task due to their relative character. Since the Bayes space methodology enables expressing these distributional data in the L_0^2 space, it is possible to apply standard methods of functional data analysis (FDA) by taking into account the zero integral constraint of clr transformed densities. This also holds for the respective spline representation which forms a cornerstone of many FDA methods. The functional linear discriminant analysis procedure, adapted here from [19] for the case of densities, avoids problems resulting from singularity of the clr transformed observations. By imposing a dimension reduction, the classification rule can be developed.

A proper treatment of probability density functions using the Bayes space methodology does not necessarily lead to better classification results, similarly as in case of discrete distributions (or, more in general, multivariate compositional data) using the logratio approach [14, 15]. Nevertheless, there is a guarantee that the densities are appropriately processed, by taking their scale invariance property into account and, accordingly, that *reasonable* outputs are produced. In any case, we aim to continue with the development of efficient classification procedures for probability density functions in the near future.

Acknowledgments

Karel Hron and Ivana Pavlů gratefully acknowledge the support of the grant No. GA19-01768S of the Grant Agency of the Czech Republic and the grant IGA_PrF_2021_008 Mathematical Models of the Internal Grant Agency of the Palacký University in Olomouc.

Bibliography

[1] J. Aitchison. *The Statistical Analysis of Compositional Data.* Chapman & Hall, London, 1986.

[2] L. Billard and E. Diday. *Symbolic Data Analysis.* Wiley, Chichester, 2006.

[3] F. Burba, F. Ferraty, and P. Vieu. *k*-nearest neighbors method in functional nonparametric regression. *Journal of Nonparametric Statistics,* 21:453–469, 2009.

[4] F. Chamroukhi and H.D. Nguyen. Model-based clustering and classification of functional data. *WIREs,* 9(4):e1298, 2019.

[5] L.-H. Chen and C.-R. Jiang. Sensible functional linear discriminant analysis. *Computational Statistics & Data Analysis,* 126:39–52, 2018.

[6] C. De Boor. *A Practical Guide to Splines.* Springer-Verlag, New York, 1978.

[7] P. Delicado. Dimensionality reduction when data are density functions. *Computational Statistics & Data Analysis,* 55(1):401–420, 2011.

[8] A. Dempster, N. Laird, and D. Rubin. Maximum likelihood from incomplete data via the EM algorithm. *Journal of the Royal Statistical Society: Series B (Statistical Methodology),* 39:1–38, 1977.

[9] S. Dias and P. Brito. Linear regression model with histogram-valued variables. *Statistical Analysis and Data Mining,* 8:75–113, 2015.

[10] S. Dias, P. Brito, and P. Amaral. Discriminant analysis of distributional data via fractional programming. *European Journal of Operational Research,* 294(1):206–218, DOI: 10.1016/j.ejor.2021.01.025, 2021.

[11] P. Dierckx. *Curve and Surface Fitting with Splines.* Oxford University Press, Oxford, 1995.

[12] J.J. Egozcue, J.L. Díaz-Barrero, and V. Pawlowsky-Glahn. Hilbert space of probability density functions based on Aitchison geometry. *Acta Mathematica Sinica, English Series,* 22(4):1175–1182, 2006.

[13] J.J. Egozcue, V. Pawlowsky-Glahn, G. Mateu-Figueras, and C. Barceló-Vidal. Isometric logratio transformations for compositional data analysis. *Mathematical Geology,* 35(3):279–300, 2003.

[14] P. Filzmoser, K. Hron, and M. Templ. Discriminant analysis for compositional data and robust parameter estimation. *Computational Statistics,* 27:585–604, 2012.

[15] P. Filzmoser, K. Hron, and M. Templ. *Applied Compositional Data Analysis*. Springer, Cham, 2018.

[16] M. Guo, L. Zhou, J.Z. Huang, and W.K. Härdle. Functional data analysis of generalized regression quantiles. *Statistics and Computing*, 25:189–202, 2013.

[17] K. Hron, A. Menafoglio, M. Templ, K. Hrůzová, and P. Filzmoser. Simplicial principal component analysis for density functions in Bayes spaces. *Computational Statistics & Data Analysis*, 94:330–350, 2016.

[18] M. Ichino. Symbolic PCA for histogram-valued data. In *Proceedings of the IASC 2008*, pages 1–10, Yokohama, Japan, 2008.

[19] G.M. James and T.J. Hastie. Functional linear discriminant analysis for irregularly sampled curves. *Journal of the Royal Statistical Society: Series B (Statistical Methodology)*, 63:533–550, 2001.

[20] R.A. Johnson and D.W. Wichern. *Applied Multivariate Statistical Analysis, 6th ed.* Prentice Hall, London, 2007.

[21] K. S. Kwong and B. Iglewicz. On singular multivariate normal distribution and its applications. *Computational Statistics and Data Analysis*, 22:271–285, 1996.

[22] J. Machalová, K. Hron, and J.S. Monti. Preprocessing of centred logratio transformed density functions using smoothing splines. *Journal of Applied Statistics*, 43:1419–1435, 2016.

[23] J. Machalová, R. Talská, K. Hron, and A. Gába. Compositional splines for representation of density functions. *Computational Statistics*, 36:1031–1064, 2020.

[24] S. Makosso-Kallyth and E. Diday. Adaptation of interval PCA to symbolic histogram variables. *Advances in Data Analysis and Classification*, 6(2):147–159, 2012.

[25] A. Menafoglio, A. Guadagnini, and P. Secchi. A kriging approach based on aitchison geometry for the characterization of particle-size curves in heterogeneous aquifers. *Stochastic Environmental Research and Risk Assessment*, 28(7):1835–1851, 2014.

[26] A. Menafoglio, A. Guadagnini, and P. Secchi. Stochastic simulation of soil particle-size curves in heterogeneous aquifer systems through a Bayes space approach. *Water Resources Research*, 52(8):5708–5726, 2016.

[27] A. Menafoglio, P. Secchi, and A. Guadagnini. A class-kriging predictor for functional compositions with application to particle-size curves in heterogeneous aquifers. *Mathematical Geosciences*, 48(4):463–485, 2016.

[28] K. Mosler and P. Mozharovskyi. Fast DD-classification of functional data. *Statistical Papers*, 58:1055–1089, 2017.

[29] M. Noirhomme-Fraiture and P. Brito. Far beyond the classical data models: Symbolic data analysis. *Statistical Analysis and Data Mining*, 4(2):157–170, 2011.

[30] V. Pawlowsky-Glahn, J.J. Egozcue, and R. Tolosana-Delgado. *Modeling and Analysis of Compositional Data*. Wiley, Chichester, 2015.

[31] R Core Team. *R: A Language and Environment for Statistical Computing*. R Foundation for Statistical Computing, Vienna, Austria, 2021.

[32] J. Ramsay and B.W. Silverman. *Functional Data Analysis, 2nd ed.* Springer, New York, 2005.

[33] D. Šimíček, O. Bábek, K. Hron, I. Pavlů, and J. Kapusta. Separating provenance and palaeoclimatic signals from geochemistry of loess-paleosol sequences using advance statistical tools: Central European loess belt, Czechia. *Sedimentary Geology*, 419, 2021.

[34] R. Talská, A. Menafoglio, K. Hron, J. J. Egozcue, and J. Palarea-Albaladejo. Weighting the domain of probability densities in functional data analysis. *Stat*, 9:e283, 2020.

[35] R. Talská, A. Menafoglio, J. Machalová, K. Hron, and E. Fišerová. Compositional regression with functional response. *Computational Statistics & Data Analysis*, 123:66–85, 2018.

[36] M. Templ, K. Hron, and P. Filzmoser. robCompositions: An R-package for robust statistical analysis of compositional data. In V. Pawlowsky-Glahn and A. Buccianti, editors, *Compositional Data Analysis: Theory and Applications*, pages 341–355. Wiley, Chichester, 2011.

[37] K.G. van den Boogaart, J.J. Egozcue, and V. Pawlowsky-Glahn. Bayes Hilbert spaces. *Australian & New Zealand Journal of Statistics*, 56(2):171–194, 2014.

[38] R. Verde and A. Irpino. Dynamic clustering of histogram data: using the right metric. In *Selected Contributions in Data Analysis and Classification*, pages 123–134. Springer, Berlin, Heidelberg, 2007.

[39] R. Verde, A. Irpino, and A. Balzanella. Dimension reduction techniques for distributional symbolic data. *IEEE Transactions on Cybernetics*, 46:344–355, 2015.

Part III

Dimension Reduction

10

Principal Component Analysis of Distributional Data

Sun Makosso-Kallyth

SM Analytic Canada, Richmond Hill, Canada

Edwin Diday

Université Paris Dauphine, Paris, France

CONTENTS

DOI: 10.1201/9781315370545-10

In this era where we get more and more data, new data driven strategies for getting insights from data are more than needed. For example, one might have one or several datasets where each observation and variable are described by frequency distributions. Such data are more and more common and are called symbolic histogram variables. Symbolic histogram variables belong to the family of distributional data. Unsupervised learning methods like symbolic Histogram Principal Component Analysis (HPCA) have proven to be extremely efficient when it comes to derive insights from that type of data. In this chapter, we compare two extensions of principal component analysis to distributional variables and introduce a generalization strategy when the knowledge experts want to perform histogram PCA on several distributional datasets. Indeed, most of proposed approaches only consider the situation where users have one dataset and from a technical standpoint, the proposed solutions are based on either the first order moments or the quantiles. In this chapter, we present two Histogram PCA's respectively based on the barycenters of distributions and on the average correlation matrix induced by the quantiles of distributions. The PCA method based on barycenters requires variables' coding, and several coding methods such as the Ridit scores and the metabin coding will be discussed. Through the coding, a classical principal component analysis of the barycenters tables is applied. Then, histograms are transformed into intervals and the hypercubes induced by those intervals are projected as supplementary elements. By contrast, Histogram PCA based on the average correlation matrix induced by quantiles does not require a coding method but users should specify the number of quantiles and their location. We review the benefits and the flaws of using the barycenters versus the quantiles and we present a generalization framework when there are more than one histogram dataset.

10.1 Introduction

Imagine you have for example a macroeconomic dataset about countries and those countries are aggregated in groups and characterized by numerical features like the Gross Domestic Product per habitant GDP and the mortality rate, and your desire is to analyse or to profile those groups rather than describing each country. What type of analyses can you use and how to ensure that the outputs analysis of those groups of countries will preserve their nature? Several types of analyses are actually possible among which principal component analysis of symbolic data. In this chapter we introduce principal component analysis of distributional data. The recent advances in computer sciences and information technology have led us to the issue of data analysis for complex and massive data. Symbolic Data Analysis (SDA) paradigm and dimension reduction methods are very helpful in exploratory data analysis for at least two reasons. First of all, Symbolic Data Analysis provides powerful tools (see [1], [4], [5], [17]) when it comes to analyse units with a high level of generality. Indeed, SDA allows the transition from individual observations (e.g. patients with Corona in a hospital, soccer players in a team) described by standard categorical and numerical variables, to the analysis of groups of observations (e.g. a group of patients from the same hospital, a group of soccer players belonging to the same team) described by variables whose values consider the variability inside these groups: interval values, sets of numbers, distributional data, etc. Overview on SDA can be found in [16] and [43]. Secondly, principal component analysis (PCA) is one the most widespread dimensionality reduction methods ([30]) and distributional variables (histogram variable) are commonly used in SDA.

The main extensions and adaptations of PCA to symbolic histogram variables have been proposed in [11], [13], [27], [35], [40], [44], [51], [52]. Most of these methods focus on the relationship between some characteristic elements of histogram, especially the quantiles and the moments. Quantiles do not required assumptions on the underlying distributions and some authors such as in [26], [27], [29], [44]. use the quantiles. The approach proposed in [29] is based on the squared Wasserstein distance. It is also possible to consider other distance measures such as the Hellinger, Kolmogorov distances and the symmetric Kullback-Leiber measure. But the specification of the distance could influence the outputs. Meanwhile, other authors, such as in [40], use the moments, especially the first-order moments (the mean) as active elements. When the underlying distribution is well known, the use of moments could be more informative than the use of quantiles [38]. For example if the underlying distribution is Normal, the amount of information that lies within one standard deviation of the mean is 68% while the interquartile range (IQR) covers 50%. In [40], for example, the authors performed a PCA on histograms means, transformed histograms into intervals and project the transformed data interval onto the space spanned by the principal components of the means. Alternative techniques to the moments and the quantiles have been used as well. For example, in order to take into account the underlying dependencies of data, the authors in [15] consider the "Copular PCA" for bar charts. The approach proposed in [15] is more general in that compared to histograms, bar chart bins are not necessarily ordered. There are in fact two kinds of bar chart variables: the nominal ones, for which the bins are not ordered, such as nationality; and the numerical ones, for which bins are ordered, such as age and salary. In this chapter, the focus will be first done by considering the two complementary approaches proposed in [40] and [15]. We present the approach proposed in [40] and its extension to three-way data. Then, we compare the previous approach with the one proposed in [15] for distributional variables. Afterwards, we present the approach proposed in [38] that uses quantiles. We also present a generalization framework when users have more than one histogram-valued dataset. Finally, in the last section, we apply the described methodologies to two macroeconomic datasets.

10.2   Approach based on mean of histograms

In this section, we present the approach proposed in [40]. For a given histogram-valued variable Y_j, we consider $Y = (Y_{ij})$, $\forall i = 1, \ldots, n$; $j = 1, \ldots, p$; n is the number of units, p the number of variables and let m_j be the number of bins for Y_j. Table 10.1 gives an example of two symbolic histogram-valued variables: the Gross Domestic Product (GDP)/hab, Y_1, and the mortality rate, Y_2, for four regions (Africa, NAFTA, Oriental Asian, Europe). We call by histogram variable Y_j, a n-uplet element $Y_{ij} = \{c_j, p_{ij}\}$ for $i = 1, \ldots, n$ where $c_j = (c_j^{(1)}, \ldots, c_j^{(m_j)})$ are the histogram bins of Y_j and $p_{ij} = (p_{ij}^{(1)}, \ldots, p_{ij}^{(m_j)})$ represents a m_j-dimensional vector of relative frequencies. In Table 10.1, histogram bins are $c_1 = (\le 1,]1, 20], > 20)$ and $c_2 = (\le 0.1, > 0.1)$. p_{ij} are compositional variables (see [3]) because they satisfy the relationship $\sum_{\ell=1}^{m_j} p_{ij}^{(\ell)} = 1$. The use of Histogram PCA based on barycenters requires a coding method. We suggest four different codings in that regard: the parametric coding, the non-parametric coding, the Ridit scores and the metabin coding. The three first codings are numerical codings in that they assign numerical values $w_j^{(\ell)}, j = 1, \ldots, p$; $\ell = 1, \ldots, m_j$ to histogram bins. However, the metabin coding, which is applicable to ordered and non-ordered bins, does not assign numerical values to histogram bins.

10.2.1 Parametric coding

We present the parametric coding which is applicable to ordered bins (see [40]). Let $\mathcal{D}_j = (\alpha_j, \beta_j)$ be a domain of all possible values of histogram bins c_j. For the first variable $Y_1 = (GDP)$, we have $\alpha_1 = 0$, $\beta_1 = +\infty$; for $Y_2 =$(mortality rate) by contrast: $\alpha_2 = 0$, $\beta_2 = 1(100\%)$. Let $\delta_j = \inf_{\ell=1,\dots,m_j} L_\ell$, where L_ℓ is the length of interval $c_j^{(\ell)}$. For the parametric coding, [40] suggested the following steps:

1.   •If $c_j^{(\ell)} =] - \infty, a_j]$ then $c_j^{(\ell)} \longrightarrow c_j^{(\ell)} =]e, a_j]$ where

$$e_j = \begin{cases} \alpha_j & \text{if } a_j - \delta_j < \alpha_j \\ a_j - \delta_j & \text{else} \end{cases}.$$

•If $c_j^{(\ell)} =]b_j, +\infty[$, then $c_j^{(\ell)} \longrightarrow c_j^{(\ell)} =]b_j, f_j]$ with

$$f_j = \begin{cases} \beta_j & \text{if } b_j + \delta_j > \beta_j \\ b_j + \delta_j & \text{else} \end{cases}.$$

2.   If histogram bins do not have the same unit, each interval $]a', b']$ should be replaced by an adjusted interval $]a'/(b' - a'); b'/(b' - a')]$.

3.   The parametric coding assigns to each bin a vector of scores $\omega_j = (\omega_j^{(1)}, \dots, \omega_j^{(m_j)})$, where $\omega_j^{(\ell)}$ is the center of the adjusted interval, for $\ell = 1, \dots, m_j$.

10.2.2 Non parametric codings

Parametric coding implementation requires a thorough knowledge of the field of expertise. Indeed, the choice of interval centers should not be arbitrary when interval lengths are infinity. For example, GDP per habitant does not have a known maximum value. So assuming a maximum finite value for GDP just for the purpose of the coding can be a problem. In order to overcome

TABLE 10.1

Example of symbolic histogram variables.

Region	GDP in k$/hab			Mortality rate	
	≤ 1	$]1, 20]$	> 20	≤ 0.1	> 0.1
Africa	0.34	0.66	0.00	0.25	0.75
Nafta	0.00	0.33	0.67	1.00	0.00
Oriental Asia	0.07	0.80	0.13	1.00	0.00
Europe	0.00	0.32	0.68	0.74	0.26

this problem, the authors in [40] also proposed the non-parametric coding. The non-parametric coding proposed in [40] uses the rank associated with the histogram bins. In Table 10.1 for example, bin scores are $w_1^{(1)} = 1$, $w_1^{(2)} = 2$, $w_1^{(3)} = 3$; and $w_2^{(1)} = 1$, $w_2^{(2)} = 2$.

Non parametric coding is easy to use but only suitable for equidistant bins. In the case of GDP variable from Table 10.1 for example, bin scores are $w_1^{(1)} = 1$, $w_1^{(2)} = 2$, $w_1^{(3)} = 3$. Histogram bins for GDP in kilo dollars are $[0, 1[$ for the first bin, $]1, 20]$ for the second bin, and $]20, +\infty]$ for the third bin. Gap levels between rich countries from the third bin and developing countries for the second bin are not necessarily equidistant with gap levels between the first bin (poor countries) and the second bin (developing countries). By using the non-parametric coding here, users could ignore the fact that from a macroeconomic perspective, poor and developing countries are generally more similar than developing countries and rich countries. To overcome the limitations of the parametric and non-parametric codings, we present in the next section a coding based on the Ridit scores.

10.2.3   Ridit scores

Ridit scores have been introduced in [9]. Initially, they had a probabilistic meaning (probability that a random variable is less than a reference value). In categorical data analysis, Ridit scores are also used for ordinal variables (see [2], [21], [42]) and in insurance industry, Ridit scores have been used in fraud detection (see [8]). If we consider the frequency vector p_{ij}, Ridit scores are

$$r_\ell = 0.5 p_{ij}^{(\ell)} + \sum_{m<\ell} p_{ij}^{(m)}. \tag{10.1}$$

To adapt the application of Ridit scores to the histogram variables bins, [39] first proposed the determination of the mean vector

$$\overline{p_{\cdot j}} = \frac{1}{n} \sum_{i=1}^{n} p_{ij}.$$

Then the raw Ridit scores of histogram variables are defined as follows (see [39]):

$$w_j^{(\ell)} = 0.5 \overline{p_{\cdot j}}^{(\ell)} + \sum_{m<\ell} p_{\cdot j}^{(m)}. \tag{10.2}$$

From Table 10.1 for example, the corresponding average vectors associated with this table are $\overline{p_{\cdot 1}} = (0.103, 0.528, 0.370)$ and $\overline{p_{\cdot 2}} = (0.748, 0.253)$. Raw Ridit scores for GDP are $w_1^{(1)} = 0.051$, $w_1^{(2)} = 0.102 + \frac{0.529}{2} = 0.366$, $w_1^{(3)} = 0.102 + 0.509 + \frac{0.369}{2} = 0.816$; for mortality rate raw Ridit scores are $w_2^{(1)} = 0.373$ and $w_2^{(2)} = 0.873$. In order to consider the scale differences between variables, [39] also suggested the use of the standardized and normalized Ridit scores such as:

$$w_j^{*(\ell)} = \frac{w_j^{(\ell)} - \mu_{s_j}}{\sigma_{s_j}},\tag{10.3}$$

where μ_{s_j} and σ_{s_j} represent the empirical mean and variance of the raw Ridit scores $w_j^{(\ell)}$. Standardization nevertheless tends to assign high scores in absolute values to the extreme bins and values close to zero to the intermediate bins. Given this, [39] defined the normalized Ridit scores such as:

$$w_j^{**(\ell)} = \frac{w_j^{(\ell)}}{\sum_{\ell=1}^{m_j} w_j^{(\ell)}}.\tag{10.4}$$

Parametric, non-parametric and Ridit scores codings assign numeric values to histogram bins. Then, histogram barycenters are computed.

10.2.4 Metabin coding

The scope of metabin coding is broader than the one from the three previous coding methods in that metabin coding is applicable to histogram variables and bar charts. Histogram variables have ordered bins and bar charts have nominal (non-ordered) bins. Metabin coding serves as vehicle when it comes to construct an optimal $(n \times m) \times p$ table (where $m = \max\{m_j, \ j = 1, \ldots, p\}$) that will be used as an active input table in a standard PCA. The use of metabins has been originally proposed in [15] for nominal bar chart variables and can be extended to histogram variables. Metabins join together bins taken from each variable.

The goal of metabin coding is not to assign numerical values to bins but to determine instead a $nm \times p$ standard input table that will be representative of the input distributional data. Every table cell induced by metabins contains the relative frequencies associated with the histogram bins. A metabin is a set of p-ranked bins, one for each of the p histogram variables. In practice, each metabin is defined by a set of p bins which are at the same position in the ranking of the bins associated with each barchart or histogram variable. Notice that for histogram variables, the bins ranking that induces metabins is not necessary the one from histograms. It is important to notice that there are no more than m metabins. From Table 10.1 for example, $p = 2$ and $m = \max\{3, 2\} = 3$ and a metabin table $\mathcal{H}$ can be defined as shown in Table 10.2.

TABLE 10.2

A Metabin table example induced by Table 10.1.

Row Metabin Name	GDP in k$/hab	Mortality rate
Africa-1	0.34	0.25
Africa-2	0.66	0.75
Africa-3	0.00	-
Nafta-1	0.00	1.00
Nafta-2	0.33	0.00
Nafta-3	0.67	-
Oriental Asia-1	0.07	1.00
Oriental Asia-2	0.80	0.00
Oriental Asia-3	0.13	-
Europe-1	0.00	0.74
Europe-2	0.32	0.26
Europe-3	0.68	-

10.2.5 PCA of means and representation of dispersion of concepts on individual map.

Histogram PCA based on barycenters is only applicable when a numerical coding method is selected (parametric, non-parametric, Ridit scores).

10.2.5.1 The use of a numerical coding

After the numerical coding of the distributional variables, [40] computes histogram means (barycenters) $g_{ij} = \sum_\ell \omega_j^{(\ell)} p_{ij}^{(\ell)}$ associated with each histogram Y_{ij}. Then, an ordinary PCA of the classic $n \times p$ table G is performed (where $G = (g_{ij})$, $i = 1, \ldots, n; j = 1, \ldots, p$). If $v_{\alpha, \alpha=1,\ldots,q}$, are the q first eigenvectors of the covariance matrix G, the Pearson correlations between the ordinary principal components $z_\alpha = G v_\alpha$ and the variables' means $(g_{.j})_{j=1,\ldots,p}$ allow the interpretation of factorial plans.

For the representation of units on the factorial plans, [40] transform histogram variables into intervals via the Tchebychev's inequality. According to this inequality, for any random variable X_j and any real number $t > 0$, the amount of information in $[\overline{X_j} - ts, \overline{X_j} + ts]$ (where $\overline{X_j}$ is the empirical mean, s is the empirical standard deviation) is larger than or equal to $1 - \frac{1}{t^2}$ i.e $P(X_j \in [\overline{X_j} - ts, \overline{X_j} + ts]) \geq 1 - \frac{1}{t^2}$. For a given value of t, [40] transform every p_{ij} into $[\alpha_{ij}, \beta_{ij}]$ ($\alpha_{ij} = \overline{X_j} - ts; \beta_{ij} = \overline{X_j} + ts$) via the Tchebychev's inequality. Afterwards, [40] determine the hypercubes associated with the units as in [12] and [11]. For p variables, each hypercube has 2^p vertices. The hypercube associated with the ith unit (or concept) is $\mathcal{M}_i$. If we have for example $p = 2$ interval variables built after transforming histograms into intervals, the unit i is described by the intervals $([a_{i1}, b_{i1}], [a_{i2}, b_{i2}])$. The corresponding hypercube

$\mathcal{M}_i$ is:

$$\mathcal{M}_i = \begin{bmatrix} a_{i1} & a_{i2} \\ a_{i1} & b_{i2} \\ b_{i1} & a_{i2} \\ b_{i1} & b_{i2} \end{bmatrix}.$$

10.2.5.2 Use of interval metabins as supplementary elements

Besides, by considering intervals as bins, possible metabins can be $(a_{i1}, a_{i2}); (b_{i1}, b_{i2})$ or $(a_{i1}, b_{i2}); (b_{i1}, a_{i2})$. Consequently, if the optimal metabins are $(a_{i1}, a_{i2}); (b_{i1}, b_{i2})$, then the corresponding metabin matrix can be

$$\mathcal{M}'_i = \begin{bmatrix} a_{i1} & a_{i2} \\ a_{i1} & b_{i2} \end{bmatrix}.$$

Then, these hypercubes (or metabin matrices) induced by intervals are projected as supplementary elements onto principal axes of the PCA of the histogram means (g_{ij}). The main steps of the Histogram PCA based on barycenters are:

- Selection of the coding method

- Computation of $G = (g_{ij})$ a $n \times p$ table of barycenters [1]

- PCA of $G = (g_{ij})_{i=1,...,n;\ j=1,...,p}$

- Transformation of data into intervals via Tchebytchev's inequality

- Construction of hypercubes or metabins matrices from intervals obtained by the Tchebytchev's inequality

- Projection of hypercubes or metabins matrices onto factorial axes.

The author in [40] also proposed the use of the angular transformation $arsin\sqrt{p_{ij}}$ as preprocessing method for relative frequencies p_{ij}.

10.3 Approach based on metabin coding

In this section, we present a coding approach that can be extendable to non-numerical bins. We suggest the use of the metabin coding in order to optimize the correlation of input variables. Indeed, if we suppose for example that we have a sample of regions described by two histogram variables $X_1=$ "Number of accidents" and $X_2=$ "Age", and those two variables have three bins (few,

---

[1]If parametric, non-parametric or Ridit scores coding are used.

medium, much), we can have as metabin example the one defined in Table 10.3. These metabins are (young, much), (old, medium) and (adult, few). Another metabin example coming from the same table could be (young, much), (old, few) and (adult, medium) but in this case the correlation between age and number of accidents might be lower since the category old has in general more accidents than the category medium.

TABLE 10.3
Example of metabins from two histogram variables.

(nm) × p table of metabins		
Regions	Age	Number of accidents
	Young	Much
Region 1	Old	Medium
	Adult	Few
	Young	Much
Region 2	Old	Medium
	Adult	Few

10.3.1   Determination of optimal metabins

There are many possible permutations from which metabins can be constructed. Therefore, finding the optimal set of metabins is needed. It is for example possible to find the optimal metabins that maximize the correlation between variables. Accordingly, we use the quality criterion defined in [15]. The best bin order is therefore the one that maximizes the squared correlation coefficients induced by the bins in each of the m metabins $\varpi_1, \ldots, \varpi_m$. This can be expressed by the following criterion:

$$\max \sum_{\ell_1=1}^{m} \sum_{\ell_2=1}^{m} \sum_{j_2=1}^{p} \sum_{j_1=1}^{p} \left\{ cor^2 \left(p_{j_1}^{(\ell_1)}, p_{j_2}^{(\ell_2)} \right) \right\}. \tag{10.5}$$

However, given the fact that the number of bins permutations is $m_j!$, there are many metabins choice options simply because metabins can be found in two ways. The first one is to select each observation (the region in Table 10.1) and map the bins combination (see Table 10.2 for example). The second one is to select every bin and map the corresponding observations. Finding optimal metabins can be computationally fastidious when m_j is large. Diday [15] has provided some solutions in that regard. One example could be the random selection of a given metabin as starting point; then, it is possible to change the bins selected for metabins and find the one that gives the best improvement.

10.3.2 Numerical and graphical tools provided by metabin coding

Let us consider M the $n \times (pm)$ table of metabins. After standardizing M, this table is used as active table for standard PCA. This standard PCA induces eigenvalues u_α and principal components z_α. In order to represent input variables as supplementary elements onto principal components, Diday [15] suggests the use of the mean of the projections when metabins are optimal. The main advantage of the metabin coding is the possibility that users have to define a new data table that [15] called "pathways data table". Each column of this pathways data table is associated with each of the p initial variables Y_j $j = 1, \ldots, p$ and each unit is multiplied by the number m pathways data table. For given fixed metabins, we could obtain a pathway of units. Indeed, by giving a direction to a pathway, we can obtain a trajectory. Accordingly, we can obtain a pathway data table of $n \times m$ rows and p columns on which a standard PCA can be applied. The pathways can then be represented onto the factorial axes by a line connecting the $n \times m$ sub-units based on their ranking. The best ranking of the bins is the one that maximizes the squared correlation of the bins in each of the m metabins. The steps of the proposed approach are as follows:

- Determination of the table M of metabins $nm \times p$;

- Standardized PCA of M;

- Determination of principal component z_α.

10.4 The use of quantiles as an alternative to the coding

Given the fact that there are multiple options for coding histogram bins, we present in this section some possible alternative methods that could be applied in the context of Histogram PCA. These alternative methods are based on the use of the quantiles and cannot be used for bar charts. Rather than coding histogram bins, it is also possible to use the quantiles derived from the histogram as suggested in [11] and [27]. But, users should specify the number of quantiles m, and their location.

10.4.1 Representation of histograms by common number of quantiles

Let us consider for $i = 1, \ldots, n$, $Y_{ij} = \{c_{ij}, p_{ij}\}$ where $c_{ij} = [a_{ij}^{(\ell)}, a_{ij}^{(\ell+1)}[$ and Y_{ij} has m_j bins. Then, under the assumption that bins have uniform distributions, we define the cumulative distribution function $F_{ij}(x)$ of the histogram as:

$$
\begin{cases}
F_{ij}(x) = 0 & \text{for } x \le a_{ij}^{(1)} \\
F_{ij}(x) = p_{ij}^{(1)}(x - a_{ij}^{(1)})/(a_{ij}^{(2)} - a_{ij}^{(1)}) & \text{for } a_{ij}^{(1)} \le x \le a_{ij}^{(2)} \\
F_{ij}(x) = F(a_{ij})^1 + p_{ij}^{(2)}(x - a_{ij}^{(2)})/(a_{ij}^{(3)} - a_{ij}^{(2)}) & \text{for } a^{(2)} \le x \le a_{ij}^{(3)} \\
\;\;\vdots & \;\;\vdots \\
F_{ij}(x) = F(a_{ij})^{m_j-1}(x) + p_{ij}^{(m_j)}(x - a_{ij}^{(m_j)})/(a_{ij}^{(m_j+1)} - a_{ij}^{(m_j)}) & \text{for } a_{ij}^{(m_j)} \le a_{ij}^{(m_j+1)} \\
F_{ij}(x) = 1 & \text{for } a_{ij}^{(m_j+1)} \le x
\end{cases}
$$

If we select the number $m_j = 4$ and the three cut points $c1 = 1/4$, $c2 = 2/4$, and $c3 = 3/4$, we can obtain three quantile values from the equations $p1 = F_{ij}(q1)$, $p2 = F_{ij}(q2)$, and $p3 = F_{ij}(q3)$. Finally, we obtain four bins $[a_{ij}^{(1)}, q_1), [q_1, q_2), [q_2, q_3)$, and $[q_3, a_{ij}^{(m_j+1)})$ and their bin probabilities $(p1 - 0)$, $(p2 - p1)$, $(p3 - p2)$, and $(1 - p3)$ with the same value $1/4$.

$$
\begin{cases}
F_{11}(x) = 0 & \text{for } x \le 0 \\
F_{11}(x) = 0.34(x - 0)/(1 - 0) & \text{for } 0 \le x \le 1 \\
F_{11}(x) = F(1) + 0.66(x - 1)/(20 - 1) & \text{for } 1 \le x \le 20 \\
F_{11}(x) = 1 & \text{for } 20 \le x
\end{cases}
$$

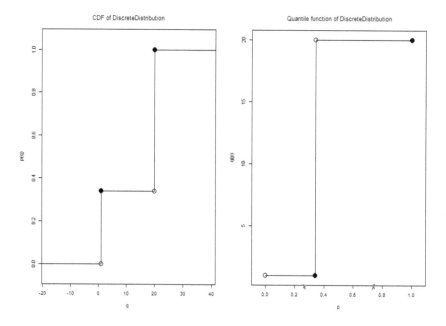

FIGURE 10.1
CDF and quantiles for *GDP* and Africa from Table 10.1.

We represent the CDF (Cumulative Density Function) and the quantiles in Figure 10.1. The quantiles induced by Table 10.1 are represented in Table 10.4.

Once the quantiles have been chosen, [27], for example, performed a standard PCA on the $mn \times p$ table induced by the quantiles. Makosso-Kallyth [38] also uses a similar approach but rather than using a unique and standard $mn \times p$ table of quantiles, the author derives m different $n \times p$ tables from which m distinct $p \times p$ correlation matrices are built. If for example, one uses $m = 3$ quantiles, then $Q_{(k)}$, $k = 1, \ldots, 3$ classical $n \times p$ tables of quantiles or CDF associated with these quantiles will be built. Each $Q_{(k)}$ will contain the quantiles from the same order. The initial table of relative frequencies H is replaced here by the tables of corresponding quantiles that we have called Q. Each of these tables will induce a $p \times p$ correlation matrix $R_{(k)}$. From those correlation matrices, [38] determines an average correlation matrix $W = \overline{R}$ using the Fisher Z score for correlations (see [20]). A SVD (Singular Value Decomposition) of W is then performed and the m tables $Q_{(k)}$, $k = 1, \ldots, m$ are projected as supplementary elements onto the eigenvectors obtained from the Singular Value Decomposition of W. Here are the main steps of the PCA based on the average correlation matrix of $Q_{(k)}$ (see also [38]):

- Choice of m, the common number of quantiles and their location.

- Extraction of $Q_{(k)}$ the tables induced by quantiles from the same order.

- Computation of $R_{(k)}$ the correlation matrices induced by $Q_{(k)}$.

- Computation of W, the average correlation matrix.

- Computation of the eigenvalues λ_α and eigenvectors u_α such that $W u_\alpha = \lambda_\alpha u_\alpha$.

- Projection of $Q_{(1)}, \ldots, Q_{(m)}$ onto principal axes u_α.

- For visualization, representation of the rectangles or $2d$ convex hulls induced by the supplementary elements.

TABLE 10.4
Table of quantiles.

| Quantiles | GDP in k$/hab | | Mortality rate ‰ | |
	$p_1 = 0.25$	$p_3 = 0.75$	$p_1 = 0.25$	$p_3 = 0.75$
Regions				
Africa	1	20	9.99	10.00
Nafta	20	20	9.99	9.99
Oriental Asia	20	20	9.99	9.99
Europe	20	20	9.99	10.00

10.5   Framework extension for three-way data

In this era of massive data, it is possible to have more than one distributional dataset. What if you have more than one dataset of distributional data and you want to implement a data driven approach similar to what we have described above? If so, your data are in three-way format or are structured in blocks. In this section, we propose how to apply the above presented methods to three-way distributional datasets. Three-way data analysis (see [22]) is an area associated with the Gifi school and includes generalized canonical correlation, STATIS method, Multiple co-inertia analysis, Multiple Factor Analysis, Procrustes Analysis, the analysis of several dissimilarity matrices, ... In the Anglo-saxon school, the first work in three-way data analysis (see [32]) started with [48] who extended principal component analysis and Spearman factor analysis (see [45], [48], [49]). The authors of [24] and [25] also extended PCA to three-way data. In [10] the authors have developed three-way data methods but their approach is more in the context of multidimensional scaling. In the French school most of three-way methods extend PCA, the most relevant ones are the double principal component analysis (DPCA) of [6], the STATIS method (see [33], [36]), MFA (Multiple Factorial Analysis) of [19] (which is different from Thurstone Multiple Factor Analysis developed in [47] that focuses on Spearman factor analysis), ACOM algorithms from [14] and [23], the regularized generalized canonical correlation analysis (RGCCA) from [46] ... We extend the approach proposed in [40] to three-way data containing symbolic histogram variables.

We now suppose that we have $q >= 1$ histogram-valued datasets and for $r = 1, \ldots, q$ n is the number of units, $p_r = p$ is the number of variables and m_{jr} is the number of bins of Y_{jr}, $j = 1, \ldots, p$. Many strategies are actually possible, the framework of canonical correlation analysis proposes several strategies in that regard. Let us suppose that we have for example two datasets of distributional data. The first one Y_1 is on economic performances (business environment, agriculture, infrastructure, economic growth) and the second one Y_2 is on political environment (leadership, democracy, corruption, transparency, safety and stability). Our extension strategies of histogram principal component suppose that the two datasets of distributional data have a symmetric importance i.e. the economic dataset explains the political dataset and vice versa. In this section we present two possible strategies when it comes to extend principal component analysis to those two datasets. Those strategies are based on the determination of a compromise matrix. The first approach is called three-way "average" Histogram PCA and supposes that the histogram-valued datasets have the same number of rows and columns. When histogram datasets do not have the same number of variables, we overcome the limitations of the previous method by applying PCA on each histogram dataset and then procrustean analysis.

10.5.1 Average three-way Histogram PCA

In this section, we first suppose that we have several histogram-valued datasets and those datasets have the same number of units (in rows) and variables (three-way data). Let us consider $Y_{ij1} = \{c_{j1}, p_{ij1}\}; \ldots; Y_{ijq} = \{c_{jq}, p_{ijq}\}$, q $n \times p$ histogram tables, the compromise $\overline{Y}_{ij}$ is a $n \times p$ table which is representative of the q tables Y_{ijr}, for $r = 1, \ldots, q$.

10.5.1.1 Transform three-way Histogram into classical three-way data

When it comes to define a three-way strategy for histogram-valued data, the first step is to decide how to characterize such data. First, we present a three-way strategy using Histogram PCA of barycenters. The proposed strategy is extendable to Histogram PCA based on the quantiles. In doing so, users should first assimilate each table of distributional data to the average table induced by the barycenters of each variable (if those barycenters are representative of the distributions). In order to determine the barycenters, one should choose a coding method and for that we suggest the use of the raw Ridit scores (for example). We get therefore q $n \times p$ tables $\overline{X}_1, \ldots, \overline{X}_q$ that each contains the barycenters of distributions.

10.5.1.2 Determination of a compromise table

Once we get a series of q standard tables of barycenters, a compromise matrix that would represent a sort of average table that will be representative of the following tables $\overline{X}_r$, $r = 1, \ldots, q$ is computed.

$$\overline{X}_{compromise} = \sum_{r=1}^{q} \pi_r \overline{X}_r \tag{10.6}$$

In expression (10.6), the π_r are such as

$$\pi_r = \sum_{i=1}^{n} \sum_{j=1}^{p} \sum_{\ell=1}^{m_j} p_{ijr}^{(\ell)} \log(p_{ijr}^{(\ell)} + 1) \tag{10.7}$$

which somehow correspond to the entropy of the Ridit scores.

10.5.2 PCA of the compromise and projection of supplementary elements

After the determination of the compromise, we apply the barycenters Histogram PCA on $\overline{X}_{compromise}$, whose entries are $g_{ij} = \sum_{r=1}^{q} \pi_r \overline{X}_{rij}$. Then, an ordinary PCA of the table $G = (g_{ij})_{i=1,\ldots,n;j=1,\ldots,p}$ is performed. Let us call $v_{\alpha,\alpha=1,\ldots,t}$, the t first eigenvectors of the covariance matrix G. Once we get those eigenvectors, histograms from each dataset are transformed into intervals and projected as supplementary elements onto $v_{\alpha,\alpha=1,\ldots,t}$.

These are the main steps of the average three-way PCA:

- Choose the coding method between the Ridit scores, metabin codings, non-parametric coding

- Compute the tables of barycenters $\overline{X}_r$, $r = 1,\ldots,q$.

- Compute the compromise matrix using expression (10.6)

- Perform SVD of the compromise matrix $\overline{X}_{compromise} v_\alpha = \lambda_\alpha v_\alpha$.

- Transform $Y_{ijr} = \{c_{jr}, p_{ijr}\}$ into intervals via Tchebytchev's inequality

- Construct hypercubes (or metabins matrices) induced by p_{ijr}

- Project hypercubes or metabins matrices onto factorial axes.

The three-way strategy described here is extendable to the quantiles tables. When histogram datasets do not have the same number of variables, we propose in the next section an alternative method similar to the one proposed in [6] that we call repeated Histogram PCA.

10.5.3   Repeated Histogram PCA

Histogram datasets have now the same number of rows but not necessarily the same number of variables. The repeated Histogram PCA overcomes this issue by performing two series of PCA. In the first one, users perform q distinct PCA's on each series of datasets $\overline{X}_r$ from which a common minimum number of principal components is determined. Let us call that minimum number of components by $t_{\min}$. Then, we get a series of $\overline{Z}_r$ tables of principal components on which we can apply once again the average three-way PCA (described above) or the generalized procrustes analysis.

10.6   Application of Histogram PCA

In this section we are going to show how histogram PCA works from a practical standpoint. To do this, we will be using two macroeconomic datasets. The first one is from [37] and the second one is OECDGrowth data from [18]. In the first example, histogram data are already built while in the second one, we will be building histogram data as we will be exploring the data.

TABLE 10.5

Selected variables.

Symbolic variables	Category Label
GDP per capita$	< 1 k$, [1k$;5k$]; [5k$;10k$]; [10k$,20k$]; > 20k$
Undernourishement	< 3%; [3%, 10 %]; [10 %;25 %]; [25%;35 %]; > 35%
Growth Population$	< 0.0%; [0;1]; [1;2] ;[2; 4] ; >= 4
HDI	< 0.5; [0.5; 0.6]; [0.6;0.7]; [0.7;0.8]; >= 0.8
Mortality Rate	< 0.5%; [0.5%; 1%]; [1%;1.5%]; [1.5%;2.0%]; >= 2.0%
Population in Millions	< 1 M; [1; 5] M; [5;10] M; [10;100] M; >= 100 M
Investment	< 500M; [500; 1000]; [1000; 30000]; [30 × 10³; 70000] ;>= 70000
Electricity kwh/pers	< 500 ; [500; 1000]; [1000;5000] ; [5000;15000]; >= 15000
GAZ kt/hab	< 1 ; [1; 2]; [2;5] ; [5;10]; >= 10
Mobil phone /1000 hab	< 10 ; [10; 100]; [100;150] ; [150;500]; >= 500
Health spent % GDP	< 3% ; [3%; 4%]; [4%;6%] ; [6%;8%]; >= 8%

10.6.1 Histogram PCA on a macroeconomical dataset

The original data, at country level, were aggregated by geographic regions ($n = 7$ regions in total) that we also call concepts (see [37]). Those concepts or regions are: Africa, South and Central America, NAFTA (Canada, US and Mexico), Oriental Asia, Europe, Middle East, the ex USSR countries. The regions are described by $p = 11$ features (see Table 10.5). Which regions are similar, which regions are the more at risk in terms of GDP or Pollution? Are those regions homogeneous? This is what histogram PCA is going to tell us.

10.6.1.1 Histogram PCA based on barycenters

We apply the approach proposed in [40] using the R package GraphPCA (see [7]). The histogram-valued data we are exploring are gathered in 11 distinct Excel files, one for each variable. In the R code below (Listing 10.1), we show how users can load those histogram data into R software. In Table 10.6, we show how the population variability looks like.

TABLE 10.6

Population

Region	Pop. 1	Pop. 2	Pop. 3	Pop. 4	Pop. 5
Africa	0.082	0.245	0.143	0.510	0.020
South Center America	0.048	0.238	0.333	0.333	0.048
Nafta	0.000	0.000	0.000	0.333	0.667
Asia Oriental	0.056	0.111	0.056	0.500	0.278
exUSSR	0.000	0.467	0.267	0.200	0.067
Europe	0.097	0.226	0.226	0.452	0.000
MiddleEast	0.091	0.273	0.182	0.364	0.091

Data Importation using R

```
1 Croissance<-read.table("Croissance.txt",h=T)
2 Electricity<-read.table("Electricity.txt",h=T)
3 GAZ<-read.table("Gaz.txt",h=T)
4 GDP<-read.table("GDP.txt",h=T)
5 HDI<-read.table("HDI.txt",h=T)
6 Health<-read.table("Health.txt",h=T)
7 Investissement<-read.table("Investissement.txt",h=T)
8 MobilPhone<-read.table("MobilPhone.txt",h=T)
9 Mortality<-read.table("Mortality.txt",h=T)
10 Population<-read.table("Population.txt",h=T)
11 Undernourishement<-read.table("Undernourishement.txt"
     ,h=T)
```

Listing 10.1
Import Histogram PCA using R.

Data preparation in R

Once datasets have been imported in R, we need to process them. For example, in the datasets imported, the first column contains regions names. We need to prepare those data in a way that only numerical data will be kept or considered (unless users ignore them as they perform Histogram PCA). In the code below (see Listing 10.2), we remove the first column and we create a dataframe with the name of the regions (it is also possible to use the *rowname* R function).

```
1 #
2 #Create data frame containing region name
3 RegionDf<-data.frame(Region=Population$Region)
4 # Remove first columns
5 Croissance2<-Croissance2[,-1]
6 Electricity2<-Electricity2[,-1]
7 GAZ2<-GAZ2[,-1]
8 GDP2<-GDP2[,-1]
9 HDI2<-HDI2[,-1]
10 Health2<-Health2[,-1]
11 Investissement2<-Investissement2[,-1]
12 MobilPhone2<-MobilPhone2[,-1]
13 Mortality2<-Mortality2[,-1]
14 Population2<-Population2[,-1]
15 Undernourishement2<-Undernourishement2[,-1]
```

Listing 10.2
Data Preparation for Histogram PCA using R.

Scoring and Histogram PCA

```
1 library(GraphPCA)
2 # Use of Ridit scores
3 # for Ridit score standardize use Ridi2(Croissance2)$
    Ridit2
4 # for Ridit score normalized use Ridi3(Croissance2)$
    Ridit3
5 # but we will be applying raw Ridit score using Ridi
    function
6 s1=Ridi(Croissance2)$Ridit
7 s2=Ridi(Electricity2)$Ridit
8 s3=Ridi(GAZ2)$Ridit
9 s4=Ridi(GDP2)$Ridit
10 s5=Ridi(HDI2)$Ridit
11 s6=Ridi(Health2)$Ridit
12 s7=Ridi(Investissement2)$Ridit
13 s8=Ridi(MobilPhone2)$Ridit
14 s9=Ridi(Mortality2)$Ridit
15 s10=Ridi(Population2)$Ridit
16 s11=Ridi(Undernourishement2)$Ridit
17
18 HistPCA(Variable=list(Croissance2,Electricity2,GAZ2,
    GDP2,HDI2,Health2,Investissement2,MobilPhone2,
    Mortality2,Population2,Undernourishement2),
19 score=list(s1,s2,s3, s4,s5,s6,s7,s8, s9, s10, s11),
20   Col.names=c("Population Growth", "Electricity", "
      Gaz", "GDP","HDI", 'Health expenses',"Investment
      ","MobilPhone","Mortality","Population","
      Undernourishement"), Row.names=RegionDf\$Region,
      transformation = 1)
```

Listing 10.3
Histogram PCA using R.

Given the limitations of the parametric and non-parametric codings mentioned above, we choose the coding method based on the raw Ridit scores (see Listing 10.3). In the factorial axes from Figure 10.2, the larger is the rectangle, the higher is the region variability. Figure 10.2 represents regions' profiles. From the first axis in this figure, we can notice three groups of regions. The first one includes Nafta and Europe, two very similar regions. The second group is the one that only includes Africa. Those two groups oppose each other. There is third group formed by countries from the ex-USSR, Middle East, South and Central America and Oriental Asia. In addition, we notice that the second axis opposes Middle East to former USSR countries.

Figure 10.3 explains and provides more details regarding the findings from Figure 10.2.

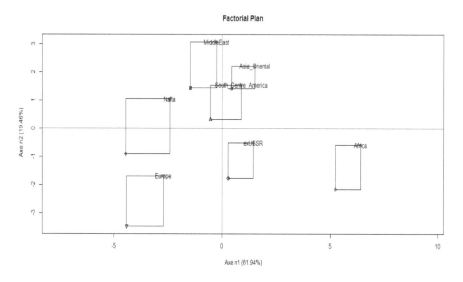

FIGURE 10.2
Factorial Plan of Histogram PCA based on barycenters.

Figure 10.3 tells us why we do see similarities and differences between the regions we are describing. The percentage of variability of the two first axes is respectively 61.94% and 19.48%. The well represented variables are GDP, Electricity, CO2/Gaz production, Undernourishment, Investment, Health (expenses percentage for health care), Fertility rate and Population growth. From the first axis of those two figures, we notice that Africa has a high level of Undernourishment rate, Fertility rate, Population growth and a low level of GDP, Electricity, GAZ production. We notice the contrary for Europe and NAFTA. Regarding the second axis of those two figures, Population growth and Mortality rate are variables that explain the opposition between Middle East and ex-USSR. In the Middle East, Population growth is high but in most of countries from the ex-USSR, the overall Population growth is weak. In the next section we are going to apply Histogram PCA based on quantiles. In the PCA performed here, the underlying assumption made is that barycenters are representative of distributions.

What if we prefer to focus on quantiles and not on barycenters? Are we going to observe the same trends and patterns? In the next section, we are going to apply Histogram PCA based on quantiles and we will see whether findings will be different or similar.

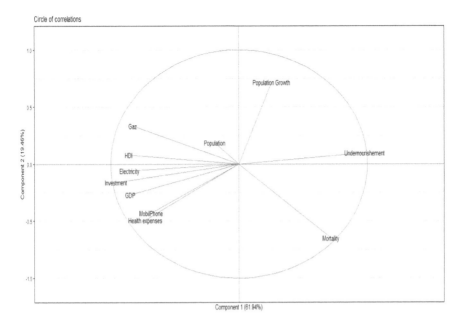

FIGURE 10.3

Correlation Map of histogram PCA based on barycenters.

10.6.1.2 PCA based on the average correlation matrix of quantiles tables

Since there is no need of coding here, users need to choose the number of quantiles and their location. We first select four quantiles: the 0.1, 0.3, 0.6, 0.9 percentiles (a combination of low and high quantiles). Other choices are also possible. Then, we apply Histogram PCA based on quantiles. Overall, the results we got are similar to those obtained by using Histogram PCA based on the barycenters. Figure 10.4 is a correlation map from the Histogram PCA based on quantiles. Results from this figure are similar to the results we get from the histogram PCA based on barycenters in Figure 10.3.

From the first PCA axis (with 50.99% of variability), we notice that in terms of profile (see Figure 10.5), Africa on the one hand and NAFTA and Europe on the other hand, oppose each other. The second axis (with 16.79% of variability) opposes Middle East and Asia, and the former USSR countries. Rectangles from the picture are reflective of the dispersion among the quantiles. However, in the PCA based on barycenters, rectangles are reflective of the dispersion around the barycenters. From the two approaches, we notice how countries from the ex-USSR seem to be the more homogeneous from a macroeconomic perspective.

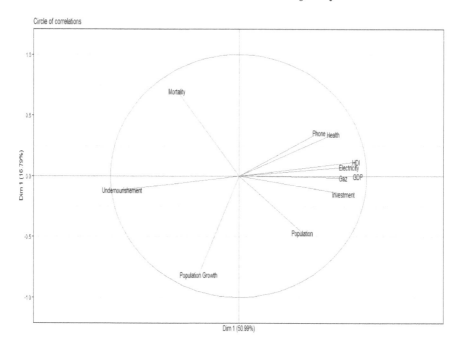

FIGURE 10.4
Correlation Map of histogram PCA based on quantiles.

In [38] the author also represents histogram PCA outputs using a 2d-convex hull representation. From Figure 10.6 (see [28]), we could notice how Middle East has the larger dispersion. Middle East seems to be a region where the countries' differences are very important.

10.6.2   Histogram PCA using OECDGrowth data

In this application, we are using data from scratch and showing users how they can start with classical data, derive distributional data and perform Histogram PCA. In that, we are using a Growth regression dataset. This dataset is provided in [18] and is available in the R package AER (Applied Econometrics with R, see [31]). GrowthDJ is a dataframe with 121 observations and 10 variables. We will be focusing on seven variables, namelye: gdp60 (Per capita GDP in 1960), gdp85 (Per capita GDP in 1985), gdpgrowth (average growth rate of per capita GDP from 1960 to 1985 in percent), popgrowthHist (Average growth rate of working-age population 1960 to 1985 (in percent)),

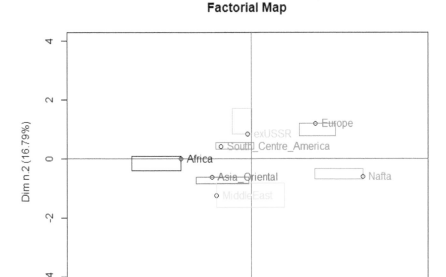

FIGURE 10.5
Histogram PCA based on quantiles: Factorial Plan using rectangles.

invest (Average ratio of investment (including Government Investment) to GDP from 1960 to 1985 (in percent)), school (Average fraction of working-age population enrolled in secondary school from 1960 to 1985 (in percent)), literacy (Fraction of the population over 15 years old that is able to read and write in 1960 (in percent)). Most of those data come from [41], except literacy60 that is from the World Bank's World Development Report.

What is the profile of those 121 countries? How many clusters could we get from those countries? What makes those clusters different or similar? As before, we are going to use Histogram PCA in order to deal with those questions. We first explored the data distributions and we noticed that there were few fields with missing data. A complete case analysis (systematic deletion of missing) would lead to 100 observations instead of 121 observations. To avoid this loss of data due to systematic deletion of missing values, we have decided to impute the data using the R package mice (Multiple Imputation using Chain Equation) (see [50]) and the R code in Listing 10.4.

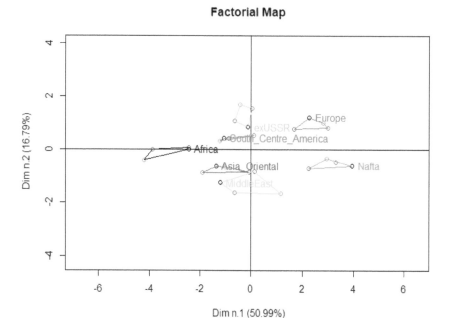

FIGURE 10.6

Histogram PCA based on quantiles: Factorial Map using convex hull.

Once data have been imputed, we explored the data through a classical PCA using the R package FactoMineR from [34]. We noticed that observation 56 (Kuwait) is an outlier country (see Figure 10.7). Indeed, gdp60 and gdp85 values from observation 56 are respectively gdp60 = 77881 and gdp85 = 25635 but the median and IQR (Inter Quartile Range) for those two variables are respectively $median(gdp60) = 1962$, $IQR = [973; 1209]$ for gdp60 and $median(gdp85) = 3484$ and $IQR = [1209; 7719]$ for $gdp85$.

We decided to remove the outlier country (observation 56, i.e. Kuwait) from our analyses and we performed a new classical PCA that was followed by a clustering. All this was done using the R package FactoMineR (see R code Listing 10.4). From the clustering, we obtained five clusters or groups of countries. Once we get those clusters of countries, the next question might be what is the profile of those clusters? What makes them close, similar, unique or different? How could we label those clusters? Is there a cluster at risk when it comes to macroeconomic performances?

```
1  # load AER library
2  library("AER")
3  # load GrowthDJ
4  data(GrowthDJ)
5  # remove the 3 first columns
6  dataGrowth<-GrowthDJ[,c(-1,-2,-3)]
7  #load library mice
8  library(mice)
9  # apply imputation, m=1 time
10 imp <- mice(dataGrowth,m=1,seed=1981,meth='pmm')
11 # imputation of missing data
12 dataGrowth3<- complete(imp,1)
13 # call of FactomineR
14 library(FactoMineR)
15 # deletion of observation 56
16 dataGrowth4<-dataGrowth3[-56,]
17 res1=PCA(dataGrowth4[,-8])
18 # clustering k=5 clusters
19 res2=HCPC(res1,nb.clust=5)
20 #
21 dataGrowth3<-res2$data.clust
22 n<-nrow(dataGrowth3)
23 # add column cluster in the data
24 dataGrowth3$clust<-paste("Cluster",dataGrowth3$clust,
       sep="_")
25 # describe the clusters
26 table(dataGrowth3$clust)
```

Listing 10.4
DataGrowth Prep imputation and determination of clusters.

From Table 10.7, we notice that Cluster 3 has 33 countries and Cluster 4 has 19 countries. In order to have a better understanding of those clusters, we build seven histogram-valued datasets with five bins from the seven input variables. For the GDP variable, for example, the five histogram bins are $[\min(GDP), a_1]$, $[a_1, a_2]$, $[a_2, a_3]$, $[a_3, a_4]$, $[a_4, \max(GDP)]$ such that:

$$\max(GDP) - a_4 = a_4 - a_3 = a_3 - a_2 = a_2 - a_1 = a_1 - \min(GDP).$$

Those five bins represent respectively the very low, low, medium, high and very high values of GDP. All seven input variables have been transformed into histograms with five bins. From those histogram-valued data, we performed barycenters Histogram PCA using R. The R implementation details can be seen in the listing 10.5 below.

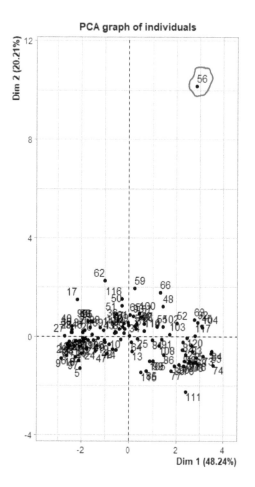

FIGURE 10.7
Classical PCA outputs of OECDGrowth data.

TABLE 10.7
Cluster Distribution.

Cluster	Number of countries
Cluster 1	23
Cluster 2	31
Cluster 3	33
Cluster 4	14
Cluster 5	19

```
1  library(GraphPCA)
2  #Z represent the column position of the reference
      categorical
3  # variable thas is Cluster ("name") in our case
4  gdp60Hist=PrepHistogram(X=sapply(dataGrowth3[,1],
      unlist),Z=dataGrowth3[,8],k=5)$Vhistogram
5  gdp85Hist=PrepHistogram(X=sapply(dataGrowth3[,2],
      unlist),Z=dataGrowth3[,8],k=5)$Vhistogram
6  gdpgrowthHist=PrepHistogram(X=sapply(dataGrowth3[,3],
      unlist),Z=dataGrowth3[,8],k=5)$Vhistogram
7  popgrowthHist=PrepHistogram(X=sapply(dataGrowth3[,4],
      unlist),Z=dataGrowth3[,8],k=5)$Vhistogram
8  investHist=PrepHistogram(X=sapply(dataGrowth3[,5],
      unlist),Z=dataGrowth3[,8],k=5)$Vhistogram
9  schoolHist=PrepHistogram(X=sapply(dataGrowth3[,6],
      unlist),Z=dataGrowth3[,8],k=5)$Vhistogram
10 literacy60Hist=PrepHistogram(X=sapply(dataGrowth3
      [,7],unlist),Z=dataGrowth3[,8],k=5)$Vhistogram
11
12 # Raw Ridit scores computation
13 ss1=Ridi(gdp60Hist)$Ridit
14 ss2=Ridi(gdp85Hist)$Ridit
15 ss3=Ridi(gdpgrowthHist)$Ridit
16 ss4=Ridi(popgrowthHist)$Ridit
17 ss5=Ridi(investHist)$Ridit
18 ss6=Ridi(schoolHist)$Ridit
19 ss7=Ridi(literacy60Hist)$Ridit
20 # Histogram PCA
21 # if transformation=2 then there is a use of angular
22 # transformation; if not, no transformation
23 res_pca=HistPCA(list(gdp60Hist, gdp85Hist,
      gdpgrowthHist,popgrowthHist,investHist,schoolHist,
      literacy60Hist),
24             score=list(ss1,ss2,ss3,ss4,ss5, ss6,ss7)
                 ,
25             Col.names=c("gdp60","gdp85","gdpgrowth",
                 "popgrowth","invest","school","
                 literacy"),
26             transformation = 1 )
27
28 # Visualization of Clusters
29 library(ggplot2)
30  Visu(res_pca$PCinterval)+theme_bw()+labs(title = "
      Country Profiles",
```

```
31|          subtitle = "Histogram PCA- Barycentre method",
32|          caption = "Principal Components",
33|          tag = "Figure 1",
34|          x = "Component 1",
35|          y = "Component 2")
```

Listing 10.5
Transforming classical data into histograms and running Histogram PCA.

We represent the PCA outputs using the two first components (see Figure 10.8). From the first component, we notice that Cluster 1 and Cluster 2 are similar and both are very different from Cluster 5 and Cluster 4. From component 2, we notice that Cluster 1 opposes Cluster 3. Variability percentage of those two first components are respectively 66.34% and 25.75%.

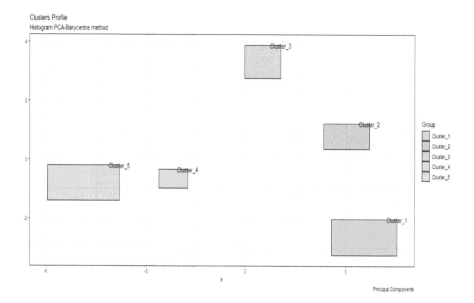

FIGURE 10.8
Histogram PCA outputs OECDGrowth data: Clusters Profile.

From Figure 10.9, we could deduce that Cluster 3, Cluster 2 and Cluster 1 have a strong Population growth (see Table 10.8). When it comes to Literacy and GDP growth, values from Cluster 3 are high. Cluster 4 and Cluster 5 have the highest values when it comes to School, Investment, GDP but low values for Population growth (see Table 10.9). Cluster 4 and Cluster 5 are composed by developed countries, as opposed to Cluster 1. Cluster 3 is the one where most of countries have a high Population and GDP growth. Results from Histogram PCA based on the quantiles method are similar to those from Histogram PCA based on the barycenters (see Appendix 2). However, when distributions are very skewed, as it is the case for the Literacy variable and Cluster 4 (see Table 10.10), the quantiles method would be preferred.

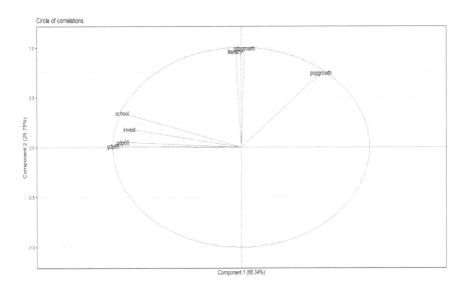

FIGURE 10.9
Histogram PCA outputs OECDGrowth data: Correlation Circle.

TABLE 10.8
OECDGrowth data: Population Growth.

Cluster	Very Low [0.3; 1.1[	Low [1.1; 1.9[	Medium [1.9; 2.7[	High [2.7; 3.5[	Very High [3.5; 4.3]
Cluster 1	0.043	0.261	0.609	0.087	0.000
Cluster 2	0.000	0.129	0.516	0.323	0.032
Cluster 3	0.000	0.000	0.333	0.606	0.061
Cluster 4	0.571	0.286	0.143	0.000	0.000
Cluster 5	0.579	0.263	0.158	0.000	0.000

TABLE 10.9
OECDGrowth data: School.

Cluster	Very Low [0.40; 2.74[	Low [2.74; 5.08[	Medium [5.08; 7.42[	High [7.42; 9.76[	Very High [9.76; 12.10]
Cluster 1	0.913	0.087	0.000	0.000	0.000
Cluster 2	0.452	0.484	0.065	0.000	0.000
Cluster 3	0.030	0.212	0.394	0.212	0.152
Cluster 4	0.000	0.071	0.357	0.286	0.286
Cluster 5	0.000	0.105	0.000	0.368	0.526

The barycenters method focus on barycenters and is more convenient when distributions are Normal or symmetric. A takeaway from those findings is that developed countries (Cluster 4 and Cluster 5) are somewhat "rich" but in those countries population is not growing very fast. Countries from Cluster 3 are those where population and economy are growing very fast. Cluster 2 is the one where countries have poor macro economical performances and where the growth of the population is important. However, countries from Cluster 1 are those where GDP, Investment and School performance indicators are weak and concerning. Decision makers can explore all those findings and set up decisions accordingly, for the benefit of the population living in those clusters.

TABLE 10.10
OECDGrowth data: Literacy60 Histogram.

Cluster	Very Low [1.0; 20.8[	Low [20.8; 40.6[	Medium [40.6; 60.4[	High [60.4; 80.2[	Very High [80.2; 100.0]
Cluster 1	0.826	0.130	0.043	0.000	0.000
Cluster 2	0.484	0.355	0.129	0.032	0.000
Cluster 3	0.091	0.152	0.061	0.485	0.212
Cluster 4	0.000	0.000	0.000	0.071	0.929
Cluster 5	0.000	0.053	0.000	0.000	0.947

TABLE 10.11
OECDGrowth data: GDP85 Histogram.

Cluster	Very Low [412.0; 4274.2[	Low [4274.2; 8136.4[	Medium [8136.4; 11998.6[	High [11998.6; 15860.8[	Very High [15860.8; 19723.0]
Cluster 1	1.000	0.000	0.000	0.000	0.000
Cluster 2	0.871	0.129	0.000	0.000	0.000
Cluster 3	0.273	0.515	0.091	0.121	0.000
Cluster 4	0.071	0.429	0.286	0.143	0.071
Cluster 5	0.000	0.000	0.105	0.632	0.263

TABLE 10.12
OECDGrowth data: GDPGrowth Histogram.

Cluster	Very Low [-0.90; 1.12[	Low [1.12; 3.14[	Medium [3.14; 5.16[	High [5.16; 7.18[	Very High [7.18; 9.20]
Cluster 1	0.174	0.739	0.087	0.000	0.000
Cluster 2	0.000	0.032	0.613	0.323	0.032
Cluster 3	0.000	0.030	0.394	0.394	0.182
Cluster 4	0.143	0.214	0.357	0.286	0.000
Cluster 5	0.000	0.316	0.632	0.053	0.000

10.7 Conclusion

In this chapter, we have shown how PCA can be applied to distributional data. Two approaches have been proposed in that regard. The first one is based on barycenters and requires a coding strategy. Several options are possible among which the metabin coding and the Ridit scores. Those two options overcome the limitations of the parametric and non-parametric codings. A practical strategy consists in choosing first the Ridit scores or metabin coding and using after the other coding methods in a sensitivity analysis. We have also presented another way to perform histogram PCA via the use of quantiles. In this case, users do not need to choose any coding strategy. However, they have to choose the number of quantiles and their location. The comparison of the two strategies, by using two macroeconomic datasets, shows that they lead to similar results. However, since Histogram PCA based on barycenters uses distribution averages, this approach is more suitable when distributions are Gaussian (Normal) or symmetric. In contrast, Histogram PCA based on quantiles works no matter of the underlying specification of distributions. We have also presented a framework extension to three-way histogram data. This strategy is based on the determination of a compromise matrix in which weights of histogram datasets are determined using histogram data entropy. All methods described here have a lot of practical applications in many domains like workforce analytics, insurance, risk and control management, internal audit, anti money laundering, healthcare analytics, sports analytics, etc. In workforce analytics for example, enterprises have data (performances, turnover, engagement, satisfaction, income, expenses) at employee level and those data are usually aggregated by team or regions for confidentiality or ethical reasons. In the insurance industry, companies have massive household or policy level data with many of numerical attributes like premiums, losses, credit score and insurance scores. In internal audit or risk management, most of enterprises have quantitative KPIs (Key Performance Metrics) or KRIs (Key Risk Indicators) and those data are usually aggregated at audit/business units levels. In healthcare analytics, one might have a list of clusters with covid19 characteristics and want to profile those clusters given some numerical features. If one of these situations applies, distributional datasets can be extracted from data and the Histogram PCA methods described here can help getting meaningful insights.

Bibliography

[1] F. Afonso, E. Diday, and C. Toque. *Data Science par Analyse des Données Symboliques: Une Nouvelle Façon d'Analyser les Données Classiques, Complexes et Massives à partir des Classes.* Editions Technip, 2018.

[2] A. Agresti and M. Kateri. *Categorical Data Analysis.* Springer Berlin Heidelberg, 2011.

[3] J. Aitchison. *The Statistical Analysis of Compositional Data.* Blackburn Press, 2003.

[4] L. Billard and E. Diday. *Symbolic Data Analysis: Conceptual Statistics and Data Mining.* Wiley Series in Computational Statistics, Berlin, 2006.

[5] H.-H. Bock and E. Diday. *Analysis of Symbolic Data, Exploratory Methods for Extracting Statistical Informations from Complex Data.* Springer, Berlin, 2000.

[6] J.M. Bouroche. *Analyse des Données Ternaires: La Double Analyse en Composantes Principales.* Thèse de 3ème cycle, Université de Paris VI, 1975.

[7] B. Brahim and S. Makosso-Kallyth. GPCIV and GraphPCA: two R packages for PCA of complex data. In *JSM 2014*, Boston, USA, 2014.

[8] P. Brockett, R. Derrig, L. Golden, A. Levine, and M. Alpert. Fraud classification using principal component analysis of RIDITs. *Journal of Risk and Insurance*, 69:341–371, 2002.

[9] I.D.J. Bross. How to use Ridit analysis. *Biometrics*, 14(1):18–38, 1958.

[10] J.D. Carroll and J-J Chang. Analysis of individual differences in multi-dimensional scaling via an n-way generalization of "Eckart-Young" decomposition. *Psychometrika*, 35(3):283–319, 1970.

[11] P. Cazes. Analyse factorielle d'un tableau de lois de probabilité. *Revue de Statistique Appliquée*, 50(3):5–24, 2002.

[12] P. Cazes, A. Chouakria, E.Diday, and Y. Schektman. Extension de l'analyse en composantes principales à des données intervalles. *Revue de Statistique Appliquée*, 45(3):5–24, 1997.

[13] M. Chen, H. Wang, and Z. Qin. Principal component analysis for probabilistic symbolic data: a more generic and accurate algorithm. *Advances in Data Analysis and Classification*, pages 1–21, 2014.

[14] D. Chessel and M. Hanafi. Analyse de la co-inertie de k nuages de points. *Revue de Statistique Appliquée*, 44(2):35–60, 1996.

[15] E. Diday. Principal component analysis for bar charts and metabins tables. *Statistical Analysis and Data Mining*, 6(5):403–430, 2013.

[16] E. Diday. Thinking by classes in data science: the symbolic data analysis paradigm. *WIREs Computational Statistics*, 8(5):172–205, 2016.

[17] E. Diday and M. Noirhomme-Fraiture. *Symbolic Data Analysis and the SODAS Software*. Wiley Interscience, Chichester, 2008.

[18] S.N. Durlauf and P.A. Johnson. Multiple regimes and cross-country growth behavior. *Journal of Applied Econometrics*, 10:365–384, 1995.

[19] B. Escofier and J. Pagès. *Analyses Factorielles Simples et Multiples: Objectifs, Méthodes et Interprétation*. Dunod, Paris, 1998.

[20] R.A. Fisher. On the probable error of a coefficient of correlation deduced from a small sample. *Metron*, 1:3–12, 1921.

[21] W.D. Gary. Ridit scores for analysis and interpretation of ordinal pain data. *European Journal of Pain*, 2(3):221 – 227, 1998.

[22] J.C. Gower. Some history of algebraic canonical forms and data analysis. In Jorg Blasius and Michael Greenacre, editors, *Visualization and Verbalization of Data*, pages 17–30. Chapman and Hall/CRC, 2014.

[23] M. Hanafi and E.M. Qannari. Nouvelles propriétés de l'analyse en composantes communes et poids spécifiques. *Journal de la Société Francaise de Statistique*, 149(2):75–97, 2008.

[24] R. Harshman. Foundations of the parafac procedure: Models and conditions for an "explanatory" multi-modal factor analysis. *UCLA Working Papers in Phonetics*, 16, 1970.

[25] R.A. Harshman and M.E. Lundy. PARAFAC: Parallel factor analysis. *Computational Statistics and Data Analysis*, 18:39–72, 1994.

[26] M. Ichino. Symbolic pca for histogram-valued data. In *IASC2008, Joint Meeting of 4th World Conference of the IASC and 6th Conference of the Asian Regional Section of the IASC on Computational Statistics & Data Analysis*, Yokohama, Japan, December 2008.

[27] M. Ichino. The quantile method for symbolic principal component analysis. *Statistical Analysis and Data Mining*, 4(2):184–198, 2011.

[28] A. Irpino, C. Lauro, and R. Verde. Visualizing symbolic data by closed shapes. In Martin Schader, Wolfgang Gaul, and Maurizio Vichi, editors, *Between Data Science and Applied Data Analysis*, Studies in Classification, Data Analysis, and Knowledge Organization, pages 244–251. Springer Berlin Heidelberg, 2003.

[29] A. Irpino, R. Verde, and F. De Carvalho. Dynamic clustering of histogram data based on adaptive squared Wasserstein distances. *Expert Systems with Applications*, 41(7):3351–3366, 2014.

[30] I. Jolliffe and J. Cadima. Principal component analysis: A review and recent developments. *Philosophical Transactions of the Royal Society A: Mathematical, Physical and Engineering Sciences*, 374:20150202, 04 2016.

[31] C. Kleiber and A. Zeileis. *Applied Econometrics with R*. Springer Verglag, New York, 3 2008.

[32] P.M. Kroonenberg. History of multiway component analysis and three-way correspondence analysis. In Jorg Blasius and Michael Greenacre, editors, *Visualization and Verbalization of Data*, pages 77–94. Chapman and Hall/CRC, 2014.

[33] C. Lavit, Y. Escoufier, R. Sabatier, and P. Traissac. The ACT (STATIS method). *Computational Statistics and Data Analysis*, 18:97–119, 1994.

[34] S. Lê, L. Josse, and F. Husson. Factominer: An R package for multivariate analysis. *Journal of Statistical Software*, 25(1):1–18, 3 2008.

[35] J. Le-Rademacher and L. Billard. Principal component analysis for histogram-valued data. *Advances in Data Analysis and Classification*, 11(2):327–351, 2016.

[36] H. L'Hermier des Plantes. *Structuration des Tableaux A Trois Indices de la Statistique*. Thèse de doctorat, Université de Montpellier, 1976.

[37] S. Makosso-Kallyth. *Analyse en Composantes Principales de Variables Symboliques de Type Histogramme*. Thèse de doctorat, Université de Paris Dauphine, 2010.

[38] S. Makosso-Kallyth. Principal axes analysis of symbolic histogram variables. *Statistical Analysis and Data Mining: The ASA Data Science Journal*, 9(3):188–200, 2016.

[39] S. Makosso-Kallyth. Analyse en composante principales d'un tableau de distributions macroéconomiques. *Monde des Util. Anal. Données*, 45:55–74, 2018.

[40] S. Makosso-Kallyth and E. Diday. Adaptation of interval pca to symbolic histogram variables. *Advances in Data Analysis and Classification*, 6:147–159, 2012.

[41] N.G. Mankiw, D. Romer, and D.N. Weil. A contribution to the empirics of economic growth. *Quarterly Journal of Economics*, 107:407–437, 1992.

[42] N. Mantel. Ridit analysis and related ranking procedures-use at own risk. *American Journal of Epidemiology*, 109(1):25–29, 01 1979.

[43] M. Noirhomme-Fraiture and P. Brito. Far beyond the classical data models: symbolic data analysis. *Statistical Analysis and Data Mining*, 4(2):157–170, 2011.

[44] O. Rodrıguez, E. Diday, and S. Winsberg. Generalization of the principal components analysis to histogram data. In *Workshop on Simbolic Data Analysis of the 4th European Conference on Principles and Practice of Knowledge Discovery in Data Bases*, pages 12–16, 2000.

[45] C. Spearman. "General intelligence" objectively determined and measured. *The American Journal of Psychology*, 15(2):201–292, 1904.

[46] A. Tenenhaus and M. Tenenhaus. Regularized generalized canonical correlation analysis. *Psychometrika*, 76(2):257–284, 2011.

[47] L.L. Thurstone. Multiple factor analysis. *Psychological Review*, 38(5):406, 1931.

[48] L.R. Tucker. Implications of factor analysis of three-way matrices for measurement of change. *Problems in Measuring Change*, pages 122–137, 1963.

[49] L.R. Tucker. The extension of factor analysis to three-dimensional matrices. *Contributions to Mathematical Psychology*, pages 109–127, 1964.

[50] S. van Buuren and K. Groothuis-Oudshoorn. Mice: Multivariate imputation by chained equations in R. *Journal of Statistical Software*, 3(45):1–67, 2011.

[51] R. Verde, A. Irpino, and A. Balzanella. Dimension reduction techniques for distributional symbolic data. *IEEE Transactions on Cybernetics*, 46(2):344–355, 2016.

[52] H. Wang, M. Chen, Shi Xi., and N. Li. Principal component analysis for normal-distribution-valued symbolic data. *IEEE Transactions on Cybernetics*, 46(2):356–365, 2016.

Appendix 1: OECDGrowth Histogram

TABLE 10.13
OECDGrowth data: Invest Histogram.

Cluster	Very Low [4.10; 10.66[	Low [10.66; 17.22[	Medium [17.22; 23.78[	High [23.78; 30.34[	Very High [30.34; 36.90]
Cluster 1	0.652	0.348	0.000	0.000	0.000
Cluster 2	0.194	0.452	0.226	0.097	0.032
Cluster 3	0.000	0.394	0.455	0.121	0.030
Cluster 4	0.000	0.071	0.214	0.429	0.286
Cluster 5	0.000	0.000	0.368	0.526	0.105

TABLE 10.14
OECDGrowth data: GDP60 Histogram.

Cluster	Very Low [383.0; 2778.8[	Low [2778.8; 5174.6[	Medium [5174.6; 7570.4[	High [7570.4; 9966.2[	Very High [9966.2; 12362.0]
Cluster 1	1.000	0.000	0.000	0.000	0.000
Cluster 2	1.000	0.000	0.000	0.000	0.000
Cluster 3	0.485	0.394	0.030	0.030	0.061
Cluster 4	0.429	0.500	0.071	0.000	0.000
Cluster 5	0.000	0.000	0.211	0.632	0.158

Appendix 2: Histogram PCA based on the quantiles-outputs

The following quantiles have been used 0.1, 0.5, 0.9.

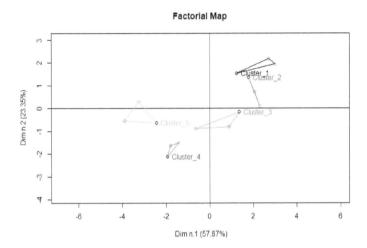

FIGURE 10.10
Histogram PCA outputs OECDGrowth data: Factorial map using convex hull.

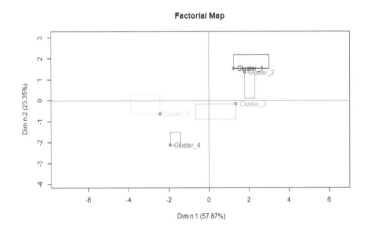

FIGURE 10.11
Histogram PCA outputs OECDGrowth data: Factorial map with rectangles.

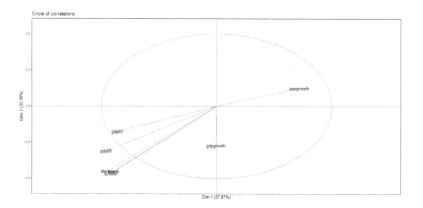

FIGURE 10.12
Histogram PCA outputs OECDGrowth data: Correlation map.

Appendix 3: Clusters and Countries details

TABLE 10.15
OECDGrowth after imputation: Cluster 1.

Obs.	gdp60	gdp85	gdpgrowth	popgrowth	invest	school	literacy60	clust	Country2
2	1588.00	1171.00	0.80	2.10	5.80	1.80	5.00	Cluster_1	Angola
3	1116.00	1071.00	2.20	2.40	10.80	1.80	5.00	Cluster_1	Benin
5	529.00	857.00	2.90	0.90	12.70	0.40	2.00	Cluster_1	Burkina Faso
6	755.00	663.00	1.20	1.70	5.10	0.40	14.00	Cluster_1	Burundi
8	838.00	789.00	1.50	1.70	10.50	1.40	7.00	Cluster_1	CentralAfr. Rep.
9	908.00	462.00	-0.90	1.90	6.90	0.40	6.00	Cluster_1	Chad
12	533.00	608.00	2.80	2.30	5.40	1.10	15.00	Cluster_1	Ethiopia
15	1009.00	727.00	1.00	2.30	9.10	4.70	27.00	Cluster_1	Ghana
16	746.00	869.00	2.20	1.60	10.90	0.60	7.00	Cluster_1	Guinea
21	1194.00	975.00	1.40	2.20	7.10	2.60	50.00	Cluster_1	Madagascar
23	737.00	710.00	2.10	2.20	7.30	1.00	3.00	Cluster_1	Mali
27	1420.00	1035.00	1.40	2.70	6.10	0.70	11.00	Cluster_1	Mozambique
29	1055.00	1186.00	2.80	2.40	12.00	2.30	15.00	Cluster_1	Nigeria
31	1392.00	1450.00	2.50	2.30	9.60	1.70	6.00	Cluster_1	Senegal
32	511.00	805.00	3.40	1.60	10.90	1.70	7.00	Cluster_1	Sierra Leone
33	901.00	657.00	1.80	3.10	13.80	1.10	2.00	Cluster_1	Somalia
35	1254.00	1038.00	1.80	2.60	13.20	2.00	13.00	Cluster_1	Sudan
40	601.00	667.00	3.50	3.10	4.10	1.10	35.00	Cluster_1	Uganda
41	594.00	412.00	0.90	2.40	6.50	3.60	31.00	Cluster_1	Zaire (DR Congo)
44	1224.00	1704.00	1.60	2.10	6.90	0.90	8.00	Cluster_1	Afghanistan
58	833.00	974.00	2.60	2.00	5.90	2.30	9.00	Cluster_1	Nepal
69	2726.00	1918.00	2.50	2.50	17.20	0.60	3.00	Cluster_1	Yemen
97	1096.00	1237.00	1.80	1.30	7.10	1.90	15.00	Cluster_1	Haiti

TABLE 10.16
OECDGrowth after imputation: Cluster 2.

Obs.	gdp60	gdp85	gdpgrowth	popgrowth	invest	school	literacy60	clust	Country2
1	2485.00	4371.00	4.80	2.60	24.10	4.50	10.00	Cluster 2	Algeria
7	889.00	2190.00	5.70	2.10	12.80	3.40	19.00	Cluster 2	Cameroon
10	1009.00	2624.00	6.20	2.40	28.80	3.80	16.00	Cluster 2	Congo (BZV)
11	907.00	2160.00	6.00	2.50	16.30	7.00	26.00	Cluster 2	Egypt
13	1307.00	5350.00	7.00	1.40	22.10	2.60	53.00	Cluster 2	Gabon
14	799.00	4371.00	3.60	2.10	18.10	1.50	68.00	Cluster 2	Gambia
17	1386.00	1704.00	5.10	4.30	12.40	2.30	5.00	Cluster 2	Ivory Coast
18	944.00	1329.00	4.80	3.40	17.40	2.40	20.00	Cluster 2	Kenya
19	431.00	1483.00	6.80	1.90	12.60	2.00	16.00	Cluster 2	Lesotho
20	863.00	944.00	3.30	3.00	21.50	2.50	9.00	Cluster 2	Liberia
22	455.00	823.00	4.80	2.40	13.20	0.60	25.00	Cluster 2	Malawi
24	777.00	1038.00	3.30	2.20	25.60	1.00	5.00	Cluster 2	Mauritania
26	1030.00	2348.00	5.80	2.50	8.30	3.60	14.00	Cluster 2	Morocco
28	539.00	841.00	4.40	2.60	10.30	0.50	1.00	Cluster 2	Niger
30	460.00	696.00	4.50	2.80	7.90	0.40	16.00	Cluster 2	Rwanda
36	817.00	7400.00	7.20	2.80	17.70	3.70	26.00	Cluster 2	Swaziland
37	383.00	710.00	5.30	2.90	18.00	0.50	10.00	Cluster 2	Tanzania
38	777.00	978.00	3.40	2.50	15.50	2.90	10.00	Cluster 2	Togo
39	1623.00	3661.00	5.60	2.40	13.80	4.30	16.00	Cluster 2	Tunisia
42	1410.00	1217.00	2.10	2.70	31.70	2.40	29.00	Cluster 2	Zambia
43	1187.00	2107.00	5.10	2.80	21.10	4.40	39.00	Cluster 2	Zimbabwe
46	846.00	1221.00	4.00	2.60	6.80	3.20	22.00	Cluster 2	Bangladesh
47	517.00	1031.00	4.50	1.70	11.40	3.50	60.00	Cluster 2	Burma
49	978.00	1339.00	3.60	2.40	16.80	5.10	28.00	Cluster 2	India
60	1077.00	2175.00	5.80	3.00	12.20	3.00	15.00	Cluster 2	Pakistan
95	2042.00	1997.00	3.30	3.30	8.00	3.90	49.00	Cluster 2	El Salvador
96	2481.00	3034.00	3.90	3.10	8.80	2.40	32.00	Cluster 2	Guatemala
98	1430.00	1822.00	4.00	3.10	13.80	3.70	45.00	Cluster 2	Honduras
106	1618.00	2055.00	3.30	2.40	13.30	4.90	39.00	Cluster 2	Bolivia
119	879.00	2159.00	5.50	1.90	13.90	4.10	39.00	Cluster 2	Indonesia
121	1781.00	2544.00	3.50	2.10	16.20	1.50	29.00	Cluster 2	Papua N. Guinea

TABLE 10.17
OECDGrowth after imputation: Cluster 3.

Obs.	gdp60	gdp85	gdpgrowth	popgrowth	invest	school	literacy60	clust	Country2
4	959.00	3671.00	8.60	3.20	28.30	2.90	90.00	Cluster.3	Botswana
25	1973.00	2967.00	4.20	2.60	17.10	7.30	91.00	Cluster.3	Mauritius
34	4768.00	7064.00	3.90	2.30	21.60	3.00	57.00	Cluster.3	South Africa
48	3085.00	13372.00	8.90	3.00	19.90	7.20	70.00	Cluster.3	Hong Kong
50	3606.00	7400.00	6.30	3.40	18.40	6.50	16.00	Cluster.3	Iran
51	4916.00	5626.00	3.80	3.20	16.20	7.40	18.00	Cluster.3	Iraq
52	4802.00	10450.00	5.90	2.80	28.50	9.50	84.00	Cluster.3	Israel
54	2183.00	4312.00	5.40	2.70	17.60	10.80	32.00	Cluster.3	Jordan
55	1285.00	4775.00	7.90	2.70	22.30	10.20	71.00	Cluster.3	Korea, Rep. Of
57	2154.00	5788.00	7.10	3.20	23.20	7.30	53.00	Cluster.3	Malaysia
59	10367.00	15584.00	6.10	3.30	15.60	2.70	38.00	Cluster.3	Oman
61	1668.00	2430.00	4.50	3.00	14.90	10.60	72.00	Cluster.3	Philippines
62	6731.00	11057.00	6.10	4.10	12.80	3.10	3.00	Cluster.3	Saudi Arabia
63	2793.00	14678.00	9.20	2.60	32.20	9.00	83.00	Cluster.3	Singapore
64	1794.00	2482.00	3.70	2.40	14.80	8.30	75.00	Cluster.3	Sri Lanka
65	2382.00	6042.00	6.70	3.00	15.90	8.80	30.00	Cluster.3	Syrian
66	8440.00	15027.00	8.00	3.30	20.70	4.80	70.00	Cluster.3	Taiwan
67	1308.00	3220.00	6.70	3.10	18.00	4.40	68.00	Cluster.3	Thailand
89	2274.00	4444.00	5.20	2.50	20.20	5.50	38.00	Cluster.3	Turkey
91	3165.00	5350.00	4.80	2.20	19.50	12.10	82.00	Cluster.3	Barbados
93	3360.00	4492.00	4.70	3.50	14.70	7.00	90.00	Cluster.3	Costa Rica
94	1939.00	3308.00	5.10	2.90	17.10	5.80	65.00	Cluster.3	Dominican Rep
100	4229.00	7380.00	5.50	3.30	19.50	6.60	65.00	Cluster.3	Mexico
101	3195.00	3978.00	4.10	3.30	14.50	5.80	90.00	Cluster.3	Nicaragua
102	2423.00	5021.00	5.90	3.00	26.10	11.60	73.00	Cluster.3	Panama
107	1842.00	5563.00	7.30	2.90	23.20	4.70	61.00	Cluster.3	Brazil
109	2672.00	4405.00	5.00	3.00	18.00	6.10	63.00	Cluster.3	Colombia
110	2198.00	4504.00	5.70	2.80	24.40	7.20	68.00	Cluster.3	Ecuador
112	1951.00	3914.00	5.50	2.70	11.70	4.40	75.00	Cluster.3	Paraguay
113	3310.00	3775.00	3.50	2.90	12.00	8.00	61.00	Cluster.3	Peru
114	3226.00	5788.00	4.50	2.70	19.40	8.10	75.00	Cluster.3	Surinam
116	10367.00	6336.00	1.90	3.80	11.40	7.00	63.00	Cluster.3	Venezuela
118	3634.00	8675.00	4.20	2.70	20.60	8.10	30.00	Cluster.3	Fiji

TABLE 10.18
OECDGrowth after imputation: Cluster 4.

Obs.	gdp60	gdp85	gdpgrowth	popgrowth	invest	school	literacy60	clust	Country2
53	3493.00	13893.00	6.80	1.20	36.00	10.90	98.00	Cluster_4	Japan
68	2761.00	18513.00	5.90	0.70	26.50	7.20	99.00	Cluster_4	U. A. Emirates.
72	2948.00	11285.00	5.20	1.70	31.20	8.20	91.00	Cluster_4	Cyprus
77	2257.00	6868.00	5.10	0.70	29.30	7.90	81.00	Cluster_4	Greece
79	4411.00	8675.00	3.80	1.10	25.90	11.40	98.00	Cluster_4	Ireland
80	4913.00	11082.00	3.80	0.60	24.90	7.10	91.00	Cluster_4	Italy
82	2293.00	13331.00	6.00	2.00	30.90	7.10	93.00	Cluster_4	Malta
85	2272.00	5827.00	4.40	0.60	22.50	5.80	62.00	Cluster_4	Portugal
86	3766.00	9903.00	4.90	1.00	17.70	8.00	87.00	Cluster_4	Spain
99	2726.00	3080.00	2.10	1.60	20.60	11.20	82.00	Cluster_4	Jamaica
105	4852.00	5533.00	2.10	1.50	25.30	5.00	91.00	Cluster_4	Argentina
108	5189.00	5533.00	2.60	2.30	29.70	7.70	84.00	Cluster_4	Chile
111	2761.00	5827.00	1.10	0.30	32.40	11.70	99.00	Cluster_4	Guyana
115	5119.00	5495.00	0.90	0.60	11.80	7.00	94.00	Cluster_4	Uruguay

TABLE 10.19
OECDGrowth after imputation: Cluster 5.

Obs.	gdp60	gdp85	gdpgrowth	popgrowth	invest	school	literacy60	clust	Country2
45	7634.00	13779.00	5.40	2.30	30.00	12.10	99.00	Cluster_5	Bahrain
70	5939.00	13327.00	3.60	0.40	23.40	8.00	99.00	Cluster_5	Austria
71	6789.00	14290.00	3.50	0.50	23.40	9.30	99.00	Cluster_5	Belgium
73	8551.00	16491.00	3.20	0.60	26.60	10.70	99.00	Cluster_5	Denmark
74	6527.00	13779.00	3.70	0.70	36.90	11.50	99.00	Cluster_5	Finland
75	7215.00	15027.00	3.90	1.00	26.20	8.90	99.00	Cluster_5	France
76	7695.00	15297.00	3.30	0.50	28.50	8.40	99.00	Cluster_5	Germany
78	8091.00	9903.00	3.90	0.60	29.00	10.20	100.00	Cluster_5	Iceland
81	9015.00	13331.00	2.80	1.40	26.90	5.00	30.00	Cluster_5	Luxembourg
83	7689.00	13177.00	3.60	1.40	25.80	10.70	99.00	Cluster_5	Netherlands
84	7938.00	19723.00	4.30	0.70	29.10	10.00	99.00	Cluster_5	Norway
87	7802.00	15237.00	3.10	0.40	24.50	7.90	99.00	Cluster_5	Sweden
88	10308.00	15881.00	2.50	0.80	29.70	4.80	99.00	Cluster_5	Switzerland
90	7634.00	13331.00	2.50	0.30	18.40	8.90	99.00	Cluster_5	UK
92	10286.00	17935.00	4.20	2.00	23.30	10.60	99.00	Cluster_5	Canada
103	9253.00	11285.00	2.70	1.90	20.40	8.80	93.00	Cluster_5	Trin & Tobago
104	12362.00	18988.00	3.20	1.50	21.10	11.90	98.00	Cluster_5	United States
117	8440.00	13409.00	3.80	2.00	31.50	9.80	100.00	Cluster_5	Australia
120	9523.00	12308.00	2.70	1.70	22.50	11.90	99.00	Cluster_5	New Zealand

11

Principal Component Analysis of Numeric Distributional Data

Meiling Chen

School of Statistics, Capital University of Economics and Business, Beijing, China

Huiwen Wang

School of Economics and Management, Beihang University, Beijing, China

CONTENTS

Numeric distributional symbolic data is known as numerical modal data according to the symbolic data analysis definitions. This chapter contributes to principal component analysis method for numeric distributional data whose realizations can be histograms, empirical distributions or empirical estimates of parametric distributions. Regarding numeric distributional data as observed random variables with a probability density function, we present an exact probability density function for each principal component by using the inversion theorem and define a covariance matrix for numeric distributional data. Furthermore, a PCA method for distributional data called DPCA based on this variance-covariance structure is established, which also considers the link between covariance matrix and projection. The effectiveness of the DPCA

DOI: 10.1201/9781315370545-11

247

method is illustrated by a simulated numerical experiment, and two real-life cases concerning evaluation of journals from Chinese Science Citation Database and innate structure of China's stock market.

11.1 Introduction

Principal Component Analysis (PCA) is a commonly used data exploration method, which aims at reducing the dimensions of a multivariate dataset by reconstructing the covariance matrix.

Supposed that the data table to be analyzed by PCA comprises p random variables $\mathbf{X}_{n \times p} = (\mathbf{X}_1, \mathbf{X}_2, \cdots, \mathbf{X}_p)$, the kth principal component (PC) $\mathbf{Y}_k$ is a linear combination of $\mathbf{X}_1, \mathbf{X}_2, \cdots, \mathbf{X}_p$, i.e.,

$$\mathbf{Y}_k = \mu_{1k}\mathbf{X}_1 + \mu_{2k}\mathbf{X}_2 + \cdots + \mu_{pk}\mathbf{X}_p \tag{11.1}$$

where the vector $\mathbf{u}_k = (\mu_{1k}, \mu_{2k}, \cdots, \mu_{pk})'$ is chosen so as to maximize the variance $var(\mathbf{Y}_k)$ of $\mathbf{Y}_k$ subject to $var(\mathbf{Y}_1) \geq var(\mathbf{Y}_2) \geq \cdots \geq var(\mathbf{Y}_k)$, $\mathbf{u}_k'\mathbf{u}_k = 1$ and $\mathbf{u}_k'\mathbf{u}_l = 0, \forall l \neq k$. It is well known that $\mathbf{u}_k$ is the eigenvector of the covariance matrix Σ of $\mathbf{X}_{n \times p}$ corresponding to the kth largest eigenvalue λ_k.

Varieties of PCA methods have been proposed for different data types. However, there are only two essential differences for different PCA methods. One is how to define a linear combination of $\mathbf{X}_1, \mathbf{X}_2, \cdots, \mathbf{X}_p$ as shown in Equation (11.1) to construct the projection of observations in that space. The other is how to construct a covariance matrix of $\mathbf{X}_1, \mathbf{X}_2, \cdots, \mathbf{X}_p$ to determine a PC space. In other words, once the two points are determined, a new PCA method is formulated.

In this chapter, we extend PCA method to numeric distributional data. As in the classical case, a symbolic variable could be distinguished as numerical or categorical. This chapter focuses on symbolic data with quantitative or numerical support, instead of categorical case. We will also use the concept of distributional data [5,7] to describe the case where each element of the observation matrix is a random variable with a general probability distribution. In this sense, intervals, histograms and continuous distributions are special cases of distributional data.

Regarding distributional data as observed random variables with a probability density function, we first present an exact probability density function for each principal component by using the inversion theorem and then define a covariance matrix for distributional data. Furthermore, a PCA method for distributional data called DPCA based on this variance-covariance structure is established, which also considers the link between covariance matrix and projection. The DPCA method may obtain the analytical results instead of approximate ones. Moreover, it is not only able to handle special data forms such as intervals or histograms, but can also apply to more complex situations, for example, the normal or other distributions.

The rest of the chapter is organized as follows. Sections 11.2–11.4 discuss the concept of distributional data, linear combination algorithm and numerical characteristics of distributional data vectors. Based on these elements, derivation of PCA on distributional data is accomplished, referred as DPCA in Sections 11.5 and 11.6. A synthetic data set will then be used to examine effectiveness of the DPCA method in Section 11.7, and two cases from real world applications are considered in Section 11.8. Finally, a conclusion is given in Section 11.9.

11.2 Numeric Distributional Data

A numeric distributional sample matrix is a symbolic data matrix $\mathbf{X}_{n \times p}$ where each element X_{ij} is a random variable with a cumulative distribution function (CDF) $F_{X_{ij}}(x)$, denoted as $X_{ij} \sim F_{X_{ij}}(x)$.

$$
\mathbf{X}_{n \times p} = \left(\begin{array}{cccc} \mathbf{X}_1 & \mathbf{X}_2 & \cdots & \mathbf{X}_p \end{array} \right) = \begin{pmatrix} \mathbf{S}'_1 \\ \mathbf{S}'_2 \\ \vdots \\ \mathbf{S}'_n \end{pmatrix} = \begin{pmatrix} X_{11} & X_{12} & \cdots & X_{1p} \\ X_{21} & X_{22} & \cdots & X_{2p} \\ \vdots & \vdots & \ddots & \vdots \\ X_{n1} & X_{n2} & \cdots & X_{np} \end{pmatrix}
$$

(11.2)

Here, $\mathbf{X}_j = (X_{1j}, X_{2j}, \ldots, X_{nj})'$ is the jth distributional variable, and $\mathbf{S}'_i = (X_{i1}, X_{i2}, \ldots, X_{ip})$ is the ith distributional observation.

Different from the classical dataset, each component in $\mathbf{X}_j$ may have different kinds of distributions. In general, the most commonly used symbolic data are as follows,

- Classical data, also called quantitative single-valued data, only if each cell X_{ij} is a real number. For this case, X_{ij} may be regarded as a special random variable with the following probability density function (PDF),

$$
f_{X_{ij}}(x) = \delta(x - x_{ij}) = \begin{cases} +\infty, & x = x_{ij} \\ 0, & x \neq x_{ij}. \end{cases}
$$

(11.3)

- Interval data, only if each element $X_{ij} = [\underline{x}_{ij}, \overline{x}_{ij}], \underline{x}_{ij}, \overline{x}_{ij} \in \mathbb{R}, \underline{x}_{ij} \leq \overline{x}_{ij}$. For this case, X_{ij} can be regarded as a uniformly distributed random variable with parameters $\underline{x}_{ij}$ and $\overline{x}_{ij}$, i.e. $X_{ij} \sim U(\underline{x}_{ij}, \overline{x}_{ij})$ with the following PDF,

$$
f_{X_{ij}}(x) = \begin{cases} \frac{1}{\overline{x}_{ij} - \underline{x}_{ij}}, & \underline{x}_{ij} \leq x \leq \overline{x}_{ij} \\ 0, & \text{others.} \end{cases}
$$

(11.4)

- Histogram data, only if each element is expressed as $X_{ij} = \{[x_{ij}^0, x_{ij}^1], p_{ij}^1;$ $[x_{ij}^1, x_{ij}^2], p_{ij}^2; \cdots ; [x_{ij}^{m-1}, x_{ij}^m], p_{ij}^m\}$, where $0 \leq p_{ij}^\ell \leq 1$, $\sum_{\ell=1}^m p_{ij}^\ell = 1$, $x_{ij}^\ell \in$ $\mathbb{R}$ $(\ell = 0, 1, \cdots, m)$, and m is the number of modalities (bins) of the histogram. It can be regarded as a random variable with an empirical density function,

$$f_{X_{ij}}(x) = \begin{cases} \frac{p_{ij}^\ell}{x_{ij}^\ell - x_{ij}^{\ell-1}}, & x_{ij}^{\ell-1} \leq x \leq x_{ij}^\ell, \quad \ell = 1, 2, \cdots, m \\ 0, & \text{others.} \end{cases} \tag{11.5}$$

- Normal-distributional data, only if each cell $X_{ij} \sim N(\mu_{ij}, \sigma_{ij}^2)$, i.e.,

$$f_{X_{ij}}(x) = \frac{1}{\sigma_{ij}\sqrt{2\pi}} e^{-\frac{1}{2}(\frac{x-\mu_{ij}}{\sigma_{ij}})^2} \tag{11.6}$$

One caveat here we need to mention is that components $X_{i1}, X_{i2}, \ldots, X_{ip}$ in each observation $\mathbf{S}_i'$ are assumed to be independent. The assumption of "orthogonal dimensions for every observation" has been widely accepted in the existing literature. For instance, [3] considered each interval-valued observation as a hypercube. An implicit assumption within this method is that the intervals in different elements are the marginal distributions of a multivariate distribution with independent components. This case also can be seen in [1], which is also consistent with the above assumption.

11.3 Linear Combination of Numeric Distributional Data

Next, we will introduce linear combination for numeric distributional data [4]. In probability theory, all random variables obey the addition and scalar-multiplication algorithm. PDF of a sum of two or more independent random variables is the convolution of their corresponding PDFs respectively, but the actual evaluation can be tedious. Based on inversion formula and the properties of characteristic function, we could give out the density function after linear combination. Because of one-to-one correspondence between characteristic function and density function, the solution is unique.

For given p numeric distributional variables $\mathbf{X}_1, \mathbf{X}_2, \cdots, \mathbf{X}_p$ with n observations, respectively, and p real numbers $\alpha_1, \alpha_2, \cdots, \alpha_p$, their linear combination is given as follows,

$$\mathbf{Y} = \sum_{j=1}^p \alpha_j \mathbf{X}_j = (Y_1, Y_2, \cdots, Y_n)' \tag{11.7}$$

where

$$Y_i = \sum_{j=1}^{p} \alpha_j X_{ij} \ for \ i = 1, 2, \cdots, n. \tag{11.8}$$

Obviously, Y_i is a random variable. This means that $\mathbf{Y}$ is a numeric distributional variable.

Since $X_{i1}, X_{i2}, \ldots, X_{ip}$ are independent, the characteristic function of Y_i is the product of their characteristic functions, i.e.,

$$\varphi_{Y_i}(t) = \prod_{j=1}^{p} \varphi_{X_{ij}}(\alpha_j t) = \prod_{j=1}^{p} \int_{-\infty}^{+\infty} e^{i(\alpha_j t)x} dF_{X_{ij}}(x). \tag{11.9}$$

where $\mathbf{i}$ is an imaginary unit with $\mathbf{i}^2 = -1$. Meanwhile, the PDF of Y_i is the convolution of the PDFs of $X_{i1}, X_{i2}, \ldots, X_{ip}$. Since the analytical form of the convolution is difficult or impossible to be obtained, the PDF of Y_i is obtained by using the inversion theorem, i.e.,

$$f_{Y_i}(x) = \frac{1}{2\pi} \int_{-\infty}^{+\infty} e^{-\mathbf{i}tx} \varphi_{Y_i}(t) dt. \tag{11.10}$$

Some distributions have simple convolutions, such as Normal distribution, Gamma distribution, Chi Square distribution and so on, which is called the additive property. For instance, if all $X_{i1}, X_{i2}, \ldots, X_{ip}$ are normally distributed [9], i.e, $X_{ij} \sim N(\mu_{ij}, \sigma_{ij}^2)$ for $j = 1, 2, \ldots, p$, then $Y_i \sim N\left(\sum_{j=1}^{p} a_j \mu_{ij}, \sum_{j=1}^{p} a_j^2 \sigma_{ij}^2 \right)$ in Equation (11.8).

The technique will be adopted to compute PCs when PC coefficients have been obtained.

11.4 Numerical Characteristics of Numeric Distributional Data

To construct the covariance matrix, we firstly propose numerical characteristics of numeric distributional variables. As proposed by [2], each symbolic observation can be pictured as a p-dimensional point group with infinitely dense points within it. By using the PDF associated with each data unit, that total information with symbolic observations could be taken into account in modeling.

Definition 1 For a distributional variable $\mathbf{X}_j = (X_{1j}, X_{2j}, \cdots, X_{nj})'$, its first moment is defined as

$$\mu_j = E_S(\mathbf{X}_j) = \frac{1}{n} \sum_{i=1}^{n} E(X_{ij}) = \frac{1}{n} \sum_{i=1}^{n} \int_{-\infty}^{+\infty} x f_{X_{ij}}(x) dx \tag{11.11}$$

and its second moment is defined as

$$E_S(\mathbf{X}_j^2) = \frac{1}{n}\sum_{i=1}^{n} E(X_{ij}^2) = \frac{1}{n}\sum_{i=1}^{n}\int_{-\infty}^{+\infty} x^2 f_{X_{ij}}(x)dx \tag{11.12}$$

where $E(X_{ij})$ and $E(X_{ij}^2)$ represent the first and second moment of the random variable X_{ij}, respectively.

Accordingly, the centralization of X_{ij} is given by $\tilde{X}_{ij} = X_{ij} - \mu_j$ for $i = 1, 2, \cdots, n$ and $j = 1, 2, \cdots, p$.

Definition 2 Given two distributional variables $\mathbf{X}_j$ and $\mathbf{X}_k(j \neq k)$, the second-order mixed moment is defined as

$$E_S(\mathbf{X}_j \cdot \mathbf{X}_k) = \frac{1}{n}\sum_{i=1}^{n} E(X_{ij}X_{ik}) = \frac{1}{n}\sum_{i=1}^{n} E(X_{ij})E(X_{ik}) \tag{11.13}$$

where X_{ij} and X_{ik} are independent according to the assumption of "orthogonal dimensions for every observation".

Using the PDFs in Equations (11.3–11.6), we could easily obtain the moments of the four types of distributional vectors that we discussed above and shown in Table 11.1.

Accordingly, the variance of $\mathbf{X}_j$ and the covariance of $\mathbf{X}_j$ and $\mathbf{X}_k$ are respectively defined as follows,

$$
\begin{aligned}
\text{var}_S(\mathbf{X}_j) &= E_S((\mathbf{X}_j - E_S(\mathbf{X}_j))^2) = E_S(\mathbf{X}_j^2) - E_S(\mathbf{X}_j)^2 & (11.14)\\
\text{cov}_S(\mathbf{X}_j, \mathbf{X}_k) &= E_S((\mathbf{X}_j - E_S(\mathbf{X}_j))(\mathbf{X}_k - E_S(\mathbf{X}_k)))\\
&= E_S(\mathbf{X}_j\mathbf{X}_k) - E_S(\mathbf{X}_j)E_S(\mathbf{X}_k). & (11.15)
\end{aligned}
$$

If distributional variables $\mathbf{X}_j$ and $\mathbf{X}_k$ are centralized, then we have

$$\text{var}_S(\mathbf{X}_j) = E_S(\mathbf{X}_j^2), \quad \text{cov}_S(\mathbf{X}_j, \mathbf{X}_k) = E_S(\mathbf{X}_j \cdot \mathbf{X}_k). \tag{11.16}$$

According to Equations (11.11) and (11.14), $\mathbf{X}_j$ can be standardized by replacing each element X_{ij} by

$$X_{ij}^* = \frac{X_{ij} - \mu_j}{\sqrt{\text{var}_S(\mathbf{X}_j)}}. \tag{11.17}$$

Combining Equations (11.14) and (11.15), the correlation coefficient of $\mathbf{X}_j$ and $\mathbf{X}_k$ can be defined as

$$\text{corr}_S(\mathbf{X}_j, \mathbf{X}_k) = \frac{\text{cov}_S(\mathbf{X}_j, \mathbf{X}_k)}{\sqrt{\text{var}_S(\mathbf{X}_j)}\sqrt{\text{var}_S(\mathbf{X}_k)}}. \tag{11.18}$$

Note that $\text{corr}_S(\mathbf{X}_j, \mathbf{X}_k)$ will be equal to $\text{cov}_S(\mathbf{X}_j, \mathbf{X}_k)$, if all the elements in $\mathbf{X}_j$ and $\mathbf{X}_k$ have been standardized.

TABLE 11.1
Moments of four types of distributional data.

Moment	Real number vectors
$E_S(\mathbf{X}_j)$	$\frac{1}{n} \sum_{i=1}^{n} x_{ij}$
$E_S(\mathbf{X}_j^2)$	$\frac{1}{n} \sum_{i=1}^{n} x_{ij}^2$
$E_S(\mathbf{X}_j \cdot \mathbf{X}_k)$	$\frac{1}{n} \sum_{i=1}^{n} x_{ij} x_{ik}$

Moment	Normal-distributional vectors
$E_S(\mathbf{X}_j)$	$\frac{1}{n} \sum_{i=1}^{n} \mu_{ij}$
$E_S(\mathbf{X}_j^2)$	$\frac{1}{n} \sum_{i=1}^{n} (\sigma_{ij}^2 + \mu_{ij}^2)$
$E_S(\mathbf{X}_j \cdot \mathbf{X}_k)$	$\frac{1}{n} \sum_{i=1}^{n} \mu_{ij} \mu_{ik}$

Moment	Interval vectors
$E_S(\mathbf{X}_j)$	$\frac{1}{2n} \sum_{i=1}^{n} (\underline{x}_{ij} + \overline{x}_{ij})$
$E_S(\mathbf{X}_j^2)$	$\frac{1}{3n} \sum_{i=1}^{n} (\underline{x}_{ij}^2 + \underline{x}_{ij}\overline{x}_{ij} + \overline{x}_{ij}^2)$
$E_S(\mathbf{X}_j \cdot \mathbf{X}_k)$	$\frac{1}{4n} \sum_{i=1}^{n} (\underline{x}_{ij} + \overline{x}_{ij})(\underline{x}_{ik} + \overline{x}_{ik})$

Moment	Histogram vectors
$E_S(\mathbf{X}_j)$	$\frac{1}{2n} \sum_{i=1}^{n} \sum_{\ell=1}^{m} p_{ij}^\ell (x_{ij}^{\ell-1} + x_{ij}^\ell)$
$E_S(\mathbf{X}_j^2)$	$\frac{1}{3n} \sum_{i=1}^{n} \sum_{\ell=1}^{m} p_{ij}^\ell [(x_{ij}^{\ell-1})^2 + x_{ij}^{\ell-1} x_{ij}^\ell + (x_{ij}^\ell)^2]$
$E_S(\mathbf{X}_j \cdot \mathbf{X}_k)$	$\frac{1}{4n} \sum_{i=1}^{n} [\sum_{\ell=1}^{m} p_{ij}^\ell (x_{ij}^{\ell-1} + x_{ij}^\ell)][\sum_{\ell=1}^{m} p_{ik}^\ell (x_{ik}^{\ell-1} + x_{ik}^\ell)]$

11.5   The PCA Algorithm on Numeric Distributional Data

Equipped with the definitions introduced in the previous sections, we begin to derive the distributional PCs [4]. Without loss of generality, we assume that all the distributional data units have been centered. Similar to the numeric case, the kth distributional PC $\mathbf{Y}_k$ is a linear combination of $\mathbf{X}_1, \mathbf{X}_2, \cdots, \mathbf{X}_p$. That is, $\mathbf{Y}_k = \mathbf{X}\mathbf{u}_k = \sum_{j=1}^{p} \mu_{kj}\mathbf{X}_j$, where $\mathbf{u}_k = (\mu_{k1}, \mu_{k2}, \cdots, \mu_{kp})' \in \mathbb{R}^p$ subject to $\mathbf{u}_k'\mathbf{u}_k = 1$ and $\mathbf{u}_k'\mathbf{u}_l = 0$, $\forall l \neq k$. The first q principal component $\mathbf{Y}_1, \mathbf{Y}_2, \cdots, \mathbf{Y}_q$ are required to maximize the total variance as much as possible to represent the original information carried by $\mathbf{X}_1, \mathbf{X}_2, \cdots, \mathbf{X}_p$.

Note that $Y_{ik} = \mathbf{S}_i'\mathbf{u}_k = \sum_{j=1}^{p} \mu_{kj}X_{ij}$ where $\mathbf{S}_i' = (X_{i1}, X_{i2}, \cdots, X_{ip})$ is the ith symbolic observation. It follows that the moments of random variable Y_{ik} can be obtained by differentiating its characteristic function $\varphi_{Y_{ik}}(t)$, for example,

$$E(Y_{ik}^2) = \mathbf{i}^{-2}\varphi_{Y_{ik}}''(0) = \mathbf{i}^{-2} \cdot \frac{d^2}{dt^2}\varphi_{Y_{ik}}(t)\,|_{t=0} \tag{11.19}$$

where $\mathbf{i}$ is an imaginary unit with $\mathbf{i}^2 = -1$. It follows from Equation (11.9) that $\varphi_{Y_{ik}}(t) = \prod_{j=1}^{p}\varphi_{X_{ij}}(\mu_{kj}t)$ where $\varphi_{X_{ij}}(\mu_{kj}t)$ is the characteristic function of $\mu_{kj}X_{ij}$. Therefore, we have

$$\frac{d^2}{dt^2}\varphi_{Y_{ik}}(t)$$

$$= \frac{d^2}{dt^2}\left[\prod_{j=1}^{p}\varphi_{X_{ij}}(\mu_{kj}t)\right]$$

$$= \frac{d}{dt}\left[\mu_{k1}\varphi_{X_{i1}}'(\mu_{k1}t)\left(\prod_{j=2}^{p}\varphi_{X_{ij}}(\mu_{kj}t)\right) + \cdots + \mu_{kp}\left(\prod_{j=1}^{p-1}\varphi_{X_{ij}}(\mu_{kj}t)\right)\varphi_{X_{ip}}'(\mu_{kp}t)\right]$$

$$= \frac{d}{dt}\left[\begin{pmatrix} \mu_{k1}, & \cdots, & \mu_{kp} \end{pmatrix}\begin{pmatrix} \varphi_{X_{i1}}'(\mu_{k1}t)\left(\prod_{j=2}^{p}\varphi_{X_{ij}}(\mu_{kj}t)\right) \\ \vdots \\ \left(\prod_{j=1}^{p-1}\varphi_{X_{ij}}(\mu_{kj}t)\right)\varphi_{X_{ip}}'(\mu_{kp}t) \end{pmatrix}\right] \tag{11.20}$$

$$= \mathbf{u}_k'\begin{pmatrix} \varphi_{X_{i1}}''(\mu_{k1}t)\left(\prod_{j=2}^{p}\varphi_{X_{ij}}(\mu_{kj}t)\right) & \cdots & \varphi_{X_{i1}}'(\mu_{k1}t)\left(\prod_{j=2}^{p-1}\varphi_{X_{ij}}(\mu_{kj}t)\right)\varphi_{X_{ip}}'(\mu_{kp}t) \\ \vdots & \ddots & \vdots \\ \varphi_{X_{i1}}'(\mu_{k1}t)\left(\prod_{j=2}^{p-1}\varphi_{X_{ij}}(\mu_{kj}t)\right)\varphi_{X_{ip}}'(\mu_{kp}t) & \cdots & \left(\prod_{j=1}^{p-1}\varphi_{X_{ij}}(\mu_{kj}t)\right)\varphi_{X_{ip}}''(\mu_{kp}t) \end{pmatrix}\mathbf{u}$$

For $i = 1, \cdots, n$, $j = 1, \cdots, p$, $k = 1, \cdots, q$, letting $t = 0$, then we have

$$\varphi_{X_{ij}}(\mu_{kj} t)|_{t=0} = \varphi_{X_{ij}}(0) = 1 \tag{11.21}$$

$$\varphi'_{X_{ij}}(\mu_{kj} t)|_{t=0} = \varphi'_{X_{ij}}(0) = \mathbf{i} \cdot \mathrm{E}(X_{ij}) \tag{11.22}$$

$$\varphi''_{X_{ij}}(\mu_{kj} t)|_{t=0} = \varphi''_{X_{ij}}(0) = \mathbf{i}^2 \cdot \mathrm{E}(X_{ij}^2). \tag{11.23}$$

Equations (11.21–11.23) are replaced in Equation (11.20) to obtain

$$\mathrm{E}(Y_{ik}^2) = \mathbf{i}^{-2} \cdot \frac{d^2}{dt^2} \varphi_{Y_{ik}}(t)|_{t=0} = \mathbf{u}'_k \begin{pmatrix} \mathrm{E}(X_{i1}^2) & \cdots & \mathrm{E}(X_{i1})\mathrm{E}(X_{ip}) \\ \vdots & \ddots & \vdots \\ \mathrm{E}(X_{i1})\mathrm{E}(X_{ip}) & \cdots & \mathrm{E}(X_{ip}^2) \end{pmatrix} \mathbf{u}_k. \tag{11.24}$$

Based on the definitions in Section 11.4 and Equation (11.24), the variance of the kth distributional PC $\mathbf{Y}_k$ can be decomposed as follows,

$$\begin{aligned}
\mathrm{var}_S(\mathbf{Y}_k) &= \mathrm{E}_S(\mathbf{Y}_k^2) = \frac{1}{n} \sum_{i=1}^{n} \mathrm{E}(Y_{ik}^2) \\
&= \frac{1}{n} \sum_{i=1}^{n} \left[\mathbf{u}'_k \begin{pmatrix} \mathrm{E}(X_{i1}^2) & \cdots & \mathrm{E}(X_{i1})\mathrm{E}(X_{ip}) \\ \vdots & \ddots & \vdots \\ \mathrm{E}(X_{i1})\mathrm{E}(X_{ip}) & \cdots & \mathrm{E}(X_{ip}^2) \end{pmatrix} \mathbf{u}_k \right] \\
&= \mathbf{u}'_k \begin{pmatrix} \mathrm{E}_S(\mathbf{X}_1^2) & \mathrm{E}_S(\mathbf{X}_1\mathbf{X}_2) & \cdots & \mathrm{E}_S(\mathbf{X}_1\mathbf{X}_p) \\ \mathrm{E}_S(\mathbf{X}_2\mathbf{X}_1) & \mathrm{E}_S(\mathbf{X}_2^2) & \cdots & \mathrm{E}_S(\mathbf{X}_2\mathbf{X}_p) \\ \vdots & \vdots & \ddots & \vdots \\ \mathrm{E}_S(\mathbf{X}_p\mathbf{X}_1) & \mathrm{E}_S(\mathbf{X}_p\mathbf{X}_2) & \cdots & \mathrm{E}_S(\mathbf{X}_p^2) \end{pmatrix} \mathbf{u}_k \\
&= \mathbf{u}'_k \Sigma_S \mathbf{u}_k \tag{11.25}
\end{aligned}$$

where Σ_S denotes the covariance matrix of distributional variables $\mathbf{X}_1, \mathbf{X}_2, \cdots, \mathbf{X}_p$.

The following process is the same as that for classical PCA. That is to look for q orthogonal vectors $\mathbf{u}_1, \mathbf{u}_2, \cdots, \mathbf{u}_q$ to maximize $\sum_{k=1}^{q} \mathrm{var}_S(\mathbf{Y}_k)$ with $\mathrm{var}_S(\mathbf{Y}_1) \geq \mathrm{var}_S(\mathbf{Y}_2) \geq \cdots \geq \mathrm{var}_S(\mathbf{Y}_q)$ by solving $\Sigma_S \mathbf{u}_k = \lambda_k \mathbf{u}_k$ for $k = 1, 2, \cdots, p$. The solutions of $\mathbf{u}_1, \mathbf{u}_2, \cdots, \mathbf{u}_p$ are the orthonormal eigenvectors of Σ_S, corresponding to the eigenvalues $\lambda_1 \geq \lambda_2 \geq \cdots \geq \lambda_p$. The accumulated contribution of the information carried by the first q PCs to the total one can be defined by

$$Q_q = \frac{\sum_{k=1}^{q} \mathrm{var}_S(\mathbf{Y}_k)}{\sum_{k=1}^{p} \mathrm{var}_S(\mathbf{Y}_k)} = \frac{\sum_{k=1}^{q} \lambda_k}{\sum_{k=1}^{p} \lambda_k} \tag{11.26}$$

which reports the percentage of the original information contained in the q-dimensional space, spanned by the reserved factor axes $\mathbf{u}_1, \mathbf{u}_2, \cdots, \mathbf{u}_q$, and helps determine q. Finally, we get the kth distributional PC $\mathbf{Y}_k = \mathbf{X}\mathbf{u}_k$, $k = 1, 2, \cdots, q$ by Equations (11.9) and (11.10).

The above derivation process can be summarized by the following algorithm, called distributional PCA (abbreviated as DPCA),

- Step 1. Compute the covariance matrix Σ_S of $\mathbf{X}_{n\times p}$, using Equations (11.14) and (11.15).

- Step 2. Solve $\Sigma_S \mathbf{u}_k = \lambda_k \mathbf{u}_k$ for $k = 1, 2, \cdots, q$ for the eigenvalues $\lambda_1 \geq \lambda_2 \geq \cdots \geq \lambda_q$ to obtain the corresponding orthonormal eigenvectors $\mathbf{u}_1, \mathbf{u}_2, \cdots, \mathbf{u}_q$ for $q \leq p$.

- Step 3. Compute the PDF of the kth PC $\mathbf{Y}_k = \mathbf{X}_{n\times p}\mathbf{u}_k$ for $k = 1, 2, \cdots, q$ by Equations (11.9) and (11.10).

11.6   Properties of DPCA

Assume that $\mathbf{Y}_1, \mathbf{Y}_2, \cdots, \mathbf{Y}_q (1 \leq q \leq p)$ are the distributional PC's, similarly to the classical PCA, we have the following properties,

1. $\mathrm{E}_S(\mathbf{Y}_k) = 0, k = 1, 2, \cdots, q$;

2. $\mathrm{D}_S(\mathbf{Y}_k) = \lambda_k, \; k = 1, 2, \cdots, q$;

3. $\mathrm{cov}_S(\mathbf{Y}_k, \mathbf{Y}_l) = 0, \forall k \neq l$.

It is worth pointing out that the DPCA method may specialize to the following special cases:

- i) Classical PCA when each element is degenerated into a real number;

- i) ND-PCA proposed in [9] for normal-distribution-valued symbolic data when each element follows a normal distribution;

- iii) CIPCA proposed in [10] for interval-valued data when each element follows a uniform distribution;

- iv) Histogram PCA in [8] when each element follows an empirical distribution.

In order to verify the above conclusions, we only need to prove that the corresponding PCA methods use the same covariance matrix since the covariance matrix determines a PC space. As mentioned in Section 11.4, the moments of the four types of symbolic vectors can be obtained. The covariance matrix defined in Section 11.4 is consistent with the existing methods when distributional data degenerate into several special cases.

For the representation problem in the PC space, the projection results of DPCA are consistent with the classical PCA and ND-PCA. However, for the situations of interval-valued and histogram-valued data, the projection results in the existing methods are all approximate, and the projection of observations in the proposed DPCA is accurate but not approximate.

11.7 Experimental Results of Synthetic Data Set

In this section, the effectiveness of DPCA is illustrated on a synthetic data set whose elements are assumed to be random variables. The synthetic dataset is a 6×4 matrix $\mathbf{X}_{6 \times 4}$ shown in Equation (11.27), whose elements are Exponential (exp), Normal (N), Lognormal ($logN$), Gamma (Γ), Beta ($beta$), Weibull (wbl), F (F), Uniform (U) Distributions, respectively. The graphs of PDFs of distributional data units are shown in Fig. 11.1.

$$\mathbf{X}_{6 \times 4} = (\ \mathbf{X}_1, \ \ \mathbf{X}_2, \ \ \mathbf{X}_3, \ \ \mathbf{X}_4 \ ) = \begin{pmatrix} \mathbf{S}'_1 \\ \mathbf{S}'_2 \\ \vdots \\ \mathbf{S}'_6 \end{pmatrix}$$

$$= \begin{pmatrix} U(0,2) & N(1.5,0.5^2) & beta(2,2) & F(60,60) \\ U(5,7) & N(5,1^2) & wbl(5,7) & logN(2,0.2) \\ U(1,4) & N(2,0.5^2) & exp(0.7) & \Gamma(1.6,0.5) \\ U(1,2) & N(1.5,0.2^2) & logN(1.8,0.2)) & \Gamma(15,0.3) \\ U(4,8) & N(8,2^2) & logN(1,0.2) & wbl(6,3) \\ U(1,3) & N(2,0.6^2) & exp(0.6) & wbl(0.6,1) \end{pmatrix} \quad (11.27)$$

Based on the synthetic distributional data matrix $\mathbf{X}_{6 \times 4}$, we may employ the DPCA method to obtain the PC coefficients $\mathbf{u}_k$, the PC variances λ_k, $k = 1, 2, 3, 4$, and the density functions $f_{1i}(x)$ and distribution functions $F_{1i}(x)(i = 1, 2, \cdots, 6)$ of the first PC (PC1 hereafter).

To simulate real-life numeric data of large-scale, we further run random samplings for each element of the synthetic distributional matrix, and pool all sampled data to form a single-valued data matrix. Then we conduct a comparison between the analytical results of DPCA and those of classical PCA by simulation. Specifically, we randomly generate M numerical data points from the ith ($i = 1, 2, \cdots, 6$) distributional observation. In this way, $6 \times M$ numerical data points will be generated as an approximation of the distributional dataset $\mathbf{X}_{6 \times 4}$. The classical PCA method is applied to the generated dataset to obtain the simulated eigenvalues, eigenvectors, and distributions of PC1 scores so that we may gauge differences between the theoretical and simulated results.

In order to avoid the random error, we repeat the numerical experiment for H times. In the hth ($h = 1, 2, \cdots, H$) iteration, the experiment will be carried out by the following three steps:

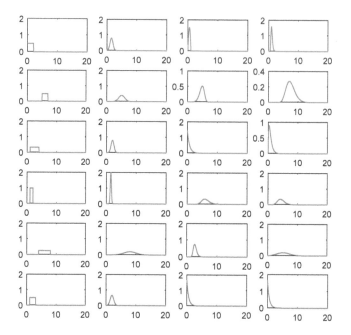

FIGURE 11.1
Graphs of probability density functions of synthetic distributional data units

Step 1. Generate M numerical data points from each distributional observation to obtain a single-valued matrix $\mathbf{K}_{(6 \times M) \times 4}(h)$.

Step 2. Apply the classical PCA on the single-valued matrix. All the eigenvalues $\lambda_k^M(h)$ and eigenvectors $\mathbf{u}_k^M(h)$ of the covariance matrix for $k = 1, 2, 3, 4$, empirical distributions $F_{1i}^M(x, h)$ for $i = 1, 2, \cdots, 6$ of PC1 scores from six observation groups could be obtained.

Step 3. Compute the following three indicators as in [10],

(a) Relative error between λ_k and $\lambda_k^M(h)$:

$$RE_k^M(h) = \left| \frac{\lambda_k^M(h) - \lambda_k}{\lambda_k} \right|, \quad k = 1, 2, 3, 4. \tag{11.28}$$

(b) Absolute cosine value (ACV hereafter) between $\mathbf{u}_k$ and $\mathbf{u}_k^M(h)$, i.e.,

$$ACV_k^M(h) = \left| \frac{\mathbf{u}_k' \mathbf{u}_k^M(h)}{\|\mathbf{u}_k\| \|\mathbf{u}_k^M(h)\|} \right|, \quad k = 1, 2, 3, 4. \tag{11.29}$$

Since cosine value measures the angle between two vectors, $ACV_j^M(h)$ characterizes the similarity between $\mathbf{u}_j$ and $\mathbf{u}_j^M(h)$. Therefore, a higher $ACV_j^M(h)$ indicates a better performance of the proposed method.

TABLE 11.2
The average value of $\overline{D}_i^M$ over 50 repeats of experiments.

M	$\overline{D}_1^M$	$\overline{D}_2^M$	$\overline{D}_3^M$	$\overline{D}_4^M$	$\overline{D}_5^M$	$\overline{D}_6^M$	Critical value of $\overline{D}_i^M$ at the significance level 0.05
50	0.1221	0.1083	0.1208	0.1350	0.1309	0.1214	0.1884
100	0.0861	0.0775	0.0795	0.0978	0.0843	0.0876	0.1340
200	0.0686	0.0597	0.0655	0.0757	0.0612	0.0679	0.0962
500	0.0424	0.0391	0.0430	0.0466	0.0412	0.0448	0.0608
1000	0.0361	0.0306	0.0349	0.0391	0.0289	0.0373	0.0430
2000	0.0285	0.0207	0.0293	0.0301	0.0230	0.0303	0.0304

(c) The value $D_i^M(h)$ in the Kolmogorov-Smirnov test to measure the similarity between the distributions of PC1:

$$D_i^M(h) = \sup_{-\infty \leq x \leq \infty} \left| F_{1i}^M(x, h) - F_{1i}(x) \right|, \quad i = 1, 2, \cdots, 6. \qquad (11.30)$$

In this part, we set $H = 50$ and the value of M gradually increases from 50 to 2000.

Finally, we calculate the average values $\overline{RE}_k^M$, $\overline{ACV}_k^M$, $\overline{D}_i^M$ of $RE_k^M(h)$, $ACV_k^M(h)$ and $D_i^M(h)$ over H replications of the experiment, respectively, i.e.,

$$\overline{RE}_k^M = \frac{1}{H} \sum_{h=1}^H RE_k^M(h), \quad \overline{ACV}_k^M = \frac{1}{H} \sum_{h=1}^H ACV_k^M(h), \quad \overline{D}_i^M = \frac{1}{H} \sum_{h=1}^H D_i^M(h).$$

These values measure the overall similarities between the theoretical and simulated results.

Fig. 11.2 shows that the average relative error $\overline{RE}_k^M$ of the experiment decreases as M increases. When M is greater than 200, each $\overline{RE}_k^M$ decreases to no more than 0.05 for $k = 1, 2, 3, 4$. Figure 11.3 shows that the directions of theoretical eigenvectors are very close to those of the classic PCA. For $M \geq 200$, the values $\overline{ACV}_k^M$ of PC coefficients are all close to 1. In fact, all of them are larger than 0.95, and increase towards 1 as M increases.

Table 11.2 reports the computational results of the values $\overline{D}_i^M$ in the Kolmogorov-Smirnov test. With the increase of M, $\overline{D}_i^M$ gradually becomes smaller for $i = 1, 2, \cdots, 6$. It implies that the theoretical distributions are more consistent with those obtained by the classical PCA. The last column of Table 11.2 gives the critical value of $\overline{D}_i^M$ for different M at the significance level 0.05. It can be seen that all $\overline{D}_i^M$ are less than the corresponding critical value. Consequently, the theoretical distribution and the empirical one may be regarded as the same in each Kolmogorov-Smirnov test.

For $i = 1, 2, \cdots, 6$, Fig. 11.4 shows the PDFs $f_{1i}(x)$ of the observations, and Fig. 11.5 shows the comparison between $F_{1i}(x)$ and $F_{1i}^M(x)$ obtained by

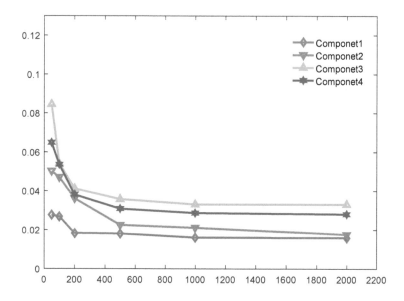

FIGURE 11.2
The trend graphs of $\overline{RE}_k^M$ as M increases.

the experiment with $M = 100$, where $F_{1i}(x)$ are displayed by dotted lines and $F_{1i}^M(x)$ are displayed by solid lines. It can be seen that all the empirical distributions are close to the theoretical ones.

In summary, the above experiment results demonstrate the effectiveness of DPCA since the theoretical PCs are very close to the classical ones as M increases.

11.8   Real-life Applications

In this section, we will examine DPCA on two real-life data sets, using histogram-valued data and normal-distributional data respectively.

11.8.1   JOURNALS Data

The first dataset comes from the Chinese Science Citation Database (CSCD) in 2007, concerning 735 Chinese scientific journals described by five variables,

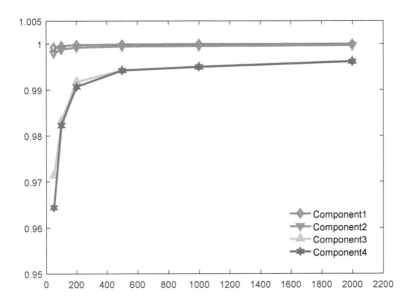

FIGURE 11.3
The trend graphs of $\overline{ACV}_k^M$ as M increases.

i.e., number of published paper ($\mathbf{Y}_1$), impact factor ($\mathbf{Y}_2$), immediacy index ($\mathbf{Y}_3$), total citation ($\mathbf{Y}_4$), and cited half-life ($\mathbf{Y}_5$).

According to the criterion from National Natural Science Foundation of China, we divide the 735 journals into 8 discipline groups: Mathematical and Physical Sciences (10%), Chemical Sciences (10%), Life Sciences (34%), Earth Sciences (10%), Engineering and Materials Sciences (15%), Information Sciences (9%), Management Sciences (4%), Comprehensive Sciences (8%). All of them are described by histograms. Therefore, an 8×5 histogram data table is obtained, which is in part shown in Table 11.3.

Since the variables are measured in different scales, all the histogram data units are first standardized according to Equation (11.17). Next DPCA is employed to obtain the first two PCs as follows:

$$\mathbf{PC}_1 = -0.21\mathbf{Y}_1 + 0.58\mathbf{Y}_2 + 0.58\mathbf{Y}_3 + 0.45\mathbf{Y}_4 + 0.28\mathbf{Y}_5,$$
$$\mathbf{PC}_2 = 0.67\mathbf{Y}_1 + 0.12\mathbf{Y}_2 + 0.01\mathbf{Y}_3 + 0.48\mathbf{Y}_4 - 0.55\mathbf{Y}_5.$$

The two PCs account for 71.3% of the total variance information in the original variables. Figure 11.6 shows the loading plots of the first and second PCs, which indicates that PC1 is strongly and positively correlated with impact

TABLE 11.3
The histogram dataset: CSCD journals.

Disciplines	Y_1	Y_2	Y_3	Y_4	Y_5
Earth Sciences	[85,103] 0.243; (103,122] 0.216; (122,141] 0.189; (141,160] 0.243; (160,179] 0.108	[0.404,0.545] 0.297; (0.545,0.685] 0.081; (0.685,0.825] 0.162; (0.825,0.966] 0.270; (0.966,1.106] 0.189	[0.035,0.059] 0.297; (0.059,0.082] 0.135; (0.082,0.106] 0.189; (0.106,0.129] 0.243; (0.129,0.153] 0.135	[427,584] 0.297; (584,741] 0.378; (741,899] 0.108; (899,1056] 0.054; (1056,1214] 0.162	[4.176,4.529] 0.189; (4.529,4.883] 0.162; (4.883,5.237] 0.297; (5.237,5.590] 0.189; (5.590,5.944] 0.162
⋮	⋮	⋮	⋮	⋮	⋮

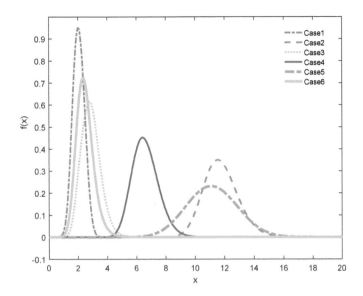

FIGURE 11.4
Graphs of the PDFs $f_{1i}(x)$ of the six observations.

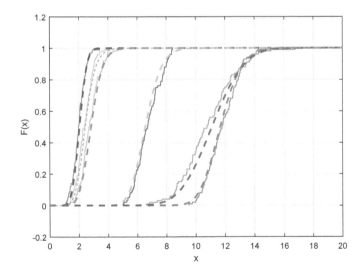

FIGURE 11.5
Graphs of the CDFs $F_{1i}(x)$ and $F_{1i}^M(x)$, when $M = 100$ of the six observations.

factor ($\mathbf{Y}_2$), immediacy index ($\mathbf{Y}_3$) and total citation ($\mathbf{Y}_4$), and moderately correlated with cited half-life ($\mathbf{Y}_5$). The second principal component (PC2) is strongly and positively correlated with number of published paper ($\mathbf{Y}_1$), however, negative correlated with cited half-life ($\mathbf{Y}_5$). As a result, we could simply name these two PCs as "citation" and "volume", respectively.

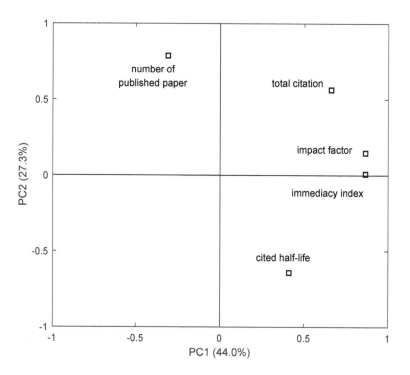

FIGURE 11.6
The loadings for the first and second principal components in JOURNALS Data.

With scores of the preserved PCs, a low-dimensional representation of journal groups could be obtained, referred to as component plots shown in Figures 11.7 and 11.8. Apparently, each of the eight journal groups featured by two distributional PCs is displayed as a probability density function.

Figure 11.7 shows the performances of the 8 groups along the PC1 ("citation"). It can be seen that journals of Earth Sciences get the highest scores on PC1. It is mainly due to their academic characteristics, i.e., more branches, frequent interdisciplinary exchanges and citations. On the other hand, variation of scores on PC1 in Earth Sciences appears larger than in other disciplines though its proportion in CSCD database is small. It is probably because journals in this field vary greatly as concerns citations. Besides, journals in Chemical Sciences, Engineering and Materials Sciences and Life Sciences also present higher scores on PC1. One of the main reasons lies in their large proportions in the statistical source, which causes higher mutual citations than for other disciplines. Compared with Earth Sciences, variations in these disciplines are much lower.

Figure 11.8 shows the performances along the PC2 ("volume"). It can be seen that journals of Information Science and Chemistry Sciences perform better. Generally speaking, a large number of published papers usually indicate rapid development and quick updates in technology. As a rising discipline, Information Science has made important breakthroughs in recent years. Nevertheless, academic communication in this discipline is expected to be further enhanced, since the journal citation is still in a relatively low level as shown in Figure 11.7. Journals of Mathematical and Physical Sciences, Management Sciences and Comprehensive Sciences perform poorly in both the two PCs. As traditional disciplines in China, Mathematical and Physical Sciences are less active, compared with other new disciplines. As for Management Sciences and Comprehensive Sciences, small proportions of journals in CSCD database may be the main reason for the low citations and paper publications.

To sum up, the component plots in Figures 11.7 and 11.8 by DPCA provide useful information for evaluating journals from different disciplines.

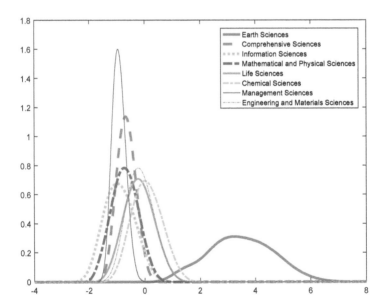

FIGURE 11.7
Performances of the 8 discipline groups in PC1.

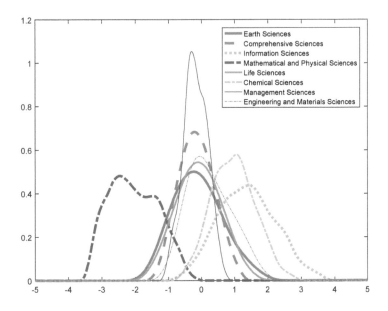

FIGURE 11.8
Performances of the 8 discipline groups in PC2.

11.8.2   STOCK Data

The second case aims at exploring the innate structure of China's stock market, by employing normal-distributional data. Taking a close look at stock records, it is not surprising to find out the high complexity in data structure, probably due to stocks' merger, restructure, rename, suspension, restoration or termination. Therefore, to avoid the difficulty of tracing stock individuals, we could employ symbolic data to describe higher-level 'concepts', such as stock styles.

Following the research in [6] [9], China's stock market can be classified into six stock styles, i.e., large-cap value stocks, large-cap growth stocks, medium-cap value stocks, medium-cap growth stocks, small-cap value stocks and small-cap stocks, constructed by the Chinese International Trust and Investment Company (CITIC) based on market size and stock style. It is valuable to compare large-cap value stocks, large-cap growth stocks, medium-cap value stocks, medium-cap growth stocks, small-cap value stocks and small-cap growth stocks along the crucial dimensions of risk and return to help fund managers to make wise investment decisions, and regulators to better understand how markets function.

More specifically, we use the year-end CITIC data for the period between 2009 and 2011 in our analysis, concerning five variables, i.e., return rates ($\mathbf{Y}_1$),

P/B ratios ($\mathbf{Y}_2$), price fluctuation swings ($\mathbf{Y}_3$), market capitalization ($\mathbf{Y}_4$) and turnover rates ($\mathbf{Y}_5$). We treat clusters of stocks as symbolic observations which are always distributed with long tails. After a normalizing transformation, these clusters could obey normal distribution. Moreover, Jarque-Bera tests do not reject that all of the variables are Normally distributed in all of the clusters. Next, we estimate mean and variance for each variable in each trimmed population and eventually we have a normal-distributional sample data as in Table 11.4.

Using DPCA, we obtain the first two principal components as follows:

$$\mathbf{PC}_1 \;=\; 0.37\mathbf{Y}_1 + 0.06\mathbf{Y}_2 + 0.59\mathbf{Y}_3 - 0.31\mathbf{Y}_4 + 0.65\mathbf{Y}_5;$$
$$\mathbf{PC}_2 \;=\; 0.37\mathbf{Y}_1 + 0.07\mathbf{Y}_2 + 0.44\mathbf{Y}_3 + 0.78\mathbf{Y}_4 - 0.25\mathbf{Y}_5.$$

These two components account for 62.3% of the variance information in the original data. Hence, they are sufficient to represent the original set of variables. Figure 11.9 plots the loadings of the first and second principal components. The first PC shows a strong association with return rates ($\mathbf{Y}_1$), price fluctuation swings ($\mathbf{Y}_3$) and turnover rates ($\mathbf{Y}_5$), all of which report transaction situations. As for the second PC, we see that it is positively correlated with market capitalization ($\mathbf{Y}_4$) and thus can be viewed as the capitalization factor. Besides, PC2 also has a negative association with turnover rates ($\mathbf{Y}_5$), which implies that Chinese investors prefer small cap stocks. Figure 11.9 also interprets the fundamental operating rules of the market. As it can be seen, return rates ($\mathbf{Y}_1$) and price fluctuation swings ($\mathbf{Y}_3$) are strongly and positively associated with each other, which illustrates the trade-off between return and risk in China's stock market in the period between 2009 and 2011. That is, the higher the return, the higher the risk. Besides, turnover rates ($\mathbf{Y}_5$) show positive correlations with the aforementioned two variables, which implies that investors prefer stocks of high risk due to their high return.

Figures 11.10 and 11.11 plot the performances of the six stock styles for each year along the first ("transaction situation") and second ("capitalization factor") PCs, respectively. From the transaction situation perspective, stock styles in each plot align from left to right according to their capitalization sizes in Figure 11.10. China's stock market ranks small-caps as the riskiest category and the large-caps as the least risky one with the medium-caps in between. All these are consistent with the standard theories in the literature. Hence China's stock market evaluates risks and return quite reasonably. For each category, value stocks are perceived less risky than the growth ones. But the value stocks (represented by solid lines) all position in the right side of the growth stocks (represented by dashed lines) regardless of the capitalization size in Figures 11.10, which may be due to their active transaction performances and the high expectations from investors.

From the capitalization factor, stock styles in each plot align from right to left according to their capitalization sizes in Figure 11.11, which implies that

TABLE 11.4
Normal-distributional dataset: stock.

	Y_1	Y_2	Y_3	Y_4	Y_5
2009 L-G	N(3.775, 0.062)	N(4.835, 0.006)	N(4.840, 0.558)	N(23.772, 1.006)	N(6.227, 0.523)
2009 L-V	N(3.785, 0.066)	N(4.868, 0.030)	N(5.127, 0.586)	N(23.551, 0.684)	N(6.338, 0.590)
2009 M-G	N(3.791, 0.046)	N(4.837, 0.006)	N(5.026, 0.480)	N(22.269, 0.252)	N(6.710, 0.344)
...					
2011 M-V	N(3.581, 0.070)	N(4.848, 0.021)	N(4.087, 0.402)	N(21.868, 0.261)	N(5.934, 0.645)
2011 S-G	N(3.501, 0.136)	N(4.827, 0.006)	N(4.070, 0.253)	N(20.620, 0.562)	N(6.258, 0.531)
2011 S-V	N(3.542, 0.370)	N(4.846, 0.046)	N(4.130, 0.362)	N(20.759, 0.515)	N(6.390, 0.615)

L-G represents large-cap growth stocks segment; L-V represents large-cap value stocks segment; M-G represents medium-cap growth stocks segment; M-V represents medium-cap value stocks segment; S-G represents small-cap growth stocks segment; S-V represents small-cap value stocks segment. The second number in each distribution is standard deviation but not variance.

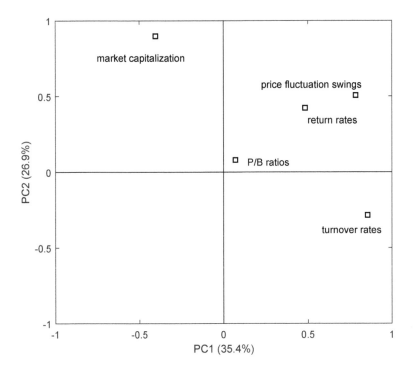

FIGURE 11.9
The loadings for the first and second principal components in STOCK Data.

small-cap stocks are the most active, followed by mid-cap stocks, and large-cap stocks are the least active. If we focus on the location change of each style from 2009 to 2011, the trading intention is gradually getting lower, showing a trend from a bull market to a bear market. The specific performances are that the return rates of stocks are gradually decreasing and the risks of stocks are also reducing. In addition, it can be found that within variations of each category increase from 2009 to 2011.

Furthermore, the above results using the NPCA method provide extra information on the variations within each category rather than only points of sample projections along the principal components as if classic PCA were applied. Within variations of each category are hard to obtain using classic PCA, because sample points projections from different categories are mixed in this case. Our approach is advantageous in this respect because normal-distributional PCs by themselves have the parameters on dispersion within each category.

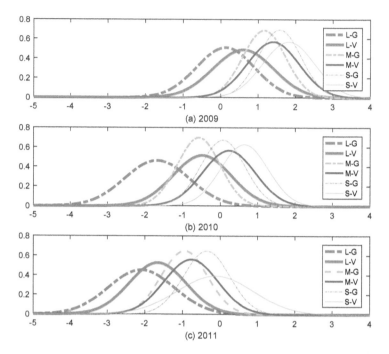

L-G represents large-cap growth stocks segment; L-V represents large-cap value stocks segment;
M-G represents medium-cap growth stocks segment; M-V represents medium-cap value stocks
segment; S-G represents small-cap growth stocks segment; S-V represents small-cap value stocks
segment.

FIGURE 11.10
Performances of 6 segments along the first principal component ("transaction
situation") in 2009 (a), 2010 (b), 2011 (c).

    In addition, our results also have implications for behavioral finance. The
within-dispersion information could be applied to measure heterogeneity of
beliefs in a certain category. For instance, Figure 11.10 indicates that among
all the segments, investors within the L-G category hold the most diversified
perceptions of rish and return. This finding suggests that a viable behavioral
pricing model should be able to factor in the heterogeneity of the beliefs
towards risk. This is particularly valuable for the fund managers who run
the business using style investments in a specific niche market, say, large-cap
growth stocks.

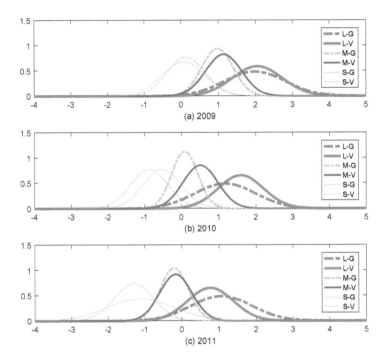

(a) 2009

(b) 2010

(c) 2011

L-G represents large-cap growth stocks segment; L-V represents large-cap value stocks segment; M-G represents medium-cap growth stocks segment; M-V represents medium-cap value stocks segment; S-G represents small-cap growth stocks segment; S-V represents small-cap value stocks segment.

FIGURE 11.11
Performances of 6 segments along the second principal component ("capitalization factor") in 2009 (a), 2010 (b), 2011 (c).

11.9   Conclusions

We put forward an analytical approach for principal components analysis of distributional data characterized by random variables. Our approach is featured by an analytical variance-covariance structure to determine the principal component space and an accurate linear operation for projecting the observations onto a PC subspace for visualization. Our method is not only able to handle special distributed data like interval data or histogram data, but can also apply to more complex data forms, for example, the situation where each element in the data table is assumed to be differently distributed.

Acknowledgment

This work was supported by the National Natural Science Foundation of China (Grant Nos. 71801162, 71420107025).

Bibliography

[1] L. Billard and E. Diday. *Symbolic Data Analysis: Conceptual Statistics and Data Mining*. Wiley, Chichester, 2006.

[2] P. Cazes. Analyse factorielle d'un tableau de lois de probabilité. *Revue de Statistique Appliquée*, 50(3):5–24, 2002.

[3] P. Cazes, A. Chouakria, E. Diday, and Y. Schektman. Extension de l'analyse en composantes principales à des données de type intervalle. *Revue de Statistique Appliquée*, 45(3):5–24, 1997.

[4] M. Chen, H. Wang, and Z. Qin. Principal component analysis for probabilistic symbolic data: a more generic and accurate algorithm. *Advances in Data Analysis and Classification*, 9(1):59–79, 2015.

[5] A. Irpino and R. Verde. Basic statistics for distributional symbolic variables: a new metric-based approach. *Advances in Data Analysis and Classification*, 9(2):143–175, 2015.

[6] W. Long, H. Mok, Y. Hu, and H. Wang. The style and innate structure of the stock markets in China. *Pacific-Basin Finance Journal*, 17(2):224–242, 2009.

[7] R. Verde and A. Irpino. New statistics for new data: a proposal for comparing multivalued numerical data. *Statistica Applicata*, 21(2):185–206, 2009.

[8] H. Wang, M. Chen, N. Li, and L. Wang. Principal component analysis of modal interval-valued data with constant numerical characteristics. In *The 58th World Statistics Congress of the International Statistical Institute. Ireland, Dublin*, 2011.

[9] H. Wang, M. Chen, X. Shi, and N. Li. Principal component analysis for normal-distribution-valued symbolic data. *IEEE Transactions on Cybernetics*, 46(2):356–365, 2016.

[10] H. Wang, R. Guan, and J. Wu. CIPCA: Complete-information-based principal component analysis for interval-valued data. *Neurocomputing*, 86:158–169, 2012.

12

Multidimensional Scaling of Distributional Data

Yoshikazu Terada

Division of Mathematical Science, Graduate School of Engineering Science, Osaka University, Japan

Patrick J.F. Groenen

Econometric Institute, Erasmus University Rotterdam, The Netherlands

CONTENTS

Multidimensional scaling (MDS) is a technique that visualizes dissimilarities between pairs of objects as distances between points in a low dimensional space. Standard MDS assumes that a single value of the dissimilarity is given for each pair of objects. To handle distributions of dissimilarities, histogram MDS has been proposed by Groenen and Winsberg [3]. They build on histograms of the empirical distributions of the dissimilarity of each of the pairs of objects. They also make use of so-called symbolic MDS for interval dissimilarities. The third ingredient of histogram MDS is the use of three-way MDS for several nested intervals that together describe the histograms. This chapter provides an overview of models for distributional dissimilarities by histogram MDS. Instead of representing an object by a single point in a low dimensional space, histogram MDS depicts objects as (concentric) circles or rectangles. The various models and their ingredients are illustrated by empirical examples.

DOI: 10.1201/9781315370545-12

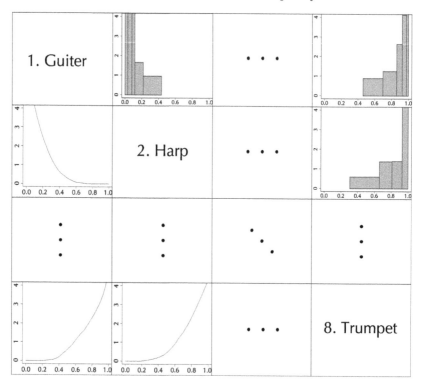

FIGURE 12.1
Illustration of a matrix of distributions of dissimilarities between pairs of musical instruments (lower triangular elements) and their empirical histograms (upper triangular elements).

12.1   Introduction

Standard multidimensional scaling (MDS) is a commonly used visualization technique to construct a configuration of points in a low-dimensional space that preserves given dissimilarities among a set of objects as distances between these points. Here, we present an overview of MDS-type models for dissimilarities that are not a single value but for which an empirical distribution is given. A graphical representation of the type of data we are dealing with is given in Figure 12.1, where the lower triangular elements show the distribution of the dissimilarity for each pair of numerical instruments and the upper triangular elements show their histogram representation. Such data can be modeled by histogram MDS where each object is represented as an appropriate geometric form such that a histogram of distances can be obtained.

There are two special cases for histograms. The most simple case of representing a distribution is to do so by a single value (e.g., by the mean, median, or mode). Then, ordinary MDS can be used, and each object is visualized by a point. In this chapter, we shall ignore this possibility. Another special case occurs when each distribution is described by a single interval (e.g., the range, the interquartile range, or the interval described by the mean plus or minus the standard deviation). In this case, only an interval of the dissimilarities needs to be described, which we will see can be done through interval MDS.

To model the histogram (or the interval) of a dissimilarity, one needs a geometric form for visualizing the objects such that for each pair of objects, a distribution (or interval) of the distances can be derived. For interval dissimilarities, two shapes for representing objects were proposed by Denœux and Masson [1]: the circle and the rectangle in two dimensions, or the hypersphere and hyperbox in higher dimensionalities. The intervals for a dissimilarity are simply defined by the smallest and largest Euclidean distances between the pairs of shapes for the two objects. For the histogram, objects are described by a set of circles or rectangles that can be chosen to be concentric or not.

In the remainder of this chapter, we present an overview of histogram MDS for distributional dissimilarities. In Section 12.2, we introduce interval-valued dissimilarity data and describe two important models, called the hypersphere model and the hyperbox model, for MDS of interval-valued dissimilarities. We also discuss the extension of interval MDS for three-way interval-valued MDS in Section 12.3. Next, we deal with histogram-valued (or percentile-valued) dissimilarity data and introduce some natural extensions of the hypersphere model and the hyperbox model for histogram-valued dissimilarity data. Finally, we conclude this chapter and discuss some future works.

12.2 Interval MDS

The main characteristic of interval MDS is that there is not a single observed value of the dissimilarities but that they are represented by an interval, that is,

$$
\Delta_{\text{Int}} := \begin{bmatrix}
- & [\delta_{12}^{(L)}, \delta_{12}^{(U)}] & \cdots & [\delta_{1n}^{(L)}, \delta_{1n}^{(U)}] \\
[\delta_{21}^{(L)}, \delta_{21}^{(U)}] & - & \cdots & [\delta_{2n}^{(L)}, \delta_{2n}^{(U)}] \\
\vdots & \vdots & \ddots & \vdots \\
[\delta_{n1}^{(L)}, \delta_{n1}^{(U)}] & [\delta_{n2}^{(L)}, \delta_{n2}^{(U)}] & \cdots & -
\end{bmatrix}, \tag{12.1}
$$

where the upper and lower bounds of the interval are given by $\delta_{ij}^{(U)}$ and $\delta_{ij}^{(L)}$ with $\delta_{ij}^{(U)} \geq \delta_{ij}^{(L)} \geq 0$, $\delta_{ij}^{(L)} = \delta_{ji}^{(L)}$, and intervals are assumed to be symmetric, that is, $\delta_{ij}^{(U)} = \delta_{ji}^{(U)}$ $(i, j = 1, \ldots, n)$ with n the number of objects. Denœux and Masson [1] proposed to find regions $R_i \subset \mathbb{R}^q$ ($i = 1, \ldots, n$) of a given q dimensional space in such a way that the differences between given dissimilarity intervals $[\delta_{ij}^{(L)}, \delta_{ij}^{(U)}]$ and reconstructed distance intervals $[d_{ij}^{(L)}(\mathcal{R}), d_{ij}^{(U)}(\mathcal{R})]$ are minimized. Therefore, the largest and smallest distances between any two regions are defined by

$$d_{ij}^{(L)}(\mathcal{R}) := \min_{x_i \in R_i, \, x_j \in R_j} \|x_i - x_j\|, \text{ and}$$

$$d_{ij}^{(U)}(\mathcal{R}) := \max_{x_i \in R_i, \, x_j \in R_j} \|x_i - x_j\|.$$

Figure 12.2 gives an example of two irregularly shaped regions R_i and the corresponding maximum distance $d_{ij}^{(U)}(\mathcal{R})$ and minimum distance $d_{ij}^{(L)}(\mathcal{R})$.

Denœux and Masson [1] defined interval MDS as the minimization of the Stress function

$$\sigma_{\text{Int}}^2(\mathcal{R}) := \sum_{i=1}^{n-1} \sum_{j=i+1}^{n} w_{ij} \left(\delta_{ij}^{(L)} - d_{ij}^{(L)}(\mathcal{R}) \right)^2$$

$$+ \sum_{i=1}^{n-1} \sum_{j=i+1}^{n} w_{ij} \left(\delta_{ij}^{(U)} - d_{ij}^{(U)}(\mathcal{R}) \right)^2, \qquad (12.2)$$

where w_{ij} is a given non-negative weight. We shall assume that the matrix with w_{ij} is irreducible, that is, it is not possible to partition the objects in subsets such that all w_{ij} between objects in different subsets are zero. If so, then the problem can be split into separate interval MDS problems. Note that $\sigma_{\text{Int}}^2(\mathcal{R})$ is sometimes referred to as the least-squares approach to interval MDS. Unfortunately, it is difficult to optimize Stress function $\sigma_{\text{Int}}(\mathcal{R})$ (or the Stress function of the possibility approach) over all sets of n connected closed region since the possible shapes of the regions R_i are potentially very irregular and do not allow for a simple parametrization for which an explicit formula for $d_{ij}^{(L)}(\mathcal{R})$ and $d_{ij}^{(U)}(\mathcal{R})$ is available. Therefore, some assumptions on the admissible shapes of the regions R_i are needed to be able to optimize the Stress function and obtain a configuration that is more easily interpretable. To this extent, two main models were proposed by Denœux and Masson [1]: (a) the circle (hypersphere) model that represents the regions by spheres with different radii and (b) the rectangle (hyperbox) model representing each object by a hyperbox of varying size.

12.2.1 Circle Model

Perhaps the most simple region is the circle (for $q = 2$) and its generalization of a hypersphere in higher dimensionality. Indeed, a hypersphere can be simply parametrized by its center coordinates $x \in \mathbb{R}^q$ and its radius $r > 0$. For an example of minimal and maximal distances between circles, see Figure 12.3. The circle model has a total of $n(q + 1)$ parameters to be estimated, nq for

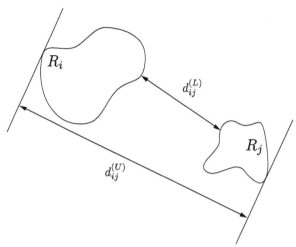

FIGURE 12.2
An illustration of the minimal and maximal distances in dimensionality $q = 2$ between the regions R_i and R_j as used in he general framework of interval MDS.

center coordinates in the $n \times q$ matrix $\mathbf{X}$ and n for the n vector $\mathbf{r}$ of radii. The lower and upper distances between hypersphere i and j are given respectively by

$$d_{ij}^{(L)}(\mathbf{X}, \mathbf{r}) \;:=\; \max\left(0, d_{ij}(\mathbf{X}) - r_i - r_j\right) \text{ and} \qquad (12.3)$$
$$d_{ij}^{(U)}(\mathbf{X}, \mathbf{r}) \;:=\; d_{ij}(\mathbf{X}) + r_i + r_j, \qquad (12.4)$$

where $d_{ij}(\mathbf{X}) := \|\mathbf{x}_i - \mathbf{x}_j\|$.

There are two approaches to obtain a hypersphere representation of objects: the least-squares and the possibility approach. In the least-squares approach, for given interval-valued dissimilarity data Δ, the Interval Stress function of the circle model is simply obtained by substituting (12.3) and (12.4) for $d_{ij}^{(L)}(\mathcal{R})$ and $d_{ij}^{(U)}(\mathcal{R})$ in the interval MDS Stress of (12.2). To find a hypersphere representation, the function $\sigma_{\text{Int}}(\mathbf{X}, \mathbf{r})$ is minimized over $\mathbf{X}$ and $\mathbf{r}$ with the constraints $r_i \geq 0$ $(i = 1, \ldots, n)$.

The possibility model provides an exact representation in some sense, whereas the least-squares approach provides an approximate representation in which lower and upper distances approximate as well as possible the given lower and upper dissimilarities, respectively. By analogy with the possibilistic fuzzy regression model of Tanaka, Uejima, and Asai [7], the goal of the possibility approach is to find the smallest hypersphere representation satisfying the following constraints:

$$[\delta_{ij}^{(L)}, \delta_{ij}^{(U)}] \subseteq [d_{ij}^{(L)}(\mathbf{X}, \mathbf{r}), d_{ij}^{(U)}(\mathbf{X}, \mathbf{r})] \quad (i, j = 1, \ldots, n).$$

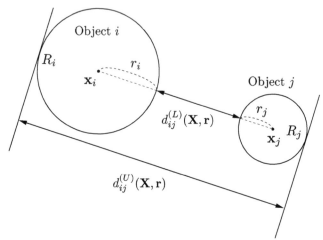

FIGURE 12.3
Example of minimal and maximal distances as defined by the circle (hyper-sphere) model in dimensionality $q = 2$.

To do so, we assume that the center coordinate matrix has already been obtained, say $\mathbf{X}^*$. For example, in Masson and Denœux [6], $\mathbf{X}^*$ is determined by the standard least-squares MDS on the centers of the interval-valued dissimilarities, that is, by minimizing the following objective function over $\mathbf{X}$:

$$\sigma(\mathbf{X}) := \sum_{i=1}^{n-1} \sum_{j=i+1}^{n} [\delta_{ij} - d_{ij}(\mathbf{X})]^2,$$

with $\delta_{ij} := (\delta_{ij}^{(L)} + \delta_{ij}^{(U)})/2$. Then, the possibility approach of the circle model is defined by the linear program

$$\min_{\mathbf{r} \in \mathbb{R}^n} \sum_{i=1}^{n} r_i$$

$$\text{subject to } \delta_{ij}^{(L)} \geq d_{ij}^{(L)}(\mathbf{X}^*, \mathbf{r}),$$
$$\delta_{ij}^{(U)} \leq d_{ij}^{(U)}(\mathbf{X}^*, \mathbf{r}), \text{ and}$$
$$r_i \geq 0 \ (i, j = 1, \ldots, n).$$

Although the two-steps in the possibility approach is attractive in that the linear programming part that fits the radii has a global solution, it has the disadvantage that the $\mathbf{X}$ and $\mathbf{r}$ are not simultaneously optimal. Therefore, in the remainder of this chapter, we shall focus on the interval MDS approach.

As an example of the interval MDS method using the circle model, the interval-valued dissimilarity data about timbre (Marozeau, de Cheveigné, McAdams, & Winsberg [5]) that we shall use throughout this chapter (see http://recherche.ircam.fr/pcm/archive/timbref0/). First, we briefly describe the details of these data. 24 subjects judged timbre dissimilarity between pairs of stimuli produced by a set of twelve musical instruments with equal fundamental frequency, duration, and loudness. There were three sessions, each at a different fundamental frequency (B3, 247Hz; C#4, 277Hz; Bb4, 466Hz). For more details about these data, see Marozeau et al. [5]. Here, we aggregate this data at fundamental frequency C#4 (277Hz) to an interval-valued dissimilarity data by lower and upper 10%-percentiles of these empirical dissimilarity distributions. Table 12.1 shows the eight instruments that we consider.

TABLE 12.1

Eight musical instruments and their abbreviation from the timbre data.

Instrument	Abbrev.	Instrument	Abbrev.
1. Guitar	Gu	5. Oboe	Ob
2. Harp	Hr	6. Clarinet	Cl
3. Violin pizzicato	Vp	7. Horn	Ho
4. Bowed Violin	Vl	8. Trumpet	Tr

The results of the circle model for the timbre data are given in Figure 12.4. For most of the instruments, the radii are similar, except for the horn (7. Ho) and clarinet (6. Cl), which are larger, and the trumpet (8. Tr), which is smaller. Therefore, the largest interval is obtained for horn and clarinet, and in general, these two instruments will have larger intervals with the other instruments, for example, with the bowed violin (4. Vl) and oboe (5. Ob). The most similar instruments are the guitar (1. Gu) and (2. Hr) and being perceived only slightly different from the violin pizzicato (3. Vp), all three being string instruments. Another cluster of instruments is formed by the bowed violin (4. Vl) and oboe (5. Ob). Finally, the trumpet (8. Tr) is considered both similar to the group of bowed violin (4. Vl) and oboe (5. Ob) and the group of horn (7. Ho) and clarinet (6. Cl).

12.2.2 Rectangle Model

In the circle model, the radii of each hypersphere represent the variability of each object. However, from a circle representation, we cannot interpret the variability of the underlying dimensions. In addition, it is difficult to interpret the dimensions due to the freedom of rotation of the circle model. As noted by Groenen, Winsberg, Rodríguez, and Diday [4], one of the important aims of MDS is to discover the relationships among the objects in terms of the underlying dimensions. Thus, it is most useful for interval-valued dissimilarity data to express the location and the variability of each object in terms of its underlying dimensions.

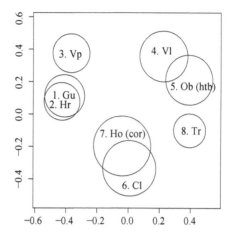

FIGURE 12.4
The result of the circle model with $q = 2$ for the timbre dissimilarity data ($\sigma_C^2 = 0.47$).

From this point of view, the rectangle representation seems to be appropriate since the sides of the rectangles correspond directly to the dimensions. The parametrization of a rectangle is slightly more complicated than that of a circle. A rectangle can be parametrized by a pair of center coordinate $\mathbf{x} \in \mathbb{R}^q$ and lengths of each dimension $\mathbf{r} \in \mathbb{R}_+^q$. Denote $\mathbf{X}$ as the matrix with rows $\mathbf{x}_i$ $(i = 1, \ldots, n)$ representing the center coordinates of the hyperbox for each object and $\mathbf{R}$ be a matrix whose rows $\mathbf{r}_i$ $(i = 1, \ldots, n)$ represent the side length vectors of the hyperbox with each object. In the rectangle model, the smallest distance between two rectangles representing objects i and j is defined by

$$d_{ij}^{(L)}(\mathbf{X}, \mathbf{R}) = \left(\sum_{s=1}^{q} \max[0, |x_{is} - x_{js}| - (r_{is} + r_{js})]^2 \right)^{1/2}, \qquad (12.5)$$

and the maximum distance by

$$d_{ij}^{(U)}(\mathbf{X}, \mathbf{R}) = \left(\sum_{s=1}^{q} [|x_{is} - x_{js}| + (r_{is} + r_{js})]^2 \right)^{1/2}. \qquad (12.6)$$

Figure 12.5 shows an example of the minimal and maximal distances the rectangle model with $q = 2$. In the least-squares approach, the Stress function is obtained by replacing $d_{ij}^{(L)}(\mathcal{R})$ and $d_{ij}^{(L)}(\mathcal{R})$ in (12.2) by $d_{ij}^{(L)}(\mathbf{X}, \mathbf{R})$ and $d_{ij}^{(U)}(\mathbf{X}, \mathbf{R})$ as defined in (12.5) and (12.6).

A solution of the rectangle model applied to the timbre data is presented in Figure 12.6. There, the variation in the range of distances along the first dimension is caused by the guitar (1. Gu), harp (2. Hr), and to a lesser extent by the bowed violin (4. Vl) and horn (7. Ho). Along the second dimension,

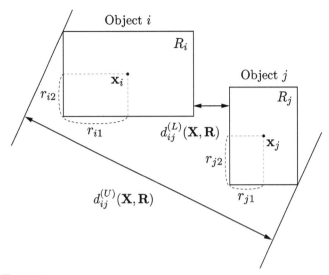

FIGURE 12.5
The rectangle model with $q = 2$.

there are large ranges for the violin pizzicato (3. Vp), the hobo (5. Ob), horn (7. Ho), clarinet (6. Cl), and trumpet (8. Tr). Note that the maximal Euclidean distances for, for example, the violin pizzicato (3. Vp) and the clarinet (6. Cl) are obtained diagonally from the upper end of the line representing the violin pizzicato (3. Vp) and the lower end of the line representing clarinet (6. Cl). It seems that the rectangle solution spreads the ranges more over the dimensions than the circle model does. It is fair to say that most of the interval information of the dissimilarities is mostly shown on the second dimension, perhaps with the exception of object pairs that involve the guitar (1. Gu), harp (2. Hr).

12.3 Histogram MDS

In most cases, interval-valued dissimilarity data consists of maximum and minimum values. However, these values are susceptible to the effect of outliers. The lower and upper α-percentiles $\alpha^{(L)}$ and $\alpha^{(U)}$ of a dissimilarity distribution are more appropriate to construct interval-valued dissimilarity data. Even if we can observe only a histogram of dissimilarity data between two objects, we can compute the percentile values from the histogram assuming uniformity within each bin. Since a histogram can be considered as a density estimator of the dissimilarity distribution, we can compute the corresponding distribution function from the histogram, and then we can get the percentile values. Alternatively, we simply consider the histogram as the discrete distribution in

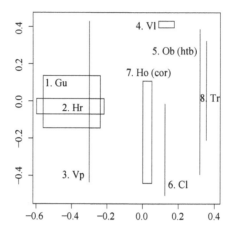

FIGURE 12.6

The result of the rectangle model with $q = 2$ for the timbre dissimilarity data $(\sigma_R^2 = 0.160)$.

which the center of each bin takes the corresponding frequency, and then we can also compute the percentile values. Consider the set A with elements A_k containing the ranges of the percentile values $[\alpha_k^{(L)}, \alpha_k^{(U)}]$ such that the ranges become increasingly larger and a smaller one contains the larger one, that is, $A_k \subset A_l$ if $k < l$. A typical example is $A_1 = [.40, .60]$, $A_2 = [.30, .70]$, and $A_3 = [.20, .80]$. Note that intervals in A_k do not need to be symmetric around .50; the only requirement is $A_k \subset A_l$ if $k < l$. We have the following type of dissimilarity data, called histogram-valued or percentile-valued dissimilarity data:

$$\Delta_{\text{Hist}} := (\Delta_{A_1}, \ldots, \Delta_{A_K}),$$

where

$$\Delta_k := \begin{bmatrix} - & [\delta_{12k}^{(L)}, \delta_{12k}^{(U)}] & \cdots & [\delta_{1nk}^{(L)}, \delta_{1nk}^{(U)}] \\ [\delta_{21k}^{(L)}, \delta_{21k}^{(U)}] & - & \cdots & [\delta_{2nk}^{(L)}, \delta_{2nk}^{(U)}] \\ \vdots & \vdots & \ddots & \vdots \\ [\delta_{n1k}^{(L)}, \delta_{n1k}^{(U)}] & [\delta_{n2k}^{(L)}, \delta_{n2k}^{(U)}] & \cdots & - \end{bmatrix},$$

with $[\delta_{ijk}^{(L)}, \delta_{ijk}^{(U)}] \subseteq [\delta_{ijl}^{(L)}, \delta_{ijl}^{(U)}]$ for $k \leq l$. Should the empirical histogram for a dissimilarity not be available for the choices of the percentiles in α, then by assuming uniformity, the required values can be interpolated.

For the histogram-valued dissimilarity data, we will use a constrained version of the standard interval MDS applied to each Δ_{A_k}. Let R_{ki} be a region of object i corresponding to percentile range A_k. It seems natural to expect that regions R_{ki} $(k = 1, \ldots, K)$ are satisfying $R_{ki} \subset R_{li}$ for $A_k \subset A_l$. Figure 12.7 shows this idea for general regions R_{ki}.

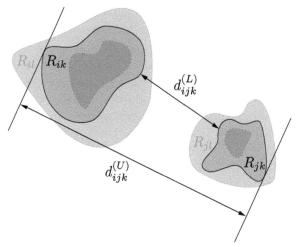

FIGURE 12.7
Histogram representation for general regions where regions R_{ik} is contained
in R_{il} if percentile range $A_k \subset A_l$.

Thus, we consider the following MDS stress function for the histogram-
valued dissimilarity data:

$$\min \sigma^2_{\text{Hist}}(\mathcal{R}_1, \ldots, \mathcal{R}_K) := \sum_{k=1}^{K} \sum_{i=1}^{n-1} \sum_{j=i+1}^{n} w_{ijk} \left(\delta^{(L)}_{ijk} - d^{(L)}_{ij}(\mathcal{R}_k) \right)^2$$

$$+ \sum_{k=1}^{K} \sum_{i=1}^{n-1} \sum_{j=i+1}^{n} w_{ijk} \left(\delta^{(U)}_{ijk} - d^{(U)}_{ij}(\mathcal{R}_k) \right)^2 \quad (12.7)$$

$$\text{subject to } R_{ik} \subset R_{il} \text{ for } A_k \subset A_l.$$

Next, four models for the histogram MDS can be thought of as shown
in Figure 12.8. They vary by the shape (circle model versus rectangle model)
and by sharing a common center (concentric) or not (non-concentric). In every
model, the restriction holds that a region representing a smaller range of α
values is contained in the region of a larger α range, thus $R_{ik} \subset R_{il}$ for
$A_k \subset A_l$.

We first provide explicit descriptions for the concentric circle and rectangle
models. As a simple extension of the circle model for interval-valued dissimilarity data, Masson and Denœux [6] proposed the concentric circle model for
the histogram-valued dissimilarity data. In the concentric circle model, the
regions R_{ik} that represent object i form nested circles that have the same
center point. Therefore, the regions R_{ik} can be described by a common center
$\mathbf{x}_i$ (with $\mathbf{x}_i$ being row i of $\mathbf{X}$) and the radii r_{ki} being ordered, that is,

$$0 \leq r_{i1} \leq r_{i2} \leq \cdots \leq r_{iK}.$$

The upper left panel of Figure 12.8 shows the regions of this model. Replacing $d_{ij}^{(L)}(\mathcal{R}_k)$ and $d_{ij}^{(U)}(\mathcal{R}_k)$ in the histogram MDS stress of (12.7) by $d_{ij}^{(L)}(\mathbf{X}, \mathbf{r}_k)$ and $d_{ij}^{(U)}(\mathbf{X}, \mathbf{r}_k)$ gives the stress function of the concentric circle model. Similarly to Groenen and Winsberg [2], an algorithm can be constructed that combines majorization minimization with monotone regression for imposing the order constraints on the r_{ik}s. Terada and Yadohisa [9] used a different approach for imposing the order constraints by defining $r_{ik} = \sum_{l=1}^{k} a_{il}$ with $a_{ik} \geq 0$. In their algorithm, the positivity restrictions on a_{il} are automatically satisfied.

Groenen and Winsberg [2] proposed the concentric rectangle model albeit not under that name. It has a common center $\mathbf{x}_i$ for the rectangles representing object i and a nested series of rectangles. The lower left panel of Figure 12.8

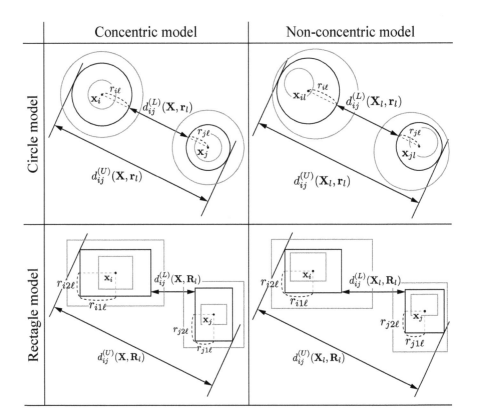

FIGURE 12.8
Four models for representing the histogram dissimilarities. The upper panels represent circle models and the lower panels the rectangle models. In the left two panels, the regions representing an object have the same center and the right two panels they have not.

shows the concentric rectangle model. The stress function of the concentric rectangle model is obtained by replacing $d_{ij}^{(L)}(\mathcal{R}_k)$ and $d_{ij}^{(U)}(\mathcal{R}_k)$ in (12.7) by the lower and upper distances $d_{ij}^{(L)}(\mathbf{X}, \mathbf{R}_k)$ and $d_{ij}^{(U)}(\mathbf{X}, \mathbf{R}_k)$ defined in (12.5) and (12.6). To ensure nestedness, the constraints

$$0 \leq r_{is1} \leq r_{is2} \leq \cdots \leq r_{isK},$$

need to be imposed. Groenen and Winsberg [2] proposed to use a majorization and monotone regression for imposing the order constraints, and Terada and Yadohisa [9] developed an algorithm that automatically satisfies these constraints by a similar alternative parametrization as they did for the concentric circle model.

The common center constraint of the concentric models is a quite restrictive condition and may not be natural in some cases. The non-concentric models relieve this constraint at the cost of being somewhat more complicated. We first consider the non-concentric circle model in which each object is represented by nested circles that do not necessarily have the same center point. The top right panel of Figure 12.8 shows this model. The nested constraint $R_{i1} \subset \cdots \subset R_{iK}$ can be rewritten by

$$\|\mathbf{x}_{ik} - \mathbf{x}_{i(k+1)}\| + r_{ik} \leq r_{i(k+1)} \quad (k = 1, \ldots, K)$$

where $\mathbf{x}_{ik} := (x_{i1k}, \ldots, x_{iqk})^T$ and r_{ik} are the center point and the radius of the circle corresponding to the α_k range of object i, respectively. Similar to the concentric model, the stress function σ_{NCC}^2 for the non-concentric circle model is obtained by substituting $d_{ij}^{(L)}(\mathbf{X}_k, \mathbf{r}_k)$ and $d_{ij}^{(U)}(\mathbf{X}_k, \mathbf{r}_k)$ for $d_{ij}^{(L)}(\mathcal{R}_k)$ and $d_{ij}^{(U)}(\mathcal{R}_k)$ in the histogram MDS stress of (12.7). Unfortunately, it is difficult to directly optimize this complex constrained optimization problem. Terada and Yadohisa [10] used unconstrained optimization with a penalty that penalizes the deviations from the constraint. The penalty term is defined by

$$g_{\mathrm{NCC}}^2(\mathbf{X}_1, \ldots, \mathbf{X}_K, \mathbf{r}_1, \ldots, \mathbf{r}_K)$$
$$:= \sum_{i=1}^{n} \sum_{k=1}^{K-1} \max \left[0, \|\mathbf{x}_{ik} - \mathbf{x}_{i(k+1)}\| + r_{ki} - r_{i(k+1)}\right]^2$$

By minimizing the extended stress function $\tilde{\sigma}_{\mathrm{NCC}}^2 := \sigma_{\mathrm{NCC}}^2 + \lambda g_{\mathrm{NCC}}^2$ for a sufficiently large $\lambda > 0$ a solution is obtained that satisfies the nestedness constraints.

In the non-concentric rectangle model, each object is represented by nested rectangles that do not necessarily have the same center point. The lower right panel of Figure 12.8 presents the non-concentric rectangle model. The stress function can be obtained by substituting $d_{ij}^{(L)}(\mathcal{R}_k)$ and $d_{ij}^{(U)}(\mathcal{R}_k)$ with $d_{ij}^{(L)}(\mathbf{X}_k, \mathbf{R}_k)$ and $d_{ij}^{(U)}(\mathbf{X}_k, \mathbf{R}_k)$ in the general histogram MDS stress (12.7). Using center coordinates and side length values, we rewrite the nestedness constraints of the stress function by the inequalities

$$r_{isk} + |x_{isk} - x_{is(k+1)}| \leq r_{is(k+1)} \tag{12.8}$$

for $(i = 1, \ldots, n, \ k = 1, \ldots, K - 1,$ and $s = 1, \ldots, p)$.

Surprisingly, there exists an efficient parametrization of the non-concentric hyperbox model in which the nested constraints are eliminated. Following Terada and Yadohisa [9], new parameters $a_{isk}^{(L)}$ and $a_{isk}^{(U)}$ are introduced so that

$$x_{isk} = x_{is1} + \frac{1}{2} \sum_{l=2}^{k} \left(a_{isl}^{(U)} - a_{isl}^{(L)} \right)$$

and

$$r_{isk} = r_{is1} + \frac{1}{2} \sum_{l=2}^{k} \left(a_{isl}^{(U)} + a_{isl}^{(L)} \right),$$

instead of using x_{isk} and r_{isk} $(k = 2, \ldots, K)$. Using this parametrization, the constraints in (12.8) can be replaced by the following non-negativity constraints

$$r_{is1} \geq 0, \ a_{isk}^{(L)} \geq 0, \ \text{and} \ a_{isk}^{(U)} \geq 0.$$

Introducing new variables $\rho_{0is}, \ \alpha_{isk}^{(L)}$ and $\alpha_{isk}^{(U)}$ $(i = 1, \ldots, n; \ k = 2, \ldots, K, \ s = 1, \ldots, p)$ so that $r_{is1} = \rho_{is1}^2, \ a_{isk}^{(L)} = \alpha_{isk}^{(L)2}$, and $a_{isk}^{(U)} = \alpha_{isk}^{(U)2}$, the non-negativity condition can be ensured. Therefore, the non-concentric hyperbox model MDS can be directly solved by unconstrained optimization.

12.3.1 Example: Timbre Data

A solution of the concentric circle model applied to the timbre data with $A_1 = (10\%, 90\%)$, $A_2 = (5\%, 95\%)$, and $A_3 = (0\%, 100\%)$ is presented in the top left panel of Figure 12.9. The radii of the nested circles of bowed violin (Vl) evenly reduce with increasing the α. This means that the variability of feeling for bowed violin (Vl) is reduced with increasing α. On the other hand, there is only a small difference between the radii of the clarinet (Cl) corresponding with A_1 and A_2. Thus, we can interpret that the variability of feeling for clarinet (Cl) is less sensitive to the increase of α.

The bottom left panel of Figure 12.9 is the resulting configuration of the concentric rectangle model for the timbre data. Compared to the concentric circle model, the variability of the different α ranges is mostly related to the first dimension for all objects. Since the brightness of sounds highly depends on the associated perceived feelings, people can more easily distinguish impulsive sounds from sustained sounds. The variability shown on the first dimension reflects this fact.

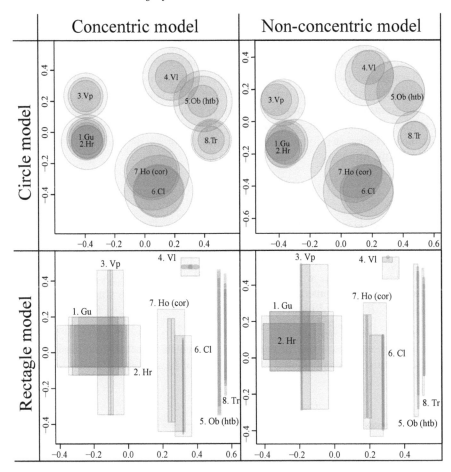

FIGURE 12.9
The results of the four models for representing the histogram timbre dissimilarities.

The top right panel of Figure 12.9 shows the result of the non-concentric circle model applied to the timbre data. With decreasing α ranges, the centers of violin pizzicato (Vp) and oboe (Ob) shift to the right and the left, respectively. This means that for a decreasing α range, the violin pizzicato (Vp) is perceived more as an impulsive sound, whereas the oboe (Ob) more as a sustained sound. The concentric circle model cannot reflect this kind of information.

A similar effect is found for the non-concentric rectangle model shown in the bottom right panel of Figure 12.9; that is, the solution is similar to the concentric rectangle model except that the rectangles are pushed to the outside from the origin.

12.3.2 Histogram MDS and Fuzzy Theory

The present formulation of histogram MDS can be considered as a similar problem for general fuzzy dissimilarities in the context of fuzzy theory as done by Masson and Denœux [6]. They assume that each object is represented by a p-dimensional fuzzy number f_i in $\mathbb{R}^p$. Let f_i be the membership function of the p-dimensional fuzzy number of object i and $R_i(\alpha)$ be the α-cut of f_i for $\alpha \in [0,1]$. In this setting, Masson and Denœux [6] mentioned that the fuzzy distance between two fuzzy numbers f_i and f_j can be defined by the following membership function:

$$d_{ij}(c) := \sup_{\mathbf{x},\mathbf{y} \in \mathbb{R}^p : c = \|\mathbf{x}-\mathbf{y}\|} \min[f_i(\mathbf{x}), f_j(\mathbf{y})].$$

Moreover, Masson and Denœux [6] show that each α-cut of d_{ij} is a closed interval $\xi_{ij}(\alpha) := [d_{ij}^{(L)}(\alpha), d_{ij}^{(U)}(\alpha)]$, where $d_{ij}^{(L)}(\alpha)$ and $d_{ij}^{(U)}(\alpha)$ are respectively the lower and upper distances between the α-cuts $R_i(\alpha)$ and $R_j(\alpha)$. For $\alpha_1 > \alpha_2$, we have $R_i(\alpha_1) \subset R_i(\alpha_2)$ ($i = 1, \ldots, n$). Histogram MDS can then be considered as the problem to estimate the hidden α-cuts $R_i(\alpha_1), \ldots, R_i(\alpha_K)$ of hidden fuzzy number f_i ($i = 1, \ldots, n$) for given α-cuts $\xi_{ij}(\alpha_k) := \left[\xi_{kij}^{(L)}, \xi_{kij}^{(U)}\right]$ ($i, j = 1, \ldots, n$; $k = 1, \ldots, K$) of fuzzy dissimilarities in least-squares sense. In particular, percentile intervals of a dissimilarity distribution can be considered as the α-cuts of fuzzy dissimilarities.

12.4 Local Minima

The stress function discussed in this chapter needs to be minimized by iterative algorithms as no analytic solution is available. For MDS and symbolic MDS, two types of algorithms are available. First, general purpose minimization algorithms such as BFGS can be applied. The latter is the steepest descent approach that needs the gradient of the stress function. If the gradient is not available, then it can estimate the gradient by multiple function evaluations near the present estimates of the parameters. For interval MDS problems with large n (say, $n > 30$), numerical gradients most likely could lead to a serious slowdown of the algorithm. The second type of algorithm is the minimization by majorization (MM) algorithm that is guaranteed to reduce the Stress function in each iteration (Groenen et al. [4]; Terada & Yadohisa [9, 10]). MM algorithms can be seen as the steepest descent algorithms with a constant step size. Here, we use three versions. The first one is the plain majorization algorithm. The second one allows in each iteration to dilate (by a single scalar) the estimates obtained from a majorization update. This scales all parameters to their size such that it minimizes stress. The third one uses a trick that doubles the step length in each iteration, the so-called relaxed update. The idea is that an ordinary MM step reduces the stress and that a step twice as long cannot increase stress. Often, the MM with step doubling decreases the number of iterations by a half. Here, we use the step doubling in combination with the optimal dilation.

The Stress function for MDS of interval dissimilarities (12.2) does not have nice properties such as convexity. Consequently, it cannot be guaranteed *a priori* that an algorithm minimizing (12.2) yields a global minimum. In fact, many local minima can occur. Here, we investigate the attraction to a (candidate) global minimum of five algorithms: (a) the majorization algorithm (MM), (b) MM with optimal dilation, (c) MM with acceleration and dilation, (d) the standard BFGS algorithm for minimizing functions in R using explicit gradients, and (e) the BFGS algorithm that uses numerically estimated gradients. We compare their performance (final Stress divided by the best Stress found over all runs by any of the algorithms) over 1000 random starts for the circle and rectangle models of interval MDS.

Figure 12.10(a) shows the results for the circle model. It presents the cumulative distribution of the performance (stress divided by overall best stress). We find that for the circle model BFGS with numeric gradients is approximately in 30% of the runs attracted to the (candidate) global minimum, whereas the three MM varieties about 5%. However, the MM algorithms and particularly MM with dilation are best able to obtain almost optimal performance in 50 to 60% of the runs. The hyperbox models seem to be much more prone to local minima. This can be seen in lower-left corner of Figure 12.10(b) that shows only a very small vertical line representing an almost 4%

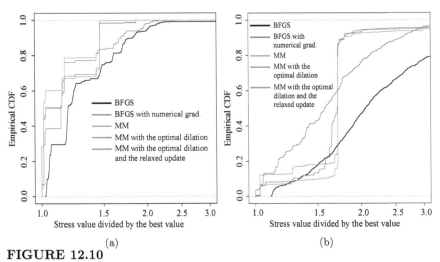

(a)   (b)

FIGURE 12.10
Empirical cumulative distributions of Stress values on the timbre data for 1000 random starts using four different minimization methods. Panel (a) contains the distributions for the circle model and Panel (b) for the rectangle model.

attraction to the (candidate) global minimum by the BFGS algorithm using numerical gradients. There is a small region of attraction (at most 10 %) of the MM methods to solutions close to the candidate global minimum. We conclude from Figure 12.10 that at least for the timbre data, the local minimum problem for σ_{Int}^2 seems less severe for the circle model than for the hyperbox model. For the circle model, the majorization approaches seem to be beneficial to increase the probability of finding the best local minimum and thus the candidate of the global minimum. For the hyperbox model, one tends to find many local minima and the BFGS method with numerical gradients seems to be better able to find good local minima.

Table 12.2 shows the computational time of these experiments for each method. For the circle model, three MM varieties are much faster than the BFGS methods, where the MM algorithm with dilation seems to be the best choice from both the computational cost and the local minimum problem perspectives. For the rectangle model, we found that the BFGS method with numerical gradients is best able to find the best local minimum. However, the three MM methods are approximately three times faster than the BFGS method with numerical gradients. Therefore, we can try 3000 random starts by using the MM methods at the same computational costs of 1000 random starts of the BFGS method with numerical gradients. Thus, considering both the computational costs and the quality of the local minimum, the MM methods (especially MM with dilation) seem to be a good choice even for the rectangle model.

TABLE 12.2

Computational time of this experiments for each method.

Circle model	time (s)
1. BFGS	35.79
2. BFGS with numerical grad.	364.37
3. MM	6.76
4. MM with the optimal dilation	6.35
5. MM with the optimal dilation and the relaxed update	5.27
Rectangle model	**time (s)**
1. BFGS	20.90
2. BFGS with numerical grad.	297.05
3. MM	103.57
4. MM with the optimal dilation	111.10
5. MM with the optimal dilation and the relaxed update	70.88

12.5 Conclusions and Discussion

In this chapter, we have provided an overview of the present state-of-the-art of histogram MDS. We discussed several cases for modeling distributions of dissimilarities. The special case of a single interval per distribution of dissimilarity can be modeled through interval MDS. The interval of the dissimilarities is then estimated by the interval of distances obtained by the minimum and maximum Euclidean distances of pairs of circles or rectangles. The standard way to model distributions of dissimilarities is through histogram MDS. It uses nested sets of quantiles of the distribution of the dissimilarity and models them by nested circles or rectangles that can be either concentric or not. The quantiles of the distribution of the dissimilarity for i and j are represented as the minimum and maximum Euclidean distances between the corresponding circles (or rectangles) of objects i and j. The nesting of the circles (or rectangles) ensures the nesting of the (nested) quantile pairs.

It is common practice in MDS to allow the dissimilarities to be optimally transformed, for example, by an ordinal transformation. We believe that it is not good to do so in the context of histogram MDS. The reason is the very fact that histogram MDS estimates quantiles. As these are numeric values, transformations do not make sense.

We believe that histogram MDS is a useful tool to model distributions of dissimilarities. Its strength lies in the simplicity of reducing $n(n-1)/2$ distributions into a much more simple visual representation of n nested shapes. As such, the interpretation of a histogram MDS solution is easier than simultaneously inspecting the distributions of all pairs of objects.

All computations in this chapter were done by the R package **smds** (Terada & Groenen [8]). This package allows fitting interval and histogram MDS, using circular or rectangle representations of the objects and, for histogram MDS, keeping the circles (or rectangles) concentric or not. In addition, it allows for choosing the type of optimization algorithm used (e.g., the MM algorithm). The core of this package is written in C++, which ensures fast computations of the solutions.

Bibliography

[1] T. Denœux and M. Masson. Multidimensional scaling of interval-valued dissimilarity data. *Pattern Recognition Letters*, 21:83–92, 2000.

[2] P.J.F. Groenen and S. Winsberg. Multidimensional scaling of histogram dissimilarities. In V. Bategelj, H.H. Bock, A. Ferligoj, and A. Žiberna,

editors, *Data Science and Classification*, pages 581–588. Springer-Verlag, 2006.

[3] P.J.F. Groenen and S. Winsberg. 3waysym-scal: Three-way symbolic multidimensional scaling. In P. Brito, G. Cucumel, P. Bertrand, and F. de Carvalho, editors, *Selected Contributions in Data Analysis and Classification*, pages 55–67. Springer-Verlag, 2007.

[4] P.J.F. Groenen, S. Winsberg, O. Rodríguez, and E. Diday. I-Scal: Multidimensional scaling of interval dissimilarities. *Computational Statistics & Data Analysis*, 51:360–378, 2006.

[5] J. Marozeau, A. de Cheveigné, S. McAdams, and S. Winsberg. The dependency of timbre on fundamental frequency. *The Journal of the Acoustical Society of America*, 114:2946–2957, 2003.

[6] M. Masson and T. Denœux. Multidimensional scaling of fuzzy dissimilarity data. *Fuzzy Sets and Systems*, 128:339–352, 2002.

[7] H. Tanaka, S. Uejima, and K. Asai. Linear regression analysis with fuzzy model. *IEEE Transactions on Systems, Man and Cybernetics*, 12:903–907, 1982.

[8] Y. Terada and P.J.F. Groenen. **smds**: Symbolic Multidimensional Scaling. http://cran.r-project.org/web/packages/smds/index.html, 2015.

[9] Y. Terada and H. Yadohisa. Multidimensional scaling with hyperbox model for percentile dissimilarities. In J. Watada, G. Phillips-Wren, L.C. Jain, and R.J. Howlett, editors, *Intelligent Decision Technologies*, pages 779–788. Springer-Verlag, 2011.

[10] Y. Terada and H. Yadohisa. Multidimensional scaling with the nested hypersphere model for percentile dissimilarities. *Procedia Computer Science*, 6:364–369, 2011.

Part IV

Regression and Forecasting

13

Regression Analysis with the *Distribution and Symmetric Distribution* Model

Sónia Dias

School of Technology and Management, Polytechnic Institute of Viana do Castelo & LIAAD-INESC TEC, Portugal

Paula Brito

Faculty of Economics, University of Porto & LIAAD-INESC TEC, Porto, Portugal

CONTENTS

The generalization of the linear regression model to histogram data is not a straightforward extension of the classical model. The semi-linearity of the space where the elements are histograms implies that the parameters in the model cannot be negative. However, a linear relation between histogram-valued variables should be allowed to be either direct or inverse. The *Distribution and Symmetric Distribution* model solves this problem and allows predicting distributions directly from other distributions. To determine the parameters of the model it is necessary to solve a quadratic optimization problem, subject to non-negativity constraints on the unknowns. It uses the Mallows Distance to evaluate the dissimilarity between the observed and predicted distributions. As in classical regression, a goodness-of-fit measure is deduced from the model, whose values range between 0 and 1.

DOI: 10.1201/9781315370545-13

13.1   Introduction

Histogram-valued variables may be considered as a generalization of interval-valued variables. For this reason, the first definitions and methods for histogram-valued variables appeared as an extension of the concepts and methods defined for interval-valued variables. The first linear regression model between distributional data was one of these cases. Billard and Diday [2,3] proposed a first model for histogram-valued variables that is a generalization of the linear regression model that they had already proposed for interval-valued variables. The model is a simple adaptation of the classical model where the parameters are estimated using the definitions of symbolic variance and symbolic covariance.

The difficulties associated with the definition of linear regression between distributions lay in the definition of linear combination between this type of elements. Histogram arithmetic is complex and the semi-linearity of the space whose elements are histograms, does not allow for the generalization of the classical definition of linear combination, and consequently of linear relation between elements of this type. In an attempt to solve this first limitation, Irpino and Verde [6] represented the histograms by the inverse of the cumulative distribution function, the quantile function. Considering this representation, instead of working with histograms, we work with piecewise linear functions and we may now define a linear regression where the involved elements are functions. However, these functions have an important property. As the subintervals within the histograms have the upper bound greater than or equal to the lower bound, the quantile functions used to represent them are always non-decreasing functions. Therefore, this kind of representation does not solve the problem of the semi-linearity of the space. In accordance with what is observed with histograms, if we multiply a quantile function by a real negative number we do not obtain a non-decreasing function, and consequently, we do not obtain a quantile function. As such, the problem of the semi-linearity of the space, whose elements are quantile functions, is still unresolved. (For more details about the behavior of the quantile functions, see Chapter 2.)

Considering the previous representation of the observations of the histogram-valued variables, two linear regression models have been proposed. The work of Irpino and Verde [7], presented in Chapter 14, proposed a regression model that decomposes the observations of the histogram-valued variables in two components. The approach of Dias and Brito [5] is based on a linear combination of couples of distributions. In this model, for each explicative variable, it is considered not only the distribution of each observation but also the respective symmetric distribution. This work will be presented in the following sections of this chapter.

13.2 The *Distribution and Symmetric Distribution* model

13.2.1 Model definition

The most straightforward approach to define a linear regression model for histogram data appeared to be a simple adaptation of the classical model. However this approach raised some problems that had to be solved, so that it would be possible to directly predict distributions from other distributions.

Consider that we want to predict the distributions of the histogram-valued variable Y from p histogram-valued variables X_j with $j \in \{1, \ldots, p\}$. Consider also that the observations of these variables are represented by the corresponding quantile functions. For each observation $i \in \{1, \ldots, n\}$, the quantile function of the predicted distribution $\widehat{Y}_i$ would then be obtained as follows:

$$\Psi_{\widehat{Y}_i}(t) = v + a_1 \Psi_{X_{i1}}(t) + a_2 \Psi_{X_{i2}}(t) + \ldots + a_p \Psi_{X_{ip}}(t). \tag{13.1}$$

The quantile functions are non-decreasing functions. However, if some parameters are negative real numbers, a decreasing function may be obtained and consequently we may be predicting a function that is not a quantile function. A possible solution to this problem is to impose non-negativity constraints on the parameters, but this would be forcing a direct linear relation between the variables. Nevertheless, it is fundamental to allow for the possibility of both direct and inverse linear relations between histogram-valued variables, i.e. it is necessary to solve the problem of the semi-linearity of the space of the quantile functions. For this reason, the linear regression model proposed by Dias and Brito [5] includes both the quantile functions $\Psi_{X_{ij}}(t)$ that represent the observed distributions of the histogram-valued variables X_j for each unit i, and the quantile functions that represent the respective symmetric histograms, i.e. $-\Psi_{X_{ij}}(1-t)$. Therefore, despite imposing non-negativity constraints on the parameters of the model, the inclusion of the quantile functions $\Psi_{X_{ij}}(t)$ and $-\Psi_{X_{ij}}(1-t)$ ensures the possibility of identifying direct and inverse linear relations between the histogram-valued variables.

Definition 1 Consider the histogram-valued variables $X_1, X_2, \ldots, X_p$. The quantile functions that represent the distribution of these histogram-valued variables for each observation i are denoted $\Psi_{X_{i1}}(t)$, $\Psi_{X_{i2}}(t)$, ..., $\Psi_{X_{ip}}(t)$ and the quantile functions that represent the respective symmetric histograms associated with each observation of the referred variables are $-\Psi_{X_{i1}}(1-t)$, $-\Psi_{X_{i2}}(1-t), \ldots, -\Psi_{X_{ip}}(1-t)$, with $t \in [0,1]$. Each quantile function $\Psi_{Y_i}(t)$ that represents the observation i of the histogram-valued variable Y may be expressed as follows:

$$\Psi_{Y_i}(t) = \Psi_{\widehat{Y}_i}(t) + e_i(t)$$

where $\Psi_{\widehat{Y}_i}(t)$ is the predicted quantile function for the observation i, obtained from

$$\Psi_{\widehat{Y}_i}(t) = v + \sum_{j=1}^{p} a_j \Psi_{X_{ij}}(t) - \sum_{j=1}^{p} b_j \Psi_{X_{ij}}(1-t)$$

with $t \in [0,1]$; $a_j, b_j \geq 0$, $j \in \{1, 2, \ldots, p\}$ and $v \in \mathbb{R}$.

It should be noted that $\Psi_{\widehat{Y}_i}(t)$ is always a quantile function since it is a linear combination of quantile functions where the coefficients are always non-negative real values. The error for each unit i, $e_i(t)$, is a piecewise linear function, but not necessarily a quantile function, given by

$$e_i(t) = \Psi_{Y_i}(t) - \Psi_{\widehat{Y}_i}(t), \qquad t \in [0,1].$$

For each unit i, the predicted distribution $\widehat{Y}_i$ may be represented by the quantile function $\Psi_{\widehat{Y}_i}$ or by the respective histogram $H_{\widehat{Y}_i}$. This linear regression model is named **Distribution and Symmetric Distribution (DSD) Regression Model** [5].

To define the *DSD* regression model, it is necessary to take into account that:

1. Not all n observations of the histogram-valued variables can be symmetric in relation to the yy-axis, otherwise the quantile functions that represent these observations and the respective symmetric would be colinear.

2. For all observations i of each variable X_j, it is assumed that the quantile functions $\Psi_{X_{ij}}(t)$, $-\Psi_{X_{ij}}(1-t)$ and $\Psi_{Y_i}(t)$ represent histograms with the same number m of subintervals and with the same set of cumulative weights $\{0, w_1, \ldots, w_{m-1}, 1\}$. For all units $i = 1, \ldots, n$, and each variable, the weights associated with corresponding intervals ℓ should be the same, p_ℓ. These weights must verify the condition $p_\ell = p_{m-\ell+1}$ with $\ell \in \{1, ..., m\}$. Note that we do not have necessarily the same weight in all subintervals of each histogram. If the histograms do not follow these conditions, it is necessary to apply the process proposed by Irpino and Verde [6]. For more details about this process see Chapter 1 of this book and Chapter 4 in [4]. As an alternative to the Irpino and Verde process and when the microdata are known, we may organize all histograms as equiprobable.

Consider the conditions enumerated above and also that $c_{X_{ij\ell}}$, $r_{X_{ij\ell}}$ represent, respectively, the center and half range of the subinterval ℓ, of the variable X_j for unit i. The quantile functions that represent the distributions of X_j, and the quantile functions that represent the respective symmetric histogram, for a given unit i, are defined respectively, by:

$$\Psi_{X_{ij}}(t) = \begin{cases} c_{X_{ij1}} + \left(\frac{2t}{w_1} - 1\right) r_{X_{ij1}} & \text{if} \quad 0 \le t < w_1 \\ c_{X_{ij2}} + \left(\frac{2(t-w_1)}{w_2 - w_1} - 1\right) r_{X_{ij2}} & \text{if} \quad w_1 \le t < w_2 \\ \vdots \\ c_{X_{ijm}} + \left(\frac{2(t-w_{m-1})}{1-w_{m-1}} - 1\right) r_{X_{ijm}} & \text{if} \quad w_{m-1} \le t \le 1 \end{cases} ; \quad (13.2)$$

$$-\Psi_{X_{ij}}(1-t) = \begin{cases} -c_{X_{ijm}} + \left(\frac{2t}{w_1} - 1\right) r_{X_{ijm}} & \text{if} \quad 0 \le t < w_1 \\ -c_{X_{ij(m-1)}} + \left(\frac{2(t-w_1)}{w_2 - w_1} - 1\right) r_{X_{ij(m-1)}} & \text{if} \quad w_1 \le t < w_2 \\ \vdots \\ -c_{X_{ij1}} + \left(\frac{2(t-w_{m-1})}{1-w_{m-1}} - 1\right) r_{X_{ij1}} & \text{if} \quad w_{m-1} \le t \le 1 \end{cases} ; \quad (13.3)$$

For each unit i, the quantile function $\Psi_{Y_i}(t)$ may be represented analogously. According to Definition 1, the quantile function $\Psi_{\widehat{Y}_i}(t)$ that predicts the histogram-valued variable Y, may be obtained as follows:

$$\Psi_{\widehat{Y}_i}(t) = \begin{cases} c_{\widehat{Y}_{i1}} + \left(\frac{2t}{w_1} - 1\right) r_{\widehat{Y}_{i1}} & \text{if} \quad 0 \le t < w_1 \\ c_{\widehat{Y}_{i2}} + \left(\frac{2(t-w_1)}{w_2 - w_1} - 1\right) r_{\widehat{Y}_{i2}} & \text{if} \quad w_1 \le t < w_2 \\ \vdots \\ c_{\widehat{Y}_{im}} + \left(\frac{2(t-w_{m-1})}{1-w_{m-1}} - 1\right) r_{\widehat{Y}_{im}} & \text{if} \quad w_{m-1} \le t \le 1 \end{cases} \quad (13.4)$$

where, for $\ell = 1, \ldots, m$ and $i = 1, \ldots, n$,

$$c_{\widehat{Y}_{i\ell}} = v + \sum_{j=1}^{p} a_j c_{X_{ij\ell}} - \sum_{j=1}^{p} b_j c_{X_{ij(m-\ell+1)}}; \quad (13.5)$$

and

$$r_{\widehat{Y}_{i\ell}} = \sum_{j=1}^{p} a_j r_{X_{ij\ell}} + \sum_{j=1}^{p} b_j r_{X_{ij(m-\ell+1)}}. \quad (13.6)$$

Alternatively, the predicted observations of the histogram-valued variable Y, may be represented by the histogram:

$$H_{\widehat{Y}_i} = \left\{ \left[l_{\widehat{Y}_{i1}}, u_{\widehat{Y}_{i1}} \right[, p_1; \ldots; \left[l_{\widehat{Y}_{im}}, u_{\widehat{Y}_{im}} \right], p_m \right\} \quad (13.7)$$

where, if $l_{X_{ij\ell}}$ and $u_{X_{ij\ell}}$ are the lower and upper bounds of the subinterval ℓ of the variable X_j for the unit i, we have:

$$l_{\widehat{Y}_{i\ell}} = \sum_{j=1}^{p} \left(a_j l_{X_{ij\ell}} - b_j u_{X_{ij(m-\ell+1)}} \right) + v$$

$$u_{\widehat{Y}_{i\ell}} = \sum_{j=1}^{p} \left(a_j u_{X_{ij\ell}} - b_j l_{X_{ij(m-\ell+1)}} \right) + v$$

for $\ell = 1, \ldots, m$ and $i = 1, \ldots, n$, with $a_j, b_j \geq 0$, $j \in \{1, 2, \ldots, p\}$ and $v \in \mathbb{R}$.

13.2.2   Interpretation of the model

Expressions (13.5) and (13.6) show that, according to the *DSD* model, for each subinterval ℓ, the predicted center $c_{\widehat{Y}_{i\ell}}$ and half range $r_{\widehat{Y}_{i\ell}}$ (or the bounds) may be obtained, respectively, by a linear relation of the centers $c_{X_{ij\ell}}$ and half ranges $r_{X_{ij\ell}}$ (or the bounds) of the explicative histogram-valued variables.

As the values of the parameters a_j, b_j and v are the same for all subintervals ℓ, we may consider the following classical linear relations between the weighted mean of the centers and between the weighted mean of the half ranges of the variables X_j and Y. That is, for each i we have:

$$\overline{Y}_i = v + \sum_{j=1}^{p} (a_j - b_j) \, \overline{X}_{ij} \tag{13.8}$$

and

$$\overline{Y}_i^r = \sum_{j=1}^{p} (a_j + b_j) \, \overline{X}_{ij}^r \tag{13.9}$$

where the weighted mean of the centers and the half ranges of the variables X_j and Y, for each i, are given by

$$\overline{X}_{ij} = \sum_{\ell=1}^{m} c_{X_{ij\ell}} p_{i\ell} \quad \text{and} \quad \overline{X}_{ij}^r = \sum_{\ell=1}^{m} r_{X_{ij\ell}} p_{i\ell}.$$

Expressions (13.8) and (13.9) show that the parameters that define the classical linear regressions between the weighted mean of the centers and between the weighted mean of the half ranges of the histogram-valued variables are not the same but are related. In spite of the fact that this model is defined between distributions and the relation between the distributions may be direct or inverse, it always induces a direct linear relation between the weighted mean of the half ranges of the variables X_j and Y, because $a_j, b_j \geq 0$. The direct or inverse relation between the histogram-valued variables is always determined by the linear relation between the weighted mean values of the observations of the histogram-valued variables. The histogram-valued variables X_j are in direct linear relation with Y when $a_j > b_j$ and the linear relation is inverse if $a_j < b_j$. The example that follows illustrates the described behavior.

Example 1 In this example, two situations will be considered, based on the data from Table 1.3 in Chapter 1. In both situations, we have simple models with one explicative variable. All observations of the histogram-valued variables were rewritten as histograms with six subintervals and for all units the weights associated with the each subinterval ℓ are the same.

Situation 1: Consider a symbolic dataset where 10 units are described by two symbolic variables: Y the response histogram-valued variable and X the explicative histogram-valued variable.

$$DSD\ Model:\ \Psi_{\widehat{Y}_i}(t) = -1.95 + 3.56\Psi_{X_i}(t) - 0.41\Psi_{X_i}(1 - t)$$

Situation 2: The observations of the explicative histogram-valued variable are the symmetric of the histogram-valued variable X. In this case, the explicative variable is denoted by $-X$ and the response variable is Y, the same as in situation 1.

$$DSD\ Model:\ \Psi_{\widehat{Y}_i}(t) = -1.95 + 0.41\Psi_{X_i}(t) - 3.56\Psi_{X_i}(1 - t)$$

Comparing the expressions of the *DSD* models, in both situations, we may observe that in the second case, as expected, the values of the parameters a and b change relatively to the first. The scatter plots corresponding to the two situations are represented in Figure 13.1 [5].

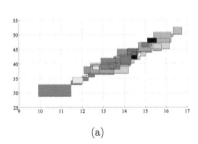
(a)

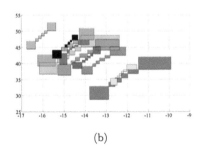
(b)

FIGURE 13.1
Scatter plots (projection in $z = 0$) of the observations of the histogram-valued variables X and Y in (a); $-X$ and Y in (b).

Observing the behavior of the scatter plots, it is important to underline that two orientations can be distinguished:

1. The orientation of the subintervals of each histogram, that is obviously always direct (when the histograms are represented by quantile functions, which are non-decreasing functions)

2. The orientation of each subinterval ℓ for all units i induces the orientation of the weighted mean values of the histograms and consequently determines the direct or inverse relation between the histogram-valued variables.

The interpretation of these behaviors is discussed in Dias and Brito [5]. In the first situation, the parameter a is larger than b so, according to expression (13.8), the classical linear relation between the weighted mean values of the distributions is direct, as illustrated in Figure 13.2(a). This behavior means that the relation between the histogram-valued variables is classified as direct. On the other hand, we consider that the linear relation between the histogram-valued variable Y and $-X$ is inverse because the parameter a is lower than b. As we can observe in Figure 13.2(b) the classical linear relation between the weighted mean values of the histograms Y_i and X_{ij} is inverse.

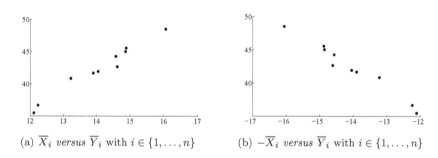

(a) $\overline{X}_i$ *versus* $\overline{Y}_i$ with $i \in \{1, \ldots, n\}$ (b) $-\overline{X}_i$ *versus* $\overline{Y}_i$ with $i \in \{1, \ldots, n\}$

FIGURE 13.2
Scatter plots considering the mean values of the observations of the histogram-valued variables X and Y in (a); $-X$ and Y in (b).

13.2.3   Estimation of the parameters

In classical statistics, the parameters of the linear regression model are estimated by minimizing the sum of the squares of the errors, i.e., solving the minimization problem $\sum_{i=1}^{n} (y_i - \widehat{y}_i)^2$, where y_i and $\widehat{y}_i$ are the observed and the predicted values for the observation i, respectively. To solve this problem the Least Squares method is used.

In this work, the error measure between distributions is calculated considering the Mallows distance [8]. Irpino and Verde [6] rewrote this distance using the center and half-range of the subintervals that compose the histograms. According to this result, the sum of the squares of the errors (SSE) between the observed and predicted distributions, Y_i and $\widehat{Y}_i$, may be defined as follows:

$$
\begin{aligned}
SSE &= \sum_{i=1}^{n} D_M^2(\Psi_{Y_i}(t), \Psi_{\widehat{Y}_i}(t)) = \\
&= \sum_{i=1}^{n} \sum_{\ell=1}^{m} p_{i\ell} \left[(c_{Y_{i\ell}} - c_{\widehat{Y}_{i\ell}})^2 + \tfrac{1}{3}(r_{Y_{i\ell}} - r_{\widehat{Y}_{i\ell}})^2 \right]
\end{aligned}
\tag{13.10}
$$

The parameters of the *DSD* model, in Definition 1, are estimated considering the Mallows distance defined as in expression (13.10) and solving a quadratic optimization problem, subject to non-negativity constraints on the unknowns.

Definition 2 Consider the quantile functions $\Psi_{Y_i}(t)$, that represent the observed distributions and $\Psi_{\widehat{Y}_i}(t)$ the predicted distributions of the reponse variable Y, obtained by the *DSD* model, according to expression (13.4). The parameters of the *DSD* model are the solution of the following constrained quadratic optimization problem:

$$
\begin{aligned}
Minimize \quad & SSE(a_j, b_j, v) = \sum_{i=1}^{n} D_M^2(\Psi_{Y_i}(t), \Psi_{\widehat{Y}_i}(t)) \\
= & \sum_{i=1}^{n} \sum_{\ell=1}^{m} p_\ell \left[\left(c_{Y_{i\ell}} - v - \sum_{j=1}^{p} \left(a_j c_{X_{ij\ell}} - b_k c_{X_{ij(m-l+1)}} \right) \right)^2 \right. \\
& \left. + \tfrac{1}{3} \left(r_{Y_{i\ell}} - \sum_{j=1}^{p} \left(a_j r_{X_{ij\ell}} + b_j r_{X_{ij(m-\ell+1)}} \right) \right)^2 \right]
\end{aligned}
$$

subject to

$$
a_j, b_j \geq 0, \quad j \in \{1, \ldots, p\}
$$

$$
v \in \mathbb{R}.
$$

The *constraint quadratic optimization problem*, that in the matricial form represents the minimization problem of the Definition 2, is the following:

$$
Minimize \quad SSE(\mathbf{b}) = \frac{1}{2}\mathbf{b}^T \mathbf{H} \mathbf{b} + \mathbf{W}^T \mathbf{b} + K
$$

subject to

$$
-a_j, -b_j \leq 0, \quad j \in \{1, \ldots, p\}
$$

$$
v \in \mathbb{R}
$$

(13.11)

where

- $\mathbf{H} = [h_{sq}]$ is the hessian matrix, a symmetric matrix of order $2p + 1$, with p the number of variables X_j. Calculating the partial derivatives and after algebric manipulation, the elements of this matrix may be obtained by:

$$h_{dq} = \begin{cases} \sum\limits_{i=1}^{n}\sum\limits_{\ell=1}^{m} 2p_\ell \left(c_{X_{i\frac{d+1}{2}\ell}} c_{X_{i\frac{q+1}{2}\ell}} + \right. \\ \qquad\qquad \left. + \frac{1}{3} r_{X_{i\frac{d+1}{2}\ell}} r_{X_{i\frac{q+1}{2}\ell}} \right) & \text{if} & \begin{array}{l} d, q \text{ odd}; \\ d, q \leq 2p \end{array} \\[4ex] \sum\limits_{i=1}^{n}\sum\limits_{\ell=1}^{m} 2p_\ell \left(c_{X_{i\frac{d}{2}(m-\ell+1)}} c_{X_{i\frac{q}{2}(m-\ell+1)}} + \right. \\ \qquad\qquad \left. + \frac{1}{3} r_{X_{i\frac{d}{2}(m-\ell+1)}} r_{X_{i\frac{q}{2}(m-\ell+1)}} \right) & \text{if} & \begin{array}{l} d, q \text{ even}; \\ d, q \leq 2p \end{array} \\[4ex] \sum\limits_{i=1}^{n}\sum\limits_{\ell=1}^{m} 2p_\ell \left(-c_{X_{i\frac{d}{2}(m-\ell+1)}} c_{X_{i\frac{q+1}{2}\ell}} + \right. \\ \qquad\qquad \left. + \frac{1}{3} r_{X_{i\frac{d}{2}(m-\ell+1)}} r_{X_{i\frac{q+1}{2}\ell}} \right) & \text{if} & \begin{array}{l} l \text{ even}, q \text{ odd}; \\ d, q \leq 2p \end{array} \\[4ex] \sum\limits_{i=1}^{n}\sum\limits_{\ell=1}^{m} 2p_\ell c_{X_{i\frac{q+1}{2}\ell}} & \text{if} & \begin{array}{l} q \text{ odd}; \\ d = 2p + 1 \end{array} \\[3ex] \sum\limits_{i=1}^{n}\sum\limits_{\ell=1}^{m} -2p_\ell c_{X_{i\frac{q}{2}(m-\ell+1)}} & \text{if} & \begin{array}{l} q \text{ even}; \\ d = 2p + 1 \end{array} \end{cases}$$

$$(13.12)$$

- $\mathbf{W} = [w_d]$ is a column vector with $2p + 1$ rows. The elements of this vector of independent terms, are defined by:

$$w_d = \begin{cases} \sum\limits_{i=1}^{n}\sum\limits_{\ell=1}^{m} p_\ell \left(-2c_{Y_{i\ell}} c_{X_{i\frac{d+1}{2}\ell}} - \right. \\ \qquad\qquad \left. - \frac{2}{3} r_{Y_{i\ell}} r_{X_{i\frac{d+1}{2}\ell}} \right) & \text{if} & d \text{ is odd}; \ d \leq 2p \\[4ex] \sum\limits_{i=1}^{n}\sum\limits_{\ell=1}^{m} p_\ell \left(2c_{Y_{i\ell}} c_{X_{i\frac{d+1}{2}(m-\ell+1)}} - \right. \\ \qquad\qquad \left. - \frac{2}{3} r_{Y_{i\ell}} r_{X_{i\frac{d+1}{2}(m-\ell+1)}} \right) & \text{if} & d \text{ is even}; \ d \leq 2p \\[4ex] \sum\limits_{i=1}^{n}\sum\limits_{\ell=1}^{m} -2p_\ell c_{Y_{i\ell}} & \text{if} & d = 2p + 1 \end{cases}$$

$$(13.13)$$

- $\mathbf{b} = (a_1, b_1, a_2, b_2, \ldots, a_p, b_p, v)^T$ is the the parameters' vector that is a column vector with $2p + 1$ rows.

- K is the real value given by $K = \sum\limits_{i=1}^{n}\sum\limits_{\ell=1}^{m} p_\ell \left(c_{Y_{i\ell}}^2 + \frac{1}{3} r_{Y_{i\ell}}^2 \right)$.

Alternatively, the minimization problem in Definition 2 may be presented in matricial form, as a *constraint least square problem*, as follows:

$$Minimize \quad SSE(\mathbf{b}) = \|\mathbf{Y} - \mathbf{Xb}\|^2$$

$$subject\ to$$

$$-a_j, -b_j \leq 0, \quad j \in \{1, \ldots, p\}$$

$$v \in \mathbb{R}.$$

(13.14)

where the matrices $\mathbf{X}$ and $\mathbf{Y}$ are built as follows:

- $\mathbf{X}$ is a matrix with $2nm$ rows and $(2p+1)$ columns defined by

$$\mathbf{X} = [\mathbf{X}_1^c \quad \mathbf{X}_1^r \quad \cdots \quad \mathbf{X}_m^c \quad \mathbf{X}_m^r]^T \tag{13.15}$$

where the odd and even rows are respectively, the $n \times (2p+1)$ matrices:

$$\mathbf{X}_\ell^c = [\mathbf{x}_{1\ell}^c \quad \mathbf{x}_{2\ell}^c \quad \cdots \quad \mathbf{x}_{n\ell}^c]^T$$

and

$$\mathbf{X}_\ell^r = [\mathbf{x}_{1\ell}^r \quad \mathbf{x}_{2\ell}^r \quad \cdots \quad \mathbf{x}_{n\ell}^r]^T.$$

The $2p+1$ dimensional vectors $\mathbf{x}_{i\ell}^c$ and $\mathbf{x}_{i\ell}^r$ that define the above matrices are:

$$\mathbf{x}_{i\ell}^c = \left(\sqrt{p_\ell} c_{X_{i1\ell}}, -\sqrt{p_\ell} c_{X_{i1(m-\ell+1)}}, \ldots, \sqrt{p_\ell} c_{X_{ip\ell}}, -\sqrt{p_\ell} c_{X_{ip(m-\ell+1)}}, \sqrt{p_\ell} \right)$$

and

$$\mathbf{x}_{i\ell}^r = \left(\sqrt{\frac{p_\ell}{3}} r_{X_{i1\ell}}, \sqrt{\frac{p_\ell}{3}} r_{X_{i1(m-\ell+1)}}, \ldots, \sqrt{\frac{p_\ell}{3}} r_{X_{ip\ell}}, \sqrt{\frac{p_\ell}{3}} r_{X_{ip(m-\ell+1)}}, 0 \right)$$

- $\mathbf{Y}$ is a the column matrix with $2nm$ rows defined by

$$\mathbf{Y} = [\mathbf{y}_1^c \quad \mathbf{y}_1^r \quad \cdots \quad \mathbf{y}_m^c \quad \mathbf{y}_m^r]^T \tag{13.16}$$

with

$$\mathbf{y}_\ell^c = \left(\sqrt{p_\ell} c_{Y_{1\ell}}, \ldots, \sqrt{p_\ell} c_{Y_{n\ell}} \right)^T$$

and

$$\mathbf{y}_\ell^r = \left(\sqrt{\frac{p_\ell}{3}} r_{Y_{1\ell}}, \ldots, \sqrt{\frac{p_\ell}{3}} r_{Y_{n\ell}} \right)^T.$$

In practical examples, to estimate the parameters of the *DSD* model it is necessary to define the respective matrices involved in the optimization problems 13.11 or 13.14. After matrix definition, these problems are solved recurring to appropriate software[2].

It is important to guarantee that the solution obtained from the optimization problem (in Definition 2) is an optimal solution, i.e., that the values obtained for the parameters are those where the *SSE* reaches the global minimum and that verify the non-negative constrains. According to [9] we can assert that:

1. the functions that define the non-negative constrains are convex, and therefore the feasible region of the optimization problem is a convex set;

2. $\mathbf{H} = 2\mathbf{X}^T\mathbf{X}$, with $\mathbf{H}$ and $\mathbf{X}$ defined in expressions (13.12) and (13.15), respectively. Since the matrix $\mathbf{H}$ is positive semi-definite, then the function to optimize is a convex function.

As the function to optimize and the functions that define the constrains are convex, it may then be ensured that the optimization problem has optimal solutions. In the cases where the objective function is strictly convex we may ensure that the optimal solution is unique. Note that this should always occur because the objective function is strictly convex when the matrix $\mathbf{H}$ is positive definite and this is verified when the columns of $\mathbf{X}$ are linearly independent, i.e. when all distributions X_j and respective symmetrics are not colinear, for all j (see Subsection 13.2.1).

13.3 Properties and results deduced from the model

13.3.1 Kuhn Tucker conditions and consequences

To prove important results concerning the predicted distribution, the Kuhn Tucker conditions must be considered. Some of these results are the counterparts of the corresponding properties in classical statistics.

Theorem 1 [9] Consider the minimization problem presented in Definition (2). If $\mathbf{b}^* = (a_1^*, b_1^*, \ldots, a_p^*, b_p^*, v^*)$ is an optimal solution of the problem, then $\mathbf{b}^*$ must satisfy the constraints of the optimization problem and the Kuhn Tucker conditions. For each $j \in \{1, 2, \ldots, p\}$ we have

- Constraints: $-a_j^* \leq 0$ and $-b_j^* \leq 0$;

---

[2]For example we may use the Matlab function *quadprog* if we treat the problem as a constraint quadratic problem (13.11) and we may use the Matlab function *lsqlin* if we write the problem as a constraint least squares problem (13.14).

- Kuhn Tucker conditions - there exist real multipliers λ_j, δ_j satisfying:

  1. $\dfrac{\partial SSE}{\partial a_j}(\mathbf{b}^*) - \lambda_j = 0$;   4. $\lambda_j a_j^* = 0$;

  2. $\dfrac{\partial SSE}{\partial b_j}(\mathbf{b}^*) - \delta_j = 0$;   5. $\delta_j b_j^* = 0$;

  3. $\dfrac{\partial SSE}{\partial v}(\mathbf{b}^*) = 0$;   6. $\lambda_j, \delta_j \geq 0$.

  Or, equivalently,

  1'. $\dfrac{\partial SSE}{\partial a_j}(\mathbf{b}^*) \geq 0$;   4'. $a_j^* \dfrac{\partial SSE}{\partial a_j}(\mathbf{b}^*) = 0$;

  2'. $\dfrac{\partial SSE}{\partial b_j}(\mathbf{b}^*) \geq 0$;   5'. $b_j^* \dfrac{\partial SSE}{\partial b_j}(\mathbf{b}^*) = 0$.

  3'. $\dfrac{\partial SSE}{\partial v}(\mathbf{b}^*) = 0$;

Considering the definition of symbolic mean for histogram-valued variables (see Chapter 2), the results previously presented in this chapter and the Kuhn Tucker conditions, we may prove the following propositions.

Proposition 1 The mean of the predicted histogram-valued variable $\overline{\widehat{Y}}$ is equal to the mean of the observed histogram-valued variable $\overline{Y}$.

Proof: See [5].

As a consequence of Proposition 1, we may prove the next result.

Proposition 2 For each observation i, the quantile function of the distribution $\widehat{Y}_i$ predicted by the *DSD* model, may be rewritten as follows:

$$\Psi_{\widehat{Y}_i}(t) - \overline{Y} = \sum_{j=1}^{p} a_j^* \left(\Psi_{X_{ij}}(t) - \overline{X}_j\right) + b_j^* \left(-\Psi_{X_{ij}}(1-t) + \overline{X}_j\right).$$

To complete the development of the linear regression model for histogram-valued variables, a goodness-of-fit measure remains to be deduced. To obtain this measure, the following two results are crucial.

Proposition 3 For the observed and predicted distributions Y_i and $\widehat{Y}_i$ of the variable Y, with $i \in \{1, \ldots, n\}$, we have

$$\sum_{i=1}^{n} \int_0^1 \left(\Psi_{Y_i}(t) - \Psi_{\widehat{Y}_i}(t)\right)\left(\Psi_{\widehat{Y}_i}(t) - \overline{Y}\right) dt = 0.$$

Proof: See [5].

Proposition 4 The sum of the square of the Mallows distance between each observed distribution i, of the histogram-valued variable Y and the symbolic mean of the histogram-valued variable Y, $\overline{Y}$, may be decomposed as follows:

$$\sum_{i=1}^{n} D_M^2\left(\Psi_{Y_i}(t), \overline{Y}\right) = \sum_{i=1}^{n} D_M^2\left(\Psi_{Y_i}(t), \Psi_{\widehat{Y}_i}(t)\right) + \sum_{i=1}^{n} D_M^2\left(\Psi_{\widehat{Y}_i}(t), \overline{Y}\right).$$

Proof: See [5].

13.3.2   Goodness-of-fit measure

In classical statistics, the coefficient of determination R^2 is defined from the decomposition of total variation into the sum of squares due to error and the sum of squares due to regression. In Proposition 4 a similar decomposition is obtained considering the Mallows distance between the observed distributions and the distributions predicted with the *DSD* model. So, similarly to the classical model, it is possible to deduce a coefficient of determination from the *DSD* model.

Definition 3 Consider the observed and predicted distributions of the histogram-valued variable Y and $\widehat{Y}$ represented, respectively, by their quantile functions $\Psi_{Y_i}(t)$ and $\Psi_{\widehat{Y}_i}(t)$. Consider also $\overline{Y}$ as the symbolic mean of the histogram-valued variable Y. The measure Ω is given by

$$\Omega = \frac{\displaystyle\sum_{i=1}^{n} D_M^2\left(\Psi_{\widehat{Y}_i}(t), \overline{Y}\right)}{\displaystyle\sum_{i=1}^{n} D_M^2\left(\Psi_{Y_i}(t), \overline{Y}\right)}.$$

In classical linear regression, the coefficient of determination R^2 ranges from 0 to 1. In this case, the measure, Ω, also verifies this condition.

Theorem 2 The goodness-of-fit measure Ω ranges from 0 to 1.

Proof: Consider the Ω measure as in Definition 3. This measure is non-negative, so $\Omega \geq 0$. From Proposition 4, we have

$$\sum_{i=1}^{n} D_M^2\left(\Psi_{Y_i}(t), \overline{Y}\right) = \sum_{i=1}^{n} \int_0^1 \left(\Psi_{Y_i}(t) - \Psi_{\widehat{Y}_i}(t)\right)^2 dt + \sum_{i=1}^{n} \int_0^1 \left(\Psi_{\widehat{Y}_i}(t) - \overline{Y}\right)^2 dt$$

The last equality may be rewritten as follows,

$$1 = \frac{\displaystyle\sum_{i=1}^{n} \int_0^1 \left(\Psi_{Y_i}(t) - \Psi_{\widehat{Y}_i}(t)\right)^2 dt}{\displaystyle\sum_{i=1}^{n} D_M^2\left(\Psi_{Y_i}(t), \overline{Y}\right)} + \frac{\displaystyle\sum_{i=1}^{n} \int_0^1 \left(\Psi_{\widehat{Y}_i}(t) - \overline{Y}\right)^2 dt}{\displaystyle\sum_{i=1}^{n} D_M^2\left(\Psi_{Y_i}(t), \overline{Y}\right)}$$

$$\Leftrightarrow \Omega = 1 - \frac{\displaystyle\sum_{i=1}^{n} \int_0^1 \left(\Psi_{Y_i}(t) - \Psi_{\widehat{Y}_i}(t)\right)^2 dt}{\displaystyle\sum_{i=1}^{n} D_M^2\left(\Psi_{Y_i}(t), \overline{Y}\right)}.$$

Since the term $\dfrac{\sum\limits_{i=1}^{n}\int\limits_0^1\left(\Psi_{Y_i}(t)-\Psi_{\widehat{Y}_i}(t)\right)^2 dt}{\sum\limits_{i=1}^{n}D_M^2\left(\Psi_{Y_i}(t),\overline{Y}\right)}$ is non-negative, the value of Ω is always lower than or equal to 1. So, we have that $0 \leq \Omega \leq 1$.

Let us now analyze the extreme situations.

- Suppose $\Omega = 0$.

In this case,

$$\sum_{i=1}^{n}D_M^2\left(\Psi_{\widehat{Y}_i}(t),\overline{Y}\right) = 0 \Leftrightarrow \sum_{i=1}^{n}\int_0^1\left(\Psi_{\widehat{Y}_i}(t)-\overline{Y}\right)^2 dt = 0.$$

So, for all $i \in \{1,\dots,n\}$, we have $\Psi_{\widehat{Y}_i}(t) - \overline{Y} = 0 \Leftrightarrow \Psi_{\widehat{Y}_i}(t) = \overline{Y}$. In this case, the predicted function for all observations i is a constant function.

- Suppose that $\Omega = 1$.

In this case,

$$\sum_{i=1}^{n}D_M^2\left(\Psi_{\widehat{Y}_i}(t),\overline{Y}\right) = \sum_{i=1}^{n}D_M^2\left(\Psi_{Y_i}(t),\overline{Y}\right).$$

From the decomposition obtained in Proposition 4 we have,

$$\sum_{i=1}^{n}D_M^2\left(\Psi_{Y_i}(t),\overline{Y}\right) = \sum_{i=1}^{n}D_M^2\left(\Psi_{\widehat{Y}_i}(t),\overline{Y}\right) + \sum_{i=1}^{n}D_M^2\left(\Psi_{\widehat{Y}_i}(t),\Psi_{Y_i}(t)\right)$$

and so,

$$\sum_{i=1}^{n}D_M^2\left(\Psi_{\widehat{Y}_i}(t),\Psi_{Y_i}(t)\right) = 0.$$

For all $i \in \{1,\dots,n\}$,

$$D_M^2\left(\Psi_{\widehat{Y}_i}(t),\Psi_{Y_i}(t)\right) = 0$$

$$\Leftrightarrow \int_0^1\left(\Psi_{\widehat{Y}_i}(t)-\Psi_{Y_i}(t)\right)^2 dt = 0 \Rightarrow \Psi_{\widehat{Y}_i}(t) = \Psi_{Y_i}(t).$$

In this case, for each observation i, the predicted and observed quantile functions are coincident.

In conclusion $0 \leq \Omega \leq 1$. If $\Omega = 0$ there is no linear relation between the explicative and response histogram-valued variables. If $\Omega = 1$, the linear relation is perfect, so the relation between the histogram-valued variable Y and histogram-valued variables X_j, with $j \in \{1,\dots,p\}$, is exactly the relation defined by the linear regression model.

□

Moreover, the dissimilarity between the observed and predicted distributions may be evaluated using the Root-Mean-Square Error ($RMSE_M$). This measure, that is also based on the Mallows distance, has been proposed by Irpino and Verde [7] and is defined as follows:

$$RMSE_M = \sqrt{\dfrac{\sum\limits_{i=1}^{n} \int_0^1 \left(\Psi_{\hat{Y}_i}(t) - \Psi_{Y_i}(t)\right)^2 dt}{n}}$$

Example 2 Consider again the Situation 1 in Example 1. For this case the value Ω is close to 1, $\Omega = 0.9631$, which allows concluding that the *DSD* relation between the histogram-valued variables is strong. In accordance, we observed a good alignment of the distributions in Figure 13.1(a).

As a consequence, and in accordance with the lower value of the $RMSE_M = 0.8946$, the observed and predicted distributions of the histogram-valued variable Y are similar. To illustrate this good prediction, Figure 13.3 represents the observed and predicted distributions for unit 1, represented by histograms and quantile functions, as the respective error function.

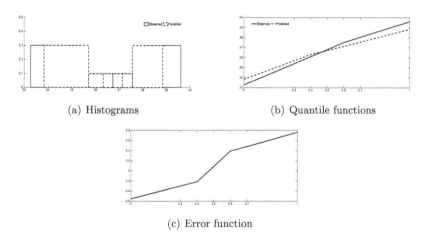

           (a) Histograms   (b) Quantile functions

(c) Error function

FIGURE 13.3
Observed and predicted distribution for observation 1 and respective error function.

13.4 An extension of the model

In this section we will present a generalization of the *DSD* model with the goal of obtaining a more flexible model. In this new approach, the linear regression model considers a quantile function instead of a real number as independent parameter. In the *DSD* model proposed in Definition 1 the independent parameter is a real number, and as such it only influences the fit of the centers of the predicted subintervals of the histogram. Moreover, this influence will be equal in all subintervals. Considering a quantile function as an independent parameter allows predicting quantile functions where the center and half range of the subintervals of each histogram may be influenced in different ways.

Definition 4 Consider the p explicative histogram-valued variables $X_1, X_2,$ $\ldots,$ X_p and the response variable Y, in the same conditions considered in Definition 1. Each quantile function Ψ_{Y_i}, may be expressed as:

$$\Psi_{Y_i}(t) = \Psi_{\widehat{Y}_i}(t) + e_i(t).$$

where $\Psi_{\widehat{Y}_i}(t)$ is the predicted quantile function for observation i, obtained from

$$\Psi_{\widehat{Y}_i}(t) = \Psi_{Constant}(t) + \sum_{j=1}^{p} a_j \Psi_{X_{ij}}(t) - \sum_{j=1}^{p} b_j \Psi_{X_{ij}}(1-t).$$

with $t \in [0,1]$ and $a_j, b_j \geq 0$, $j \in \{1, 2, \ldots, p\}$. The independent parameter $\Psi_{Constant}(t)$ is now a quantile function, that we assume to be continuous, defined by

$$\Psi_{Constant}(t) = \Psi_c(t) + \Psi_r(t) \tag{13.17}$$

where

$$\Psi_c(t) = \begin{cases} c_v & \text{if} \quad 0 \leq t < w_1 \\ c_v + r_{v_1} + r_{v_2} & \text{if} \quad w_1 \leq t < w_2 \\ c_v + r_{v_1} + 2r_{v_2} + r_{v_3} & \text{if} \quad w_2 \leq t < w_3 \\ \vdots & \\ c_v + r_{v_1} + \sum_{\ell=2}^{m-1} 2r_{v_\ell} + r_{v_m} & \text{if} \quad w_{m-1} \leq t \leq 1 \end{cases} \tag{13.18}$$

and for each subinterval ℓ,

$$\Psi_r(t) = \left(\frac{2(t - w_{\ell-1})}{w_\ell - w_{\ell-1}} - 1 \right) r_{v_\ell}, \quad w_{\ell-1} \leq t < w_\ell,$$

with $r_{v_\ell} \geq 0$, $\ell \in \{1, 2, \ldots, m\}$ and $c_v \in \mathbb{R}$.

The error, for each observation i, is the piecewise linear function given by

$$e_i(t) = \Psi_{Y_i}(t) - \Psi_{\widehat{Y}_i}(t), \qquad t \in [0,1]$$

According to the model in Definition 4, the quantile function that represents the distribution of the predicted histogram-valued variable Y, for a given observation i is:

$$\Psi_{\widehat{Y}_i}(t) = \Psi_{\widehat{Y}_{c_i}}(t) + \Psi_{\widehat{Y}_{r_i}}(t). \tag{13.19}$$

In the piecewise functions, each piece with $w_{\ell-1} \leq t \leq w_\ell$, is defined by:

$$\Psi_{\widehat{Y}_{c_i}}(t) = \sum_{j=1}^{p} \left(a_j c_{X_{ij\ell}} - b_j c_{X_j(i)_{m-\ell+1}} \right) + \Psi_c(t)$$

and

$$\Psi_{\widehat{Y}_{r_i}}(t) = \left(\frac{2(t - w_{\ell-1})}{w_\ell - w_{\ell-1}} - 1 \right) \left(\sum_{j=1}^{p} \left(a_j r_{X_{ij\ell}} + b_k r_{X_{ij(m-\ell+1)}} \right) + r_{v_\ell} \right)$$

The generalization of the *DSD* model induces also a linear relation between the centers and between the half ranges of each subinterval of the histogram. From these relations we may define classical linear regressions between the weighted mean of the centers and between the weighted mean of the half ranges of the variables X_j and Y. For each i we have:

$$\overline{\widehat{Y}}_i = \overline{c}_v + \sum_{j=1}^{p} (a_j - b_j) \overline{X}_{ij} \tag{13.20}$$

and

$$\overline{\widehat{Y}}_i^r = \overline{r}_v + \sum_{j=1}^{p} (a_j + b_j) \overline{X}_{ij}^r \tag{13.21}$$

with $\overline{X}_{ij}, \overline{Y}_i, \overline{Y}_i^r, \overline{X}_i^r$ defined as in expressions (13.8) and (13.9), $\overline{c}_v = \sum_{\ell=1}^{m} p_\ell c_{v_\ell}$ (symbolic mean of the distribution represented by $\Psi_{Constant}(t)$ with c_{v_ℓ} defined according to the expression (13.17)) and $\overline{r}_v = \sum_{\ell=1}^{m} p_\ell r_{v_\ell}$.

These relations are similar to the ones defined in expressions (13.8) and (13.9) of the first approach. Consequently, the relation between the variables for this case is also in accordance with the relation between the means of the centers of the variables.

To predict the parameters of this extension of the *DSD* model, we define an optimization problem similar to the one defined in the first approach. The main difference is that now it is necessary to impose more constraints because we must ensure that the independent parameter is a quantile function. It is necessary to impose the non-negativity of all half ranges that define the independent parameter. In this case and under the same conditions considered in Definition 2, the quadratic optimization problem is defined as follows:

$$Minimize \quad SSE(a_j, b_j, v) = \sum_{i=1}^{n} D_M^2(\Psi_{Y_i}(t), \Psi_{\widehat{Y}_i}(t))$$

subject to

$$a_j, b_j \geq 0, \quad j \in \{1, \ldots, p\} \tag{13.22}$$

$$c_v \in \mathbb{R}$$

$$r_{v_\ell} \geq 0, \quad \ell \in \{1, \ldots, m\}.$$

Results similar to the ones proved before may also be obtained when the distributions are predicted with Definition 4. Also in this case, it is possible to deduce the goodness of fit measure Ω, that also verifies the properties proved in Section 3.

13.5 Illustrative example

Consider an adaptation to histogram data of the Cardiological data set of Billard and Diday [1]. In symbolic data Tables 13.1 and 13.2, we have the distributions of the pulse rate, systolic blood pressure and diastolic blood pressure for each of eleven patients. The aim in this example is to predict the distribution of values of the pulse rate from the distributions of their systolic and diastolic blood pressures.

TABLE 13.1
Cardiological histogram data set (part I)

Patients	Pulse Rate
1	$\{[44; 60), 0.8; [60; 68], 0.2\}$
2	$\{[60; 70), 0.5; [70; 72], 0.5\}$
3	$\{[56; 80), 0.6; [80; 90], 0.4\}$
4	$\{[70; 75), 0.4; [75; 112], 0.6\}$
5	$\{[54; 56), 0.2; [56; 72], 0.8\}$
6	$\{[70; 80), 0.5; [80; 100], 0.5\}$
7	$\{[63; 73), 0.4; [73; 75], 0.6\}$
8	$\{[72; 79), 0.5; [79; 100], 0.5\}$
9	$\{[76; 80), 0.2; [80; 98], 0.8\}$
10	$\{[86; 94), 0.8; [94; 96], 0.2\}$
11	$\{[86; 89), 0.6; [89; 100], 0.4\}$

According to the conditions of the model, all observations of the histogram-valued variables were rewritten as histograms with six subintervals where the

TABLE 13.2

Cardiological histogram data set (part II)

Patients	Systolic blood pressure	Diastolic blood pressure
1	$\{[90;95),0.2;[95;100],0.8\}$	$\{[50;60),0.4;[60;70],0.6\}$
2	$\{[90;110),0.4;[110;130],0.6\}$	$\{[70;80),0.2;[80;90],0.8\}$
3	$\{[140;160),0.5;[160;180],0.5\}$	$\{[90;92),0.5;[92;100],0.5\}$
4	$\{[110;120),0.2;[120;142],0.8\}$	$\{[80;85),0.6;[85;108],0.4\}$
5	$\{[90;98),0.6;[98;100],0.4\}$	$\{[50;63),0.4;[63;70],0.6\}$
6	$\{[130;150),0.4;[150;160],0.6\}$	$\{[80;90),0.5;[90;100],0.5\}$
7	$\{[140;145),0.2;[145;150],0.8\}$	$\{[60;80),0.2;[80;100],0.8\}$
8	$\{[130;140),0.4;[140;160],0.8\}$	$\{[76;85),0.5;[85;90],0.5\}$
9	$\{[110;160),0.5;[160;190],0.5\}$	$\{[70;100),0.4;[100;110],0.6\}$
10	$\{[138;142),0.5;[142;180],0.5\}$	$\{[90;100),0.4;[110;110],0.6\}$
11	$\{[110;135),0.2;[135;150],0.8\}$	$\{[78;88),0.2;[88;100],0.8\}$

weight set is $\{0.2;0.2;0.1;0.1;0.2;0.2\}$. Table 13.3 presents the parameters and the goodness-of-fit measures for the two approaches presented in this chapter. We name *DSD I* the *Distribution and Symmetric Distribution* model where the independent parameter is a real number v and *DSD II* the extended model where the independent parameter is a constant quantile function $\Psi_{Constant}(t)$.

TABLE 13.3

Parameters and measures of the *DSD* models

Method	Parameters					Goodness-of-fit	
	$v/\Psi(t)$	a_1	b_1	a_2	b_2	Ω	$RMSE_M$
DSD I	10.0366	0.0206	0.0551	0.8397	0	0.7452	6.8484
DSD II	$\Psi_{Constant}(t)$	0	0	0.7710	0	0.7533	6.7394

The independent parameter in *DSD II* is the quantile function

$$
\Psi_{Constant}(t) = \begin{cases}
9.75 & \text{if } 0 \le t < 0.4 \\
10.00 + \left(\frac{2(t-0.4)}{0.1} - 1\right) \times 0.25 & \text{if } 0.4 \le t < 0.5 \\
10.87 + \left(\frac{2(t-0.5)}{0.1} - 1\right) \times 0.62 & \text{if } 0.5 \le t < 0.6 \\
12.33 + \left(\frac{2(t-0.6)}{0.2} - 1\right) \times 0.83 & \text{if } 0.6 \le t < 0.8 \\
14.19 + \left(\frac{2(t-0.8)}{0.2} - 1\right) \times 1.03 & \text{if } 0.8 \le t \le 1
\end{cases}
$$

Observing the parameters of the models, from expression (13.8) we may conclude that, in both models, for the set of patients to which the data refers, the symbolic mean of the pulse rate increases around 0.8 for each unit of increase of the symbolic mean of diastolic blood pressure. As this value is positive we may consider that the relation between these histogram-valued

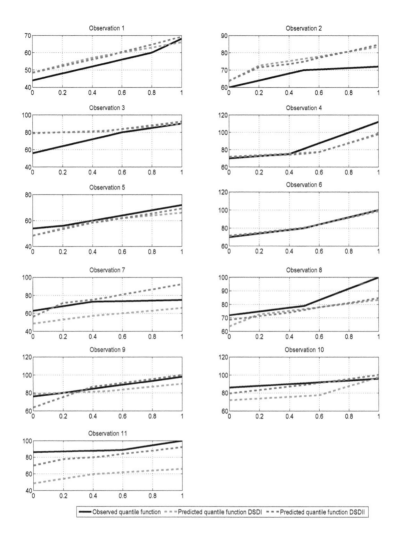

FIGURE 13.4

Observed and predicted quantile functions with the *DSD* models.

variables is direct. In the case of the symbolic mean of the systolic blood pressure we may observe that this variable has an approximately null or a null influence in the models.

The quantile functions of the observed and predicted distributions of the histogram-valued variable pulse rate, for the two approaches are represented in Figure 13.4. As it may be observed, the predicted distributions are similar in accordance with the similar values of the measures Ω and $RMSE_M$ in both cases. In the generalized model the goodness-of-fit measures are only slightly better than the ones obtained for the model with a real independent parameter. Considering a quantile function as an independent parameter we expect a more flexible model, however the prediction of this function imposes more constrains in the optimization problem. This is why *DSD II* may not provide better predictions.

As the values of Ω are not very close to one, Figure 13.4 shows that the predicted and observed distributions of the values of the pulse rate are not always close.

13.6 Conclusion

In this chapter we proposed two approaches to obtain a linear regression model for histogram-valued variables (whose observations are represented by quantile functions) which allow predicting distributions immediately and directly from the histogram-valued explicative variables. In DSD regression model I the independent parameter is a real number and because of this it only influences the fit of the centers of the predicted subintervals of the histograms. In DSD regression model II a quantile function is considered as the independent parameter, and hence it will allow predicting quantile functions where the center and half range of the subintervals of each predicted histogram may be affected in different ways. The parameters of the models are obtained by solving a quadratic optimization problem subject to non-negativity constrains on the unknowns and using the Mallows distance. This distance is also the used to deduce a goodness-of-fit measure for the model. Furthermore, the proposed method may be considered a theoretical generalization of the descriptive classical linear regression model, since when applied to degenerate variables (histograms with only one interval with the same bounds and weight equal to one) we obtain the classical linear regression model. The proposed approach is very flexible as it may also be applied to the particular case of the interval-valued variables, i.e. histograms with only one interval (where different distributions may be assumed) and weight equal to one. Other linear regression models for numerical distributional data have also been proposed. In Chapter 14 an alternative linear regression method, proposed by Irpino and Verde [7] will be presented. This method is based on a particular decomposition of the Mallows distance, also named L_2 Wasserstein distance.

Bibliography

[1] L. Billard and E. Diday. Regression analysis for interval-valued data. In *Data Analysis, Classification and Related Methods. Proceedings of the 7th Conference of the International Federation of Classification Societies*, pages 369–374. Springer Berlin Heidelberg, 2000.

[2] L. Billard and E. Diday. Symbolic regression analysis. In *Classification, Clustering, and Data Analysis. Proceedings of the 8th Conference of the International Federation of Classification Societies*, pages 281–288. Springer Berlin Heidelberg, 2002.

[3] L. Billard and E. Diday. *Symbolic Data Analysis: Conceptual Statistics and Data Mining*. John Wiley & Sons, Inc. New York, NY, USA, 2006.

[4] S. Dias. *Linear Regression with Empirical Distributions*. PhD thesis, University of Porto, Porto, Portugal, 2014.

[5] S. Dias and P. Brito. Linear regression model with histogram-valued variables. *Statistical Analysis and Data Mining*, 8(2):75–113, 2015.

[6] A. Irpino and R. Verde. A new Wasserstein based distance for the hierarchical clustering of histogram symbolic data. In *Data Science and Classification. Proceedings of the 10th Conference of the International Federation of Classification Societies*, pages 185–192. Springer Berlin Heidelberg, 2006.

[7] A. Irpino and R. Verde. Linear regression for numeric symbolic variables: a least squares approach based on Wasserstein Distance. *Advances in Data Analysis and Classification*, 9(1):81–106, 2015.

[8] C.L. Mallows. A note on asymptotic joint normality. *The Annual of Mathematical Statistics*, pages 508–515, 1972.

[9] W. Winston. *Operations Research. Applications and Algorithms*. Wadsworth, 1994.

14

Regression Analysis of Distributional Data Based on a Two-Component Model

Antonio Irpino

Department of Mathematics and Physics, University of Campania "L. Vanvitelli", Caserta, Italy

Rosanna Verde

Department of Mathematics and Physics, University of Campania "L. Vanvitelli", Caserta, Italy

CONTENTS

The ordinary least squares (OLS) model is the most known method for modeling a causal relationship between a dependent variable and a set of independent ones. This chapter presents a (population) OLS model for data described by numeric distributional variables. Indeed, well-grounded and accepted inference on distributional variables is not yet available. The model is based on a particular decomposition of the 2-norm Wasserstein distance allowing the definition of a model of two components. The two components are related to the internal and between data variability inherent to distributional-valued data. An application on a climatic dataset shows the usefulness of the approach. An analysis of residuals supports the application and suggests some implications for further developments.

DOI: 10.1201/9781315370545-14

14.1 Introduction

A main task of statistical analysis is the discovery and modeling causal relationships among variables. The multiple linear regression model is the most known method for analyzing the relationship between a dependent variable and a set of independent ones. In this chapter, a linear regression model for distributional variables is presented. In the framework of Symbolic Data Analysis (SDA) [3, 4, 6], where observations are expressed by set-valued data, several extensions of regression models have been proposed for interval-valued data, readers may refer to [10, 15–17]. For histogram-valued variables, a particular case of distributional ones, some contributions are given in [3] where the parameters of a classic linear model are estimated by using basic statistics and covariance matrices defined in [1, 3].

A second group of linear regression models [5, 13, 20] have been proposed by using the L_2 Wasserstein distance [19] (also known as Mallows distance) between distributions for computing the squared errors for the model parameters estimation.

In this chapter, we present the regression model proposed in [13, 20], based on a particular decomposition of the L_2 Wasserstein distance [11]. A linear model is estimated by using the Ordinary Least Squares method, where the squared residuals between the predicted and the observed distributional values are expressed by using the L_2 Wasserstein distance.

The chapter is organized as follows. Section 14.2 shows how to define the Ordinary Least Square problem to estimate the parameters of a linear population regression model. The error between the expected and the observed distribution-valued description of the dependent variable is obtained using the L_2 Wasserstein distance. Using a decomposition of the sum of square errors related to the L_2 Wasserstein distance, we introduce a two-component regression method. We discuss the interpretation of the obtained parameters, and we provide some indexes for assessing the goodness of fit of the estimated model. Section 14.3 shows an application on a distribution-valued dataset measuring the ozone concentration distributions observed in some US stations and how the ozone concentration is related to other climatic distribution-valued independent variables. Section 14.4 ends the chapter with some concluding remarks.

14.2 OLS linear regression for distributional data

Linear regression of distributional variables is a method for modeling a causal relationship between a (distributional) dependent variable Y and $p \geq 1$ (distributional) explanatory variables denoted $X_1, \ldots, X_j, \ldots, X_p$. The realization of each variable for the generic unit s_i $(i = 1, \ldots, n)$ is an empirical distribution with probability density function (pdf) denoted as $f_{ij}(x)$, for the explanatory variables, or $f_i(y)$ for the dependent one. Let ϕ be a linear (in the parameters) function of the dependent variables depending on a set Θ of scalar parameters, and ε an error term, the linear regression model can be written as follows:

$$Y = \phi(X_1, \ldots, X_p | \Theta) + \varepsilon. \tag{14.1}$$

We assume a generalization of the hypotheses of the classical regression model to this distributional data model: the X_j (for $j = 1, \ldots, p$) are deterministic; no-strong correlation between pairs of X_j is assumed (where the correlation between distributional variables is defined according to [12]); the error functions $\varepsilon_i : [0, 1] \to \Re$, are realizations of the same random process $\mathcal{E}(t)$, $t \in [0, 1]$, such that $E(\mathcal{E}) = 0$ (zero mean process), $VAR(\mathcal{E}) = c$ (constant variance), and such that $E(\mathcal{E}_i\mathcal{E}_h) = 0$, for $i \neq h$.

By analogy with the classical regression model, Ordinary Least Squares (OLS) approach is used for the parameters estimation of the model. OLS is based on the minimization of squared errors between the observed and expected distribution-values. We measure the squared errors through a distance between the observed and expected response. Several distances exist for distributions (see [8] for a review of the main probabilistic distances). However, not all the distances allow for a quadratic form of the OLS problem, or for the solution of the problem in a closed form. In this chapter, we present an approach that uses the L_2 Wasserstein distance between pdfs, and that is consistent with the decomposition of the distance proposed in [11, 12].

Being f and g two pdfs, and denoting by F and G the corresponding *cumulative distribution functions* (cdf), and by Ψ_f and Ψ_g the corresponding *quantile functions* (namely, the inverse of the cdfs), we recall that the L_2 Wasserstein distance between f and g is the following:

$$d_W(f, g) = \sqrt{\int_0^1 [\Psi_f(t) - \Psi_g(t)]^2 \, dt}. \tag{14.2}$$

It corresponds to the Euclidean distance between the *quantile functions* of two distributions[1]. Using this metric in the space of quantile functions, a set of operators and statistics have been proposed for distributional data [12]. Before introducing the method, we recall some basic operators and statistics for distributional data based on quantile functions. If we consider a random variable X_f having pdf f, we can compute the following statistics as:

- $\mu_f = E(X_f) = \int_{-\infty}^{+\infty} x\,f(x)dx = \int_0^1 \Psi_f(t)dt$

---

[1]About the advantages of using quantile functions instead of pdfs or cdfs, we suggest to refer to [9].

- $\sigma_f = SD(X_f) = \sqrt{\int\limits_{-\infty}^{+\infty} x^2 f(x)dx - \mu_f^2} = \sqrt{\int\limits_{0}^{1} [\Psi_f(t)]^2 \, dt - \mu_f^2}$

- Given two *pdf*s f and g and their quantile functions, the dot product that induces a L_2 Wasserstein metric between *pdf*s is defined as follows:

$$\int\limits_{0}^{1} \Psi_f(t)\Psi_g(t)dt = \sigma_f \sigma_g r_{\Psi_f \Psi_g} + \mu_f \mu_g; \tag{14.3}$$

where $r_{\Psi_f \Psi_g}$ is the correlation measure between the quantile functions Ψ_f and Ψ_g associated with the corresponding *pdf*s f and g. That is:

$$r_{\Psi_f \Psi_g} = \frac{\int_0^1 \Psi_f(t) \cdot \Psi_g(t)dt - \mu_f \cdot \mu_g}{\sigma_f \cdot \sigma_g}. \tag{14.4}$$

So far, we identified a metric for measuring the difference between two distributions. Thus, OLS minimization problem will be formulated as follows:

$$\hat{\Theta} = \arg\min_{\Theta} \|Y - \phi(X_1, \ldots, X_p|\Theta)\| =$$

$$= \arg\min_{\Theta} \sum_{i=1}^{n} [d_W (f_{Y_i}, \phi(X_{i1}, \ldots, X_{ip}|\Theta))]^2, \tag{14.5}$$

where X_{ij} denotes the $f_{X_{ij}}$ density function. We remark that quantile functions are into a one-one correspondence with the respective *pdf*s, and that the distance computation depends on the quantile functions only. Thus, the OLS problem in Equation (14.5) can be written as follows:

$$\hat{\Theta} = \arg\min_{\Theta} \sum_{i=1}^{n} \int\limits_{0}^{1} [\Psi_{Y_i} - \phi(\Psi_{X_{i1}}, \ldots \Psi_{X_{ip}}|\Theta)]^2 \, dt. \tag{14.6}$$

Considering that quantile functions are non-decreasing functions in $[0, 1]$, some restrictions are required. If ϕ is linear, only conic combinations of quantile functions (namely, linear combinations with positive coefficients) guarantee that the result is a quantile function as well [9]. Thus, such a non-negativity constraint must be taken into consideration. If we use a simple formulation of the model as follows:

$$\Psi_{Y_i} = \theta_0 + \sum_{j=1}^{p} \theta_j \Psi_{X_{ij}} + \varepsilon_i; \tag{14.7}$$

the model in Equation (14.7) is solved via a constrained version of the OLS problem in Equation (14.6) as follows:

$$\hat{\Theta} = \arg\min_{\Theta} \sum_{i=1}^{n} \int\limits_{0}^{1} \left[\Psi_{Y_i} - \theta_0 - \sum_{j=1}^{p} \theta_j \Psi_{X_{ij}}\right]^2 \, dt \tag{14.8}$$

$s.a.\ \theta_j \geq 0\ j = 1, \ldots, p.$

We remark that the θ_0 parameter has no restriction because it only shifts the response quantile function with respect to its mean. The simple model in Equation (14.7) can be improved, from a predictive point of view, in several ways. For example, in [5] a solution is proposed by introducing new distributional variables that are connected with the observed ones.

In our model, we use a particular decomposition of the L_2 Wasserstein distance.

14.2.1 A decomposition of the L_2 Wasserstein distance

Given two *pdf*s f and g, associated with the respective μ_f and μ_g means, and the quantile functions Ψ_f and Ψ_g, we denote by Ψ_f^c and Ψ_g^c the respective centered quantile functions, namely, the quantile functions minus the respective means. According to [2, lemma 8.8], the (squared) L_2 Wasserstein distance can be rewritten as

$$d_W^2(f,g) = (\mu_f - \mu_g)^2 + \int\limits_0^1 \left[\Psi_f^c(t) - \Psi_g^c(t) \right]^2 dt. \qquad (14.9)$$

This property allows the decomposition of the squared distance as the sum of two components, the first related to the location of the distributions and the second related to their variability structure. In [11] an improvement of the decomposition shows that d_W^2 can be decomposed into three quantities (related, respectively, to the parameters of position, scale and shape), as follows:

$$d_W^2(f,g) = (\mu_f - \mu_g)^2 + (\sigma_f - \sigma_g)^2 + 2\sigma_f\sigma_g \left(1 - r_{\Psi_f\Psi_g} \right), \qquad (14.10)$$

where $r_{\Psi_f\Psi_g}$ is the above defined correlation measure, that can be considered as a measure of comparison of the shapes of two distributions. Let us consider two vectors of quantile functions $\mathbf{x} = [\Psi_{X_i}]_{n\times 1}$ and $\mathbf{y} = [\Psi_{Y_i}]_{n\times 1}$. Using the inner product inducing the L_2 Wasserstein distance, it follows that the dot product of two vectors of distributions is computed as follows:

$$\mathbf{x}^T\mathbf{y} = \sum_{i=1}^n \langle \Psi_{Y_i}, \Psi_{X_i} \rangle = \sum_{i=1}^n \left[r_{\Psi_{X_i}\Psi_{Y_i}} \cdot \sigma_{X_i} \cdot \sigma_{Y_i} + \mu_{X_i} \cdot \mu_{Y_i} \right]; \qquad (14.11)$$

where ^{T} denotes the vector-matrix transposition operator. Equation (14.11) permits to extend the classical product between matrices of quantile functions.

14.2.2 The two-component linear regression model

Let $\Psi_{X_{ij}}^c = \Psi_{X_{ij}} - \mu_{X_{ij}}$ be the centered (w.r.t. the respective mean) quantile function associated with $\Psi_{X_{ij}}$. The same is valid for $\Psi_{Y_{ij}}$. Let us consider the following vectors and matrices:

- $\mathbf{Y} = [\Psi_{Y_i}]_{n \times 1}$ is the response vector with elements the quantile functions. $\mathbf{Y} = \bar{\mathbf{Y}} + \mathbf{Y}^c$ where[2] $\bar{\mathbf{Y}} = [\mu_{Y_i}]_{n \times 1}$ and $\mathbf{Y}^c = \left[\Psi^c_{Y_{ij}}\right]_{n \times 1}$ are, respectively, the vector (of scalars) of the means and the vector of the centered quantile functions of the dependent variable.

- $\mathbf{X} = [\Psi_{X_{ij}}]_{n \times p}$ is the matrix of the explanatory variables of quantile functions. $\mathbf{X} = \bar{\mathbf{X}} + \mathbf{X}^c$ where $\bar{\mathbf{X}} = [\mu_{X_{ij}}]_{n \times p}$ and $\mathbf{X}^c = \left[\Psi^c_{X_{ij}}\right]_{n \times p}$ are, respectively, the matrix (of scalars) of the means and the matrix of the centered quantile functions of the explanatory variables.

- $\bar{\mathbf{X}}_+ = [\mathbf{1}|\bar{\mathbf{X}}]$ is the juxtaposition of the a column vector $\mathbf{1} = [1]_{n \times 1}$ and the matrix $\bar{\mathbf{X}}$.

- $\boldsymbol{\Theta} = [\mathbf{B}|\boldsymbol{\Gamma}]$ is a row vector of scalars such that $\mathbf{B} = [\beta_0, \beta_1, \ldots, \beta_p]$ and $\boldsymbol{\Gamma} = [\gamma_1, \ldots, \gamma_p]$.

- $\mathbf{e} = [e_i]_{n \times 1}$ is the vector of the error functions e_i.

The two-component model assumes that each response quantile function Ψ_{Y_i} can be expressed as a linear combination of the means $\mu_{X_{ij}}$ and of the centered quantile functions $\Psi^c_{X_{ij}}$ and error term e_i[3] as follows:

$$\Psi_{Y_i} = \beta_0 + \sum_{j=1}^{p} \beta_j \mu_{X_{ij}} + \sum_{j=1}^{p} \gamma_j \Psi^c_{X_{ij}} + e_i. \tag{14.12}$$

Using the matrix notation, the model in Equation (14.12) is expressed as follows:

$$\mathbf{Y} = \bar{\mathbf{X}}_+ \mathbf{B} + \mathbf{X}^c \boldsymbol{\Gamma} + \mathbf{e}. \tag{14.13}$$

To provide an OLS parameter estimation, we define the *SSE* (Sum of Squared Errors) function using the L_2 Wasserstein measure as follows:

$$SSE(\beta_0, \beta_1, \ldots, \beta_j, \gamma_1, \ldots, \gamma_j) = \sum_{i=1}^{n} \int_0^1 e_i^2(t)dt =$$

$$= \sum_{i=1}^{n} \int_0^1 \left[\Psi_{Y_i}(t) - \left(\beta_0 + \sum_{j=1}^{p} \beta_j \mu_{X_{ij}} + \sum_{j=1}^{p} \gamma_j \Psi^c_{X_{ij}}(t)\right)\right]^2 dt. \tag{14.14}$$

Using a matrix notation, Equation (14.14) is expressed as:

$$SSE(\mathbf{B}, \boldsymbol{\Gamma}) = \mathbf{e}^T \mathbf{e} = \left[\mathbf{Y} - \bar{\mathbf{X}}_+ \mathbf{B} - \mathbf{X}^c \boldsymbol{\Gamma}\right]^T \left[\mathbf{Y} - \bar{\mathbf{X}}_+ \mathbf{B} - \mathbf{X}^c \boldsymbol{\Gamma}\right]. \tag{14.15}$$

---

[2]Note that the sum between a quantile function Ψ and a scalar value α is equal to $\alpha + \Psi(t)$ $\forall t \in [0, 1]$, while the sum of two quantile functions Ψ_x and Ψ_y is $\Psi_x(t) + \Psi_y(t)$ $\forall t \in [0, 1]$.

[3]We remark that the function $e_i : [0, 1] \to \Re$ is not a quantile function.

Recalling that $\mathbf{Y} = \bar{\mathbf{Y}} + \mathbf{Y}^c$, and using matrix algebra, Equation (14.15) can be decomposed in two additive components as follows[4]:

$$SSE(\mathbf{B}, \mathbf{\Gamma}) = SSE(\mathbf{B}) + SSE(\mathbf{\Gamma}) = \bar{\mathbf{e}}^T \bar{\mathbf{e}} + (\mathbf{e}^c)^T \mathbf{e}^c, \tag{14.16}$$

where

$$\bar{\mathbf{e}} = \bar{\mathbf{Y}} - \bar{\mathbf{X}}_+ \mathbf{B}, \tag{14.17}$$
$$\mathbf{e}^c = \mathbf{Y}^c - \mathbf{X}^c \mathbf{\Gamma}. \tag{14.18}$$

Equation (14.16) shows that the problem is the sum of two independent subproblems to minimize. The former sub-problem is related to the means of the predictor quantile functions $\mu_{X_{ij}}$'s , while the latter one is related to the variability of the centered quantile functions $\Psi^c_{X_{ij}}$'s. Thus, model in Equation (14.13) is a *two-component* model with the following independent models to estimate:

$$\bar{\mathbf{Y}} = \bar{\mathbf{X}}_+ \mathbf{B} + \bar{\mathbf{e}}, \tag{14.19}$$
$$\mathbf{Y}^c = \mathbf{X}^c \mathbf{\Gamma} + \mathbf{e}^c. \tag{14.20}$$

The first model, in Equation (14.19), involves matrices and vectors of scalars only, and is solved as a classical OLS problem for the estimation of $\mathbf{B}$:

$$\underset{\mathbf{B}}{argmin}\, SSE(\mathbf{B}) = \bar{\mathbf{e}}^T \bar{\mathbf{e}} = \left[\bar{\mathbf{Y}} - \bar{\mathbf{X}}_+ \mathbf{B}\right]^T \left[\bar{\mathbf{Y}} - \bar{\mathbf{X}}_+ \mathbf{B}\right]. \tag{14.21}$$

The OLS estimates for the β parameters are classically obtained as:

$$\hat{\mathbf{B}} = \left(\bar{\mathbf{X}}_+^T \bar{\mathbf{X}}\right)^{-1} \bar{\mathbf{X}}_+^T \bar{\mathbf{Y}}. \tag{14.22}$$

The second model, in Equation (14.20), involves quantile functions. In this case, it is necessary to impose a non-negativity constraint on the parameters. Therefore, the parameters are estimated by using the non–negative least squares (NNLS) algorithm of [14], and the product between vectors or matrices of quantile functions, as in Equation (14.11). The NNLS problem can be expressed as follows:

$$\underset{\mathbf{\Gamma}}{argmin}\, SSE(\mathbf{\Gamma}) = (\mathbf{e}^c)^T \mathbf{e}^c = [\mathbf{Y}^c - \mathbf{X}^c \mathbf{\Gamma}]^T [\mathbf{Y}^c - \mathbf{X}^c \mathbf{\Gamma}] \tag{14.23}$$
$$s.a. \qquad \gamma_j \geq 0 \quad j = 1, \dots, p.$$

If all the γ parameters are positive, the estimate of the $\hat{\mathbf{\Gamma}}$ vector is:

$$\hat{\mathbf{\Gamma}} = \left(\mathbf{X}^{cT} \mathbf{X}^c\right)^{-1} \mathbf{X}^{cT} \mathbf{Y}^c, \tag{14.24}$$

else the NNLS algorithm is used. Algorithm 1 shows the NNLS algorithm proposed in [14, Algorithm (23.10)], where products of matrices containing

---

[4]Details of this decomposition may be found in [13].

quantile functions are computed according to Equation (14.11). The algorithm returns a set of $\hat{\gamma}_j \geq 0$ consistently with the Kuhn-Tucker conditions of the NNLS problem expressed in [14, Theorem (23.4)]. Using the *two-component* model, the i-th predicted quantile function, denoted by $\Psi_{\hat{Y}_i}$, is obtained by the sum of the mean value, predicted by the model in Equation (14.19), and the centered quantile function predicted by the model in Equation (14.20), as follows:

$$\Psi_{\hat{Y}_i} = \hat{\mu}_{Y_i} + \Psi_{\hat{Y}_i}^c = \hat{\beta}_0 + \sum_{j=1}^{p} \hat{\beta}_j \mu_{X_{ij}} + \sum_{j=1}^{p} \hat{\gamma}_j \Psi_{X_{ij}}^c. \tag{14.25}$$

14.2.2.1 Interpretation of the parameters

The two-component model has the advantage of a straightforward interpretation of the estimated parameters because the two sets, the **B** and **Γ**, pertain to the two main aspects of distributional variables: the position and the internal variability structure. Basing on the decomposition of the L_2 Wasserstein distance, the two-component model can model a linear relationship between the means of the distributional data, like in the classical OLS regression. Indeed, the β_j parameters are scalars, and their interpretation is precisely the same as the classical regression model parameters. Also, their inferential aspects are the same as in classical regression.

The interpretation of the γ_j parameters, which must be non-negative because they are the coefficients of the quantile-centered functions, is more complex. The second component of the model expresses a relationship between two distributional variables whose realizations are zero-mean distributions (because quantile functions are centered with respect to their means). Thus, the variability of the two new distributional variables is related only to the different scales and shapes of the observed data. In order to suggest an interpretation, we should consider that these two aspects are strictly related. Indeed, as described in [12] and observing the decomposition in Equations (14.10) and (14.11), we consider that $r_{\Psi_f \Psi_g}$ is a correlation index that is always positive and assumes value 1 when the two quantile functions have identical shaped *pdf*s (namely, two Gaussians, two uniforms, etc.). Let us consider the following cases:

- $r_{\Psi_f \Psi_g} = 0$ for each pair of observed data. For example, that occurs when at least one variable is expressed by distributions that degenerate into points. In this case, Equation (14.11) depends on the sum of the products of the mean values of the distributions (namely $\sum_{i=1}^{n} \mu_{X_i} \cdot \mu_{Y_i}$) only, and thus the second component of the model cannot be estimated.

- $r_{\Psi_f \Psi_g} = 1$ for each couple of observed data. This happens when all the distributional data have the same shape. In this case, Equation (14.11) depends on the sum of the products of the mean values and that of the standard

---

Algorithm 1 Non–Negative Least Squares for Centered Quantile Function solving problem in Equation (14.23)

Require: $\mathbf{X}^c$ the $n \times p$ matrix of the centered quantile functions of the p predictors variables

Require: $\mathbf{Y}^c$ the n vector of the centered quantile functions of the response variable

Require: $\hat{\boldsymbol{\Gamma}}^T := 0$ the initial vector of p γ's coefficients

Require: $\mathcal{P} := \emptyset$ the set of non-active columns

Require: $\mathcal{Z} := \{1, \ldots, p\}$ the set of active columns

1: Compute the initial solution for $\mathbf{w} := \mathbf{X}^{cT}\mathbf{Y}^c - \left(\mathbf{X}^{cT}\mathbf{X}^c\right)\hat{\boldsymbol{\Gamma}}$ using Equation (14.11) for the product of quantile matrices

2: **if** set $\mathcal{Z}$ is empty **or** $w_j \geq 0 \ \forall j \in \mathcal{Z}$ **then**

3:     Go to Step 16

4: **end if**

5: Find an index $t \in \mathcal{Z}$, such that $w_t = max\{w_j : j \in \mathcal{Z}\}$

6: Move the index t from set $\mathcal{Z}$ to set $\mathcal{P}$

7: Let $\mathbf{C}$ denote the $n \times p$ matrix with generic $j - th$ column $\mathbf{C}_j$

$$\mathbf{C}_j = \begin{cases} \text{column } j \text{ of } \mathbf{X} & \text{if } j \in \mathcal{P} \\ \mathbf{0} & \text{if } j \in \mathcal{Z} \end{cases}$$

8: Compute the $\mathbf{z}$ vector as solution of the least squares problem $\mathbf{C}_j\mathbf{z} \cong \mathbf{Y}^c$ that is solved also by $(\mathbf{C}^{*T}\mathbf{C}^*)\mathbf{z}^* \cong \mathbf{C}^{*T}\mathbf{Y}^c$, i.e., $\mathbf{z}^* = (\mathbf{C}^{*T}\mathbf{C}^*)^{-1}\mathbf{C}^{*T}\mathbf{Y}^c$ where $\mathbf{C}^*$ is the matrix formed by the columns for which $j \in \mathcal{P}$ and $\mathbf{z}^*$ is the vector of z_j, $j \in \mathcal{P}$. Note that only the components z_j, $j \in \mathcal{P}$ are determined by this problem. Define $z_j := 0$ for $j \in \mathcal{Z}$.

9: **if** $z_j > 0$ for all $j \in \mathcal{P}$ **then**

10:     $\hat{\boldsymbol{\Gamma}} := \mathbf{z}$ go to step 1

11: **end if**

12: Find an index $q \in \mathcal{P}$ such that $\hat{\gamma}_q/(\hat{\gamma}_q - z_q) = min\{\hat{\gamma}_j/(\hat{\gamma}_j - z_j) : z_j \leq 0, j \in \mathcal{P}\}$

13: Set $\alpha := \hat{\gamma}_q/(\hat{\gamma}_q - z_q)$.

14: Set $\hat{\boldsymbol{\Gamma}} := \hat{\boldsymbol{\Gamma}} + \alpha(\mathbf{z} - \hat{\boldsymbol{\Gamma}})$.

15: Move from set $\mathcal{P}$ to set $\mathcal{Z}$ all indices $j \in \mathcal{P}$ for which $\hat{\gamma}_j = 0$. Go to Step: 7.

16: End of computation.

---

deviations only. The second component of the model is a constrained linear regression (passing through the origin because the constant term is not estimated in the model) between the standard deviations of the distributions observed for the response variable and the standard deviations of the distributions observed for the explanatory ones.

In general, the higher the correlation, the more the γ_j depend only on the standard deviations of the distributions. If the differences in shape are small enough, we may observe the γ_j parameters assuming the following values:

- $\gamma_j = 0$, (all the other variables fixed) there is no effect in the variation of the mean response variability (scale) when an increase of the X_j variability is observed.

- $0 < \gamma_j < 1$ (all the other variables fixed), the mean variability of the response variable increase is lower than the increase of the variability of the X_j explanatory variable.

- $\gamma_j = 1$ (all the other variables fixed), the mean variability of the response variable increases in the same proportion of an increase of the variability of the X_j explanatory variable.

- $\gamma_j > 1$ (all the other variables fixed), the mean variability of the response variable increases more than the variability of the X_j explanatory variable. In this case, we can consider the γ_j value as a *shrinking* (when it is lower than 1) or an *expanding* (when it is greater than 1) factor of the predicted distributional values with respect to the explanatory ones.

- $0 < r_{\Psi_f, \Psi_g} < 1$, both scale and shape influence the γ_j estimation. Even if the two effects are mixed together, we remark that r_{Ψ_f, Ψ_g} is always less than 1, thus, the dot product in Equation (14.11) is always lower than the dot product of equally shaped quantile functions. Let us consider the case of univariate regression. Using a few algebra, it is possible to show that γ_1 is as follows:

$$\hat{\gamma}_1 = \frac{\sum_{i=1}^{n} r_{\Psi_{Y_i} \Psi_{X_{i1}}} \cdot \sigma_{Y_i} \cdot \sigma_{X_{i1}}}{\sum_{i=1}^{n} \sigma_{X_{i1}}^2}. \tag{14.26}$$

In simple regression, the $\hat{\gamma}_1$ is maximum when all the $r_{\Psi_f \Psi_g}$ are equal to 1, namely when all data are equally–shaped distributions.

14.2.2.2   Goodness-of-fit indices

Considering the nature of the data, assessing the goodness-of-fit of the model is not straightforward. In this chapter, we consider three indices that can be used for evaluating the goodness-of-fit of a regression model of distributional variables: the Ω measure proposed in [5], the $Pseudo - R^2$ proposed in [20], and the classical root mean squared error (RMSE) using the L_2 Wasserstein distance.

Ω **measure.** A measure of goodness-of-fit of linear models on distributional data has been proposed in [5]. The authors define the Ω index through the following ratio:

$$\Omega = \frac{\sum_{i=1}^{n} d_W^2 \left(f_{\hat{Y}_i}, \bar{Y} \right)}{\sum_{i=1}^{n} d_W^2 \left(f_{Y_i}, \bar{Y} \right)}, \tag{14.27}$$

where $\bar{Y}$ is the (scalar) average of the distributional variable Y. It ranges from 0 to 1 as shown in Chapter 13.

PseudoR^2 was proposed in [20], for the simple linear regression model of histogram data. In [20], it is shown that the L_2 Wasserstein distance can be used for the definition of the total sum of squares of Y (SSY) as follows:

$$SSY = n \cdot VAR(Y) = \sum_{i=1}^{n} d_W^2 \left(f_{Y_i}, f_{\bar{Y}} \right) = \sum_{i=1}^{n} \int_0^1 [\Psi_{Y_i}(t) - \Psi_{\bar{Y}}(t)]^2 \, dt, \tag{14.28}$$

where $\Psi_{\bar{Y}}(t)$ is the average quantile function of the response variable.

A well-known index of the goodness-of-fit of a model is the R^2 *coefficient of determination*. It is based on the classical decomposition of the sum of squares of the response variable (SSY) into the regression sum of squares (SSR) plus the error sum of squares (SSE)

$$SSY = SSR + SSE. \tag{14.29}$$

We remark that in the *two-component* model, the second component is constrained to have positive solutions, and it has not a constant term. Therefore, because of the constraint and the absence of the constant term, solutions may be biased, and Equation (14.29) cannot hold as the sum of two independent amounts of variability. In [13], a detailed discussion about this index is presented. Considering that the classical R^2 assumes values in $[0; 1]$, in order to obtain a conservative measure of goodness-of-fit, we propose to adopt the following restrictions on the range of values that it can assume as follows:

$$PseudoR^2 = \min \left[\max \left[0; 1 - \frac{SSE}{SSY} \right]; 1 \right], \tag{14.30}$$

where

$$SSE = \sum_{i=1}^{n} \int_0^1 \left[\Psi_{\hat{Y}_i}(t) - \Psi_{Y_i}(t) \right]^2 \, dt.$$

Root Mean Square Error (RMSE) The RMSE is generally used as a measure of goodness-of-fit. To be consistent with the proposed model and the related SSE of the OLS problem in Equation (14.15), we propose the following measure for the RMSE:

$$RMSE = \sqrt{\frac{\sum_{i=1}^{n} \int_{0}^{1} \left[\Psi_{\hat{Y}_i}(t) - \Psi_{Y_i}(t) \right]^2 dt}{n}} = \sqrt{\frac{SSE}{n}}. \qquad (14.31)$$

RMSE is a positive number and is expressed in the same scale of measurement of the dependent variable. It is equal to zero when the model fits the data perfectly.

14.3 Application

This section presents an application of the two-component regression model on data coming from air quality monitoring. For the analysis, we used the `HistDAWass` package, developed in R [18], and the `OzoneFull` histogram-valued dataset contained in the same package. The R code for the analysis is freely available at the GitHub repository `https://github.com/Airpino/Clustering_DD_app/Regression_example`.

14.3.1 Data description

The `OzoneFull` dataset is a `MatH` object, namely, a table of histogram-valued data, representing aggregate data downloaded from the Clean Air Status and Trends Network (CASTNET)[5], which is an air-quality monitoring network of the United States designed to provide data to assess trends in air quality, atmospheric deposition and ecological effects due to changes in air pollutant emissions. The main sites monitored by CASTNET are shown in Figure 14.1.

One of the main tasks of CASTNET is the control of the level of Ozone, which is a gas that has been proven to induce respiratory diseases in the population. In the literature, several studies have found evidence of a connection between ozone concentration levels and temperature, wind speed, and solar radiation [7].

The `OzoneFull` dataset contains the histograms of the *ozone concentration* (Y) (particles per billion), *temperature* (X_1) (degrees Celsius), *solar radiation* (X_2) (Watts per square meter) and *wind speed* (X_3) (meters per second) for 78 US sites. Each histogram summarizes the hourly data during the 2010 summer and the central hours of the days (10 a.m.–5 p.m.). Each histogram contains

---

[5]`http://java.epa.gov/castnet/`

FIGURE 14.1
Ozone dataset. Map of monitored sites.

100 equi-frequent bins, i.e. the bins are not of equal width as usual, but the histogram was build such that each bin contains the 1% of raw data.

In Table 14.1, we show the first three rows of the OzoneFull.

TABLE 14.1
Ozone dataset: first three rows of the histogram-valued data table.

ID	Y Ozone.Conc. (ppb) Bin	p	X₁ Temperature (C) Bin	p	X₂ Solar Radiation (Watt/M²) Bin	p	X₃ Wind Speed (m/Sec) Bin	p
	$[8.77 - 16.62)$	0.01	$[8.45 - 11.65)$	0.01	$[25.29 - 75.88)$	0.01	$[0.10 - 0.35)$	0.01
	$[16.62 - 17.54)$	0.01	$[11.65 - 13.06)$	0.01	$[75.88 - 108.27)$	0.01	$[0.35 - 0.41)$	0.01
I1	$\cdots$	$\cdots$	$\cdots$	$\cdots$	$\cdots$	$\cdots$	$\cdots$	$\cdots$
	$[65.68 - 67.78)$	0.01	$[28.87 - 29.23)$	0.01	$[914.12 - 933.30)$	0.01	$[3.52 - 3.79)$	0.01
	$[67.78 - 89.60]$	0.01	$[29.23 - 30.18]$	0.01	$[933.30 - 942.00]$	0.01	$[3.79 - 4.48]$	0.01
	Bin	p	Bin	p1	Bin	p	Bin	p
	$[9.00 - 15.00)$	0.01	$[9.50 - 9.75)$	0.01	$[49.00 - 56.16)$	0.01	$[0.10 - 0.55)$	0.01
	$[15.00 - 17.00)$	0.01	$[9.75 - 10.38)$	0.01	$[56.16 - 71.50)$	0.01	$[0.55 - 0.80)$	0.01
I2	$\cdots$	$\cdots$	$\cdots$	$\cdots$	$\cdots$	$\cdots$	$\cdots$	$\cdots$
	$[54.24 - 58.00)$	0.01	$[29.02 - 29.60)$	0.01	$[910.00 - 916.84)$	0.01	$[7.52 - 8.37)$	0.01
	$[58.00 - 63.00]$	0.01	$[29.60 - 30.70]$	0.01	$[916.84 - 944.00]$	0.01	$[8.37 - 9.60]$	0.01
	Bin	p	Bin	p1	Bin	p	Bin	p
	$[9.25 - 17.99)$	0.01	$[17.57 - 20.13)$	0.01	$[52.57 - 78.67)$	0.01	$[0.08 - 0.26)$	0.01
	$[17.99 - 20.31)$	0.01	$[20.13 - 20.63)$	0.01	$[78.67 - 105.48)$	0.01	$[0.26 - 0.38)$	0.01
I3	$\cdots$	$\cdots$	$\cdots$	$\cdots$	$\cdots$	$\cdots$	$\cdots$	$\cdots$
	$[62.38 - 64.11)$	0.01	$[36.10 - 36.42)$	0.01	$[979.18 - 990.02)$	0.01	$[3.77 - 4.07)$	0.01
	$[64.11 - 69.45]$	0.01	$[36.42 - 37.07]$	0.01	$[990.02 - 1020.00]$	0.01	$[4.07 - 4.81]$	0.01
$\cdots$	$\cdots$	$\cdots$	$\cdots$	$\cdots$	$\cdots \; \cdots$		$\cdots$	$\cdots$

Since the histogram representation of varying bin-width histograms is not always pleasant, we plot them using only ten equi-frequent bins in Figure 14.2, where, for the sake of space, only the first ten of the 78 sites are shown.

We computed the mean distribution for each variable as the Frechét mean using the L_2 Wasserstein distance as a variability measure. The plot of the four means was computed using the WH.vec.mean() function are shown in

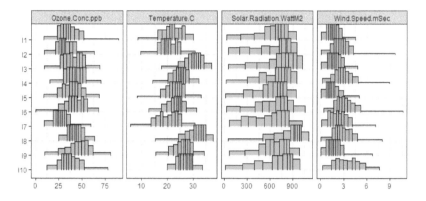

FIGURE 14.2
Ozone dataset. Plot of the data table with the first ten of 78 sites.

Figure 14.3. We also report the basic statistics of the four mean distributions in Table 14.2.

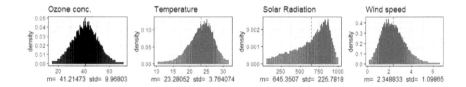

FIGURE 14.3
Ozone dataset. Plot of the mean distributions for each variable. m and std are the mean and the standard deviation of each distribution.

TABLE 14.2
Basic statistics of mean distributions.

Variable	Mean	St. dev.	First quartile	Median	Third quartile	Skewness	Kurtosis
Ozone	41.21	9.97	34.27	41.15	48.05	0.08	2.78
Temperature	23.28	3.76	21.04	23.71	25.93	-0.60	3.39
Solar Radiation	645.35	225.78	496.05	701.63	826.56	-0.71	2.58
Wind Speed	2.35	1.10	1.54	2.22	3.03	0.65	3.46

Looking at the plot and at the basic statistics table, we note that *Ozone* average distribution is approximately Gaussian. In contrast, the mean distributions of the independent variables are unimodal. However, they are differently skewed, and with a kurtosis (computed as the fourth standardized moment of each distribution) that is further from the Gaussian kurtosis (where the kurtosis index is equal to 3) than the *Ozone* one.

Using the basic statistics for distributional variables which are based on the L_2 Wasserstein distance, we computed the standard deviation of each distributional variable and the correlation matrix of the four variables, as reported in Table 14.3. All the variables are positively correlated, and the highest correlation is observed between *Ozone* and *Solar Radiation* (0.639), while mild correlations are observed between *Ozone* and *Wind Speed* (0.402), and between *Solar Radiation* and *Temperature* (0.454), or *Wind Speed* (0.439).

TABLE 14.3
Basic statistics of the distributional variables.

		Correlation matrix		
	Ozone	Temperature	Solar Rad.	Wind Speed
Ozone	1.000	0.247	0.639	0.402
Temperature		1.000	0.454	0.143
Solar Rad.			1.000	0.439
Wind Speed				1.000
		Standard deviations		
	9.530	3.842	113.431	1.314

14.3.2 The Two-component model estimation

Given the distribution of *temperature* (X_1) (degrees Celsius), the distribution of *solar radiation* (X_2) (Watts per square meter) and the distribution of *wind speed* (X_3) (meters per second), the main objective is to predict the distribution of *ozone concentration* (Y) (particles per billion) using a linear model.

The two-component model to estimate is as follows [6]:

$$\Psi_{Y_i} = \beta_0 + \sum_{j=1}^{3} \beta_j \mu_{X_{ij}} + \sum_{j=1}^{3} \gamma_j \Psi^c_{X_{ij}} + e_i \tag{14.32}$$
$$\text{subject to} \quad \gamma_j \geq 0 \quad \forall j \in \{1,2,3\}$$

In order to validate the model estimation, we performed a bootstrap analysis of the model using 1,000 bootstrap samples of the rows. In Table 14.4, we report the parameter estimates of the full model and the bootstrap results. We report the values of the Ω and $PseudoR^2$ and $RMSE$ indexes for assessing the goodness of fit of the full model.

The results reported in Table 14.4 suggest the following conclusions. We may assert that the ozone concentration distribution of a site depends on the mean *solar radiation*, where for each $\Delta Watt/m^2$ a 0.070 (*ppb*) variation of the *ozone concentration* mean level is expected, while in general we cannot say that the mean levels of *temperature* and *wind speed* induce a significant variation of the *ozone concentration* mean level (the 95% bootstrap confidence

---

[6]The model parameters estimates are obtained by the `WH.regression.two.components()` function contained in the `HistDAWass` package.

TABLE 14.4

Ozone dataset. Full and bootstrap model estimates on 1,000 samples. Goodness-of-fit indexes are reported for the full model only.

		Estimated model			

$$\Psi_{Y_i} = \hat{\beta}_0 + \sum_{j=1}^{3} \hat{\beta}_j \mu_{X_{ij}} + \sum_{j=1}^{3} \hat{\gamma}_j \Psi^c_{X_{ij}} + \hat{e}_i$$

			Bootstrap results		
	Full	Bootstrap		95% C.I.	
	model	average	bias	2.5%	97.5%
$\hat{\beta}_0$ Intercept	2.927	*3.209*	-0.282	-11.587	15.136
$\hat{\beta}_1$ (Temp.)	-0.346	*-0.352*	0.006	-0.813	0.179
$\hat{\beta}_2$ (Sol. Rad.)	0.070	*0.070*	0.000	0.051	0.090
$\hat{\beta}_3$ (Wind Sp.)	0.395	*0.378*	0.017	-1.301	1.942
$\hat{\gamma}_1$ (Temp.)	0.915	*0.911*	0.004	0.474	1.371
$\hat{\gamma}_2$ (Sol. Rad.)	0.018	*0.018*	0.000	0.012	0.024
$\hat{\gamma}_3$ (Wind Sp.)	1.887	*1.973*	-0.087	1.044	3.118
	Goodness-of-fit indexes				
Ω	0.742				
$PseudoR^2$	0.460				
$RMSE$	7.000				

intervals include the zero). Furthermore, the variability of the *ozone concentration* increases almost similarly (0.915) as the increase of the variability of *temperature*, while a unitary variation in the variability of the *solar radiation* induces a variation of 0.018 (*ppb*) and a unitary variation in the variability of the *wind speed* causes an increase in the variability of 1.958 (*ppb*) times. Consequently, it is important to consider more the variability component of the independent variables than their average values for giving a better explanation of the *Ozone* distributional variable.

We end the application performing a residual analysis. For each observation, we considered the functions $\hat{e}_i : [0,1] \to \Re$, which are not quantile functions and are computed as in Equation (14.33):

$$\hat{e}_i = \Psi_{Y_i} - \left[\hat{\beta}_0 + \sum_{j=1}^{3} \hat{\beta}_j \mu_{X_{ij}} + \sum_{j=1}^{3} \hat{\gamma}_j \Psi^c_{X_{ij}} \right]. \tag{14.33}$$

The L_2 Wasserstein distance corresponds to the Euclidean distance between functions with domain in $[0,1]$. In order to analyze the regression residuals, we standardize them by the $RMSE$ value. We plotted the standardized $\hat{e}_i$ functions in Figure 14.4.

From the plot of standardized residual functions, we can observe that the average residual function is almost close to the zero line, suggesting that the process generating the errors has zero mean for each $t \in [0,1]$. Except for

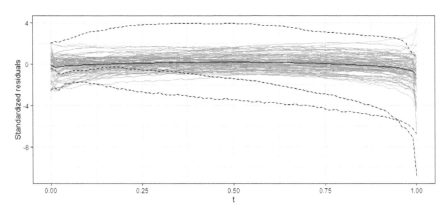

FIGURE 14.4
Ozone dataset. Standardized residuals. The solid black line is the average residual function, the dashed lines concern the three worst predictions.

three sites ($I23$, $I41$, and $I69$ of the `OzoneFull` dataset), which are the worst predicted ones, the residual functions lie in a ± 2 band almost everywhere with respect to their domain, with an increase of variability on the maximum values of the observed distributions (namely, when t is close to 1).

We end the section showing the worst and the best site predicted distributions of the *Ozone concentration distribution* obtained from the model in Figure 14.5. The worst predicted sites are identified by considering both their average position and size. At the same time, as concerns the shape, we note that the observed distribution of I41 is quite different from the predicted one as well.

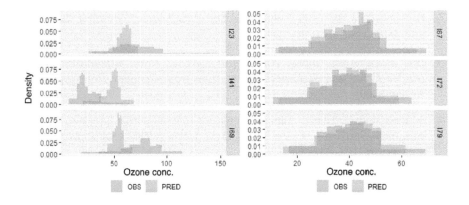

FIGURE 14.5
Ozone dataset. Observed vs predicted *Ozone concentration* distributions. The worst (on the left) and best (on the right) predicted sites.

14.4 Conclusions

This chapter presented a two-component linear regression technique for data described by probability or frequency distributions, which employs quantile functions and an OLS method based on the L_2 Wasserstein distance. Considering the nature of the data, we used a particular decomposition of the L_2 Wasserstein distance for the definition of the regression model. The model suggests a useful interpretation of the parameters in terms of the position and variability component of the distributional variables. Considering the complexity of the error term, the classical inference on parameters cannot be directly extended to the regression for distributional variables, except by using resampling techniques. However, the analysis of residual functions suggests that the process generating errors has zero mean. These results encourage new efforts about inferential properties of the estimators, a problem that has not yet found a proper, well-grounded, and widely accepted definition.

Bibliography

[1] P. Bertrand and F. Goupil. Descriptive statistics for symbolic data. In H.-H. Bock and E. Diday, editors, *Analysis of Symbolic Data: Exploratory Methods for Extracting Statistical Information from Complex Data*, pages 106–124. Springer Berlin Heidelberg, 2000.

[2] P.J. Bickel and D.A. Freedman. Some asymptotic theory for the bootstrap. *Ann. Stat.*, 9:1196–1217, 1981.

[3] L. Billard and E. Diday. *Symbolic Data Analysis: Conceptual Statistics and Data Mining*. Wiley, 2006.

[4] H.H. Bock and E. Diday. *Analysis of Symbolic Data: Exploratory Methods for Extracting Statistical Information from Complex Data*. Springer verlag, 2000.

[5] S. Dias and P. Brito. Linear regression model with histogram-valued variables. *Statistical Analysis and Data Mining: The ASA Data Science Journal*, 8(2):75–113, 2015.

[6] E. Diday and M. Noirhomme-Fraiture. *Symbolic Data Analysis and the SODAS Software*. Wiley, 2008.

[7] C. Dueñas, M.C. Fernández, S. Cañete, J. Carretero, and E. Liger. Assessment of ozone variations and meteorological effects in an urban area

in the mediterranean coast. *Science of The Total Environment*, 299(1-3):97–113, 2002.

[8] A.L. Gibbs and F.E. Su. On choosing and bounding probability metrics. *International Statistical Review*, 70(3):419–435, 2002.

[9] W. Gilchrist. *Statistical Modelling with Quantile Functions*. CRC Press, Abingdon, 2000.

[10] P. Giordani. Lasso-constrained regression analysis for interval-valued data. *Advances in Data Analysis and Classification*, 9(1):5–19, 2015.

[11] A. Irpino and E. Romano. Optimal histogram representation of large data sets: Fisher vs piecewise linear approximation. In M. Noirhomme-Fraiture and G. Venturini, editors, *EGC*, volume RNTI-E-9 of *Revue des Nouvelles Technologies de l'Information*, pages 99–110. Cépaduès-Éditions, 2007.

[12] A. Irpino and R. Verde. Basic statistics for distributional symbolic variables: a new metric-based approach. *Advances in Data Analysis and Classification*, 9(2):143–175, 2015.

[13] A. Irpino and R. Verde. Linear regression for numeric symbolic variables: a least squares approach based on Wasserstein distance. *Advances in Data Analysis and Classification*, 9(1):81–106, 2015.

[14] C.L. Lawson and R.J. Hanson. *Solving Least Square Problems*. Prentice Hall, Edgeworth Cliff, NJ, 1974.

[15] E.A. Lima Neto and F.A.T. de Carvalho. Centre and range method for fitting a linear regression model to symbolic interval data. *Computational Statistics & Data Analysis*, 52(3):1500–1515, 2008.

[16] E.A. Lima Neto, F.A.T. de Carvalho, and P.T. Camilo. Univariate and multivariate linear regression methods to predict interval-valued features. In *Australian Conference on Artificial Intelligence*, pages 526–537, 2004.

[17] E.A. Lima Neto and F.A.T. de Carvalho. Constrained linear regression models for symbolic interval-valued variables. *Computational Statistics & Data Analysis*, 54(2):333–347, 2010.

[18] R Core Team. *R: A Language and Environment for Statistical Computing*. R Foundation for Statistical Computing, Vienna, Austria, 2021. https://www.R-project.org/.

[19] L. Rüshendorff. Wasserstein metric. In *Encyclopedia of Mathematics*. Springer, 2001.

[20] R. Verde and A. Irpino. Ordinary least squares for histogram data based on Wasserstein distance. In Y. Lechevallier and G. Saporta, editors, *Proceedings of COMPSTAT 2010*, chapter 60, pages 581–588. Physica-Verlag HD, Heidelberg, 2010.

15

Forecasting Distributional Time Series

Javier Arroyo

Instituto de Tecnología del Conocimiento
Universidad Complutense de Madrid, Spain

CONTENTS

In this chapter, we introduce the concept of distributional time series and propose the fundamentals for forecasting them. We propose the use of error measures based on distances for distributions and to measure the error for a

DOI: 10.1201/9781315370545-15

specific set of quantiles of interest. We adapt the main components of classic time series, namely, trend and seasonality, to distributional time series and define the autocorrelation function for distributional time series. We explain how to forecast distributional time series using exponential smoothing, k-Nearest Neighbours and autoregressive methods. Finally, we illustrate this concepts forecasting two distributional time series: one representing rainfall data and another intra-daily returns of an equity.

15.1   Introduction

Forecasting is the process of making statements about future events. Forecasts are needed when there is uncertainty about a future outcome. Formal forecasting procedures are applied to reduce the uncertainty in order to make better decisions. They are crucial tools for effective and efficient planning. As a result, nowadays forecasting is common practice in fields such as economy, finance, business management, meteorology, environmental sciences and social sciences to mention a few.

Forecasting methods can be divided into qualitative and quantitative. On the one hand, qualitative methods are applied when there is no data available, or when data is not relevant to forecast. They are well-defined structured approaches such as the Delphi method. On the other hand, quantitative methods are applied when there is data about the past available and when it is expected that some aspects of the past patterns in the data will continue in the future. Quantitative methods usually deal with time series data, i.e. data collected at regular intervals over time. However, in some occasions, the term of cross-sectional forecasting is used to refer to the prediction of the value of a case not observed using the information on observed cases. This kind of prediction can be carried out by means of regression methods.

For the case of distributional data, only quantitative methods have been proposed so far. We can consider time series of distributions, that is, time series where a distribution of values is regularly observed over a time interval. The rest of this chapter will be devoted to explain the concept of distributional time series and review the methods and tools already proposed to forecast them.

15.2 Fundamentals of distributional time series

Definition 1 A time series of distributions $\{X_i\}$ with $i = 1, ..., n$ is a sequence of distributions successively observed at regular intervals over time. Each of these distributions X_i can be represented by a histogram as follows

$$X_i = \{(I_{i1}p_{i1}), ..., (I_{im}p_{im_i})\}, \text{ with } i = 1, ..., n, \tag{15.1}$$

where $\{p_{i\ell}\}$ with $\ell = 1, ..., m_i$ is a frequency or probability distribution over the considered domain that satisfies $p_{i\ell} \geq 0$ and $\sum_{\ell=1}^{m_i} p_{i\ell} = 1$; and where $I_{i\ell}$ $\forall i, \ell$, is an interval defined as $I_{i\ell} = [l_{i\ell}, u_{i\ell}]$ with $-\infty < l_{i\ell} \leq u_{i\ell} < \infty$ and $l_{i\ell} \leq u_{i\ell}$ $\forall i, \ell$, with $\ell \geq 2$.

The origin of an observed time series of distributions usually is the result of an aggregation process. The aggregation can be either contemporaneous or temporal.

In **contemporaneous aggregation**, we typically would have a variable measured in a group of elements through time and the interest of the analysis lies in the group behavior and not in the individual behavior of each element. This might be the case of the distribution of wind power production of the generators in a wind farm, the distribution of calls dispatched by the operators of a call center, the distribution of a network of sensors distributed in an area (for example, the sensor network of air pollution monitoring in a city). In these examples it would be possible to analyze and forecast the individual time series instead of the aggregated distribution.

However, in contemporaneous aggregation, the elements in the group may be different at each time i, as is shown in Figure 15.1. In this case, aggregation is the only way to analyse data, and a distribution provides more information than other aggregation operators such as the total, the mean or a bivariate time series with the first two moments of the distribution. This could be the case for example of the distribution of a variable measured for all the inhabitants of a region. For example, age (population pyramids), years of schooling, annual incomes, etc. Another example would be the distribution of a variable measured for all the products in successive batches in a manufacturing process, or the distribution of eggs laid by the chicken in a poultry farm.

In **temporal aggregation**, we would have a variable measured at a given frequency, but the interest would lie in the distribution of the observed values at a lower frequency. This is typically the case of sensor data recorded at continuous time, but analysed hourly or daily, or daily data aggregated quarterly as suggested in Figure 15.2. Examples of this kind of data could be solar radiation or wind speed measurements taken in renewable energy installations. Another field where this kind of time series arise naturally is finance, where

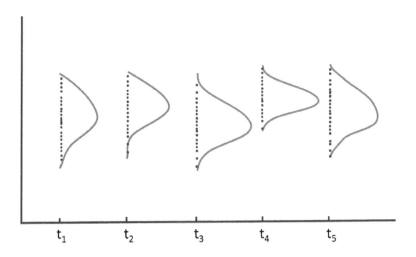

FIGURE 15.1
Stylized representation of contemporaneous aggregation.

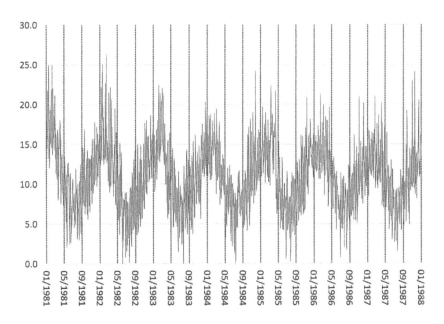

FIGURE 15.2
Daily time series that could be aggregated quarterly (between dashed lines).

prices or returns can be recorded at intra-daily frequency, but the interest of analysis and forecasting could be at lower frequencies, e.g. daily.

15.3 A theoretical approximation to distributional time series

In the classic context, a time series can be seen as a realization of a discrete time stochastic processes. The discrete time stochastic process is defined as a sequence of n scalar random variables. In this setting, an observed time series is the observation of the stochastic process. Usually only a single realization (also known as path or trajectory) is observed. As a result, in order to be able to study the properties of the stochastic process, it is required to be stationary and ergodic, which implies that the random process will not change its statistical properties with time and that its statistical properties can be deduced from a single, sufficiently long realization (observation) of the process.

In our context, the object under study is a time series of distributions. For each instant of time a distribution of values is observed. Each of these single observed values could be considered itself as the realization of a random variable. In other words, each distribution consists of a set of values each one being an observation from their respective random variable. Note that in the case of temporal aggregation, the set of random variables that produces the distribution is in fact a stochastic process. Thus, in distributional time series, each distribution is obtained as the result of observations from a set of random variables.

Another way to consider a distributional time series is shown in [17]. This work considers that each of the observed distributions is a realization of a random variable. The formal definition follows.

Let X be a distributional variable and X_i the observation for unit i with $i = 1, ..., n$. Let $(\Omega, \mathcal{F}, P)$ be a probability space, where Ω is the set of elementary events, $\mathcal{F}$ is the σ-field of events and $P : \mathcal{F} \to [0, 1]$ the σ-additive probability measure. Define a partition of Ω into sets $A_X(x)$ such that $A_X(x) = \{\omega \in \Omega | X(\omega) = x\}$, where $x \in \{X_i, i = 1, ..., n\}$.

Definition 2 A mapping $X : \mathcal{F} \to \{X_i\}$, such that, for all $x \in \{X_i, i = 1, ..., n\}$ there is a set $A_X(x) \in \mathcal{F}$, is called a distributional random variable.

Definition 3 A distribution-valued stochastic process is a collection of distributional random variables that are indexed by time, i.e. $\{X_i\}$ for $i \in T \subset \mathbb{R}$, with each X_i following Definition 2.

According to these definitions, a distributional time series is a realization of a distribution-valued stochastic process. The realized distributions at times i and j may be different as a result of the fact that their respective random variables are also different. This approximation may be more valid for the case of contemporaneous aggregation, where we observe a set of measures at each time point. For the case of temporal aggregation, it assumes that all the observed measures at the high frequency that are going to be aggregated come from the same random variable. This assumption could be correct at some occasions but could be wrong at others (e.g. imagine the case where the time series to be aggregated is subject to a structural break).

These two approximations make it clear that it is not straightforward to establish a theoretical framework for distributional time series and to study its properties. Such a thing is out of the scope of this chapter, that will be devoted to propose methods to describe and forecast distributional time series.

15.4 Error measurement for distributional time series

In the classic context the basic operation for error measurement in time series is the subtraction between the observed and the predicted value at time i. In the context of distributional time series, we need a basic operation to effectively measure the dissimilarity between the observation and the prediction when they are distributions. Below we review several candidate approaches and analyse the appropriateness of each one.

There is a definition of **arithmetic for histograms** which includes a subtraction operation, see [11] and [27]. *A priori* it would be possible to use this operation to define the concept of error. However, the result of this operation does not inform about the dissimilarity between the distributions considered. The reason is that such a histogram arithmetic is a probabilistic arithmetic. The probabilistic histogram bounds the actual value and describes the probability of finding it within the bounds. Then the probabilistic histogram arithmetic makes it possible to operate with probabilistic histograms in a way that the result is also a probabilistic histogram that encloses the whole set of possible results and describes their probability density. Such an approach is related to numerical analysis and is applied in fields such as reliability and tolerance analysis.

To illustrate the inappropriateness of histogram arithmetic, consider the case of a perfect forecast such as that $X_t = \hat{X}_t$. The histogram subtraction would be expected to produce a result such that $X_t - \hat{X}_t = 0$ or zero in histogram form $\{([0,0]1)\}$. However, this is only the case if and only if the histogram is a classic (single) value , that is $X_t = \hat{X}_t = \{([a,a]1)\}$ with $a \in \Re$.

Another approach that could be considered for measuring distributions discrepancy would be the use of **statistical tests** to determine whether the observed and the forecasted distributions are the same. The usual statistical tests assess whether two or more groups of observed values have identical distributions. However, it would not be straightforward in distributional time series, since while for the observed distribution we observe the actual (disaggregated) values, for the case of the forecasted distribution we do not have the disaggregated values. A testing approach is used in [7] in the context of

TABLE 15.1
Dissimilarity measures for distributions.

Dissimilarity measures	Definitions
Kullback-Leibler divergence	$D_{K-L}(f,g) = \int_{\Re} \log\{\frac{f(x)}{g(x)}\} f(x) dx$
Jeffrey divergence	$D_J(f,g) = D_{K-L}(f,g) + D_{K-L}(g,f)$
χ^2 divergence	$D_{\chi^2}(f,g) = \int_{\Re} \frac{|f(x)-g(x)|^2}{g(x)} dx$
Hellinger distance	$D_H(f,g) = \left[\int_{\Re}(\sqrt{f(x)} - \sqrt{g(x)})^2 dx\right]^{\frac{1}{2}}$
Total Variation distance	$D_{var}(f,g) = \int_{\Re} |f(x) - g(x)| dx$
Patrick-Fisher distance	$D_{P-F}(f,g) = \sqrt{\int_{\Re}(f(x) - g(x))^2 dx}$
Kolmogorov distance	$D_K(f,g) = \max_{\Re} |F(x) - G(x)|$
L_1 Wasserstein distance	$D_{W1}(f,g) = \int_{\Re} |F^{-1}(t) - G^{-1}(t)| dt$
L_2 Wasserstein distance	$D_{W2}(f,g) = \sqrt{\int_{\Re} |F^{-1}(t) - G^{-1}(t)|^2 dt}$

density forecasting in classic time series, where the density forecast provided by a forecasting model is compared to observed data. Such an approach will not be explored here because it does not provide an interpretable measure of the dissimilarity between the distributions.

On the contrary, the approach that will be considered here would be precisely the use of **dissimilarity measures for distributions**. This approach provides a single numerical measure of how alike two distributions are according to a specified criterion. A dissimilarity measure provides a single value that can play the role of error that we are looking for. Such an approach also has a precedent in density forecasting, in [18] the Kullback-Leibler information is used for combining density forecasts and measure the difference between the forecasted and the true (and unknown) density.

The use of dissimilarity measures for error measurement in distributional time series will be explored in detail in the next subsection.

15.4.1 Distance-based error measures for distributional time series

In [12] and [16] different dissimilarity measures for probability distributions are reviewed. Table 15.1 shows some of them considering that $f(x)$ and $g(x)$ are density functions, $F(x)$ and $G(x)$ are cumulative distribution functions for $x \in \Re$, and $F^{-1}(t)$ and $G^{-1}(t)$ are quantile factions for $t \in [0,1]$.

Kullback-Leibler, Jeffrey and χ^2 divergences all suffer from the same problem, they need that both density functions $f(x)$ and $g(x)$ have the same range (or support of the density function), which is often not the case for distributional data.

The values of Hellinger, Total Variation, Patrick-Fisher and Kolmogorov distances are restricted to the following intervals: $[0, \sqrt{2}]$, $[0, 2]$, $[0, 2]$ and $[0, 1]$, respectively. They all take the maximum value when the supports of both density functions are disjoint. This is the case regardless of the separation between the supports. This is not a desirable property for error measurement with distributional data, because in this context it is often convenient to consider as more dissimilar distributions those whose supports are more distant.

Finally, the values of L_1 Wasserstein and L_2 Wasserstein distances are restricted to the interval $[0, d]$, where d is the lenghth of the domain. In this sense, the more distant the support of the considered distributions the greater their distance.

Both distances come from the same family of distances for distribution functions that satisfies the following equation:

$$D_{Wp}(f, g) = \left(\int_0^1 |F^{-1}(t) - G^{-1}(t)|^p dt \right)^{1/p}, \qquad (15.2)$$

where $F^{-1}(t)$ and $G^{-1}(t)$ are the inverse of cumulative distribution functions, that is the quantile functions, defined for $t \in [0, 1]$. Both distances are particular cases ($p = 1$ and $p = 2$) of the general expression, as the Manhattan and Euclidean distances are particular cases of the Minkowski metric. Another point in favour is that they are metrics, which ensures some interesting properties that dissimilarity measures do not satisfy such as symmetry and subadditivity (triangle inequality).

It is interesting to note that this family of distances, also known as Monge–Kantorovich, has a discrete counterpart, the Earth Mover's distance. This family of distances roughly considers each distribution as a unit amount of earth or mass and estimates the minimum amount of work needed to turn one distribution into the other. Figure 15.3 shows a stylized representation of the idea, where histograms are composed of blocks and the arrows represent the required minimum work. This analogy explains the name of Earth Mover's distance and also provides a meaningful metaphor that helps us to understand the L_p Wasserstein distances.

Figure 15.4 shows the cumulative distribution functions of histograms $X = \{([2, 3]\ 0.5), ([3, 4]\ 0.3), ([4, 5]\ 0.2)\}$ and $Y = \{([1, 2]\ 0.1), ([2, 3]\ 0), ([3, 4]\ 0.1), ([4, 5]\ 0.3), ([5, 6],\ 0.5)\}$. The area in light green represents the discrepancy between the cumulative distribution functions (or the quantile functions) that is key in the L_p Wasserstein distances. This figure helps us to understand what is measured by Equation (15.2).

As a result, these distances can be deemed as appropriate for defining error measurement in distributional time series according to their properties, range of values, meaning and intuition. We can define the following error measures.

Definition 4 Let $\{X_i\}$ and $\{\hat{X}_i\}$ with $i = 1, ..., n$ be an observed and forecasted time series of distributions. The Mean Distance Error can be defined as

$$MDE(\{X_i\}, \{\hat{X}_i\}) = \frac{\sum_{i=1}^n \left(D(X_i, \hat{X}_i) \right)}{n}, \qquad (15.3)$$

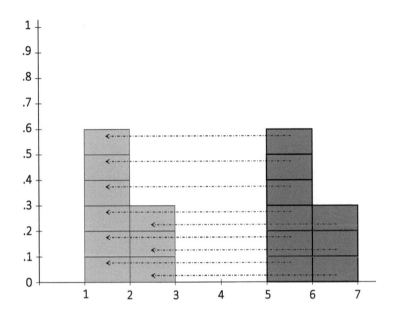

FIGURE 15.3
Stylized representation of the Earth Mover's distance between two histograms.

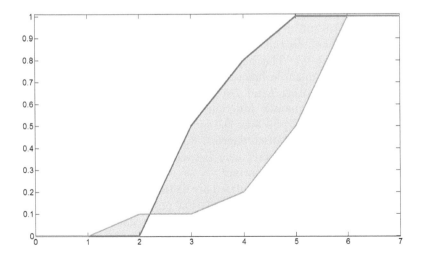

FIGURE 15.4
Cumulative distribution functions of histograms X and Y.

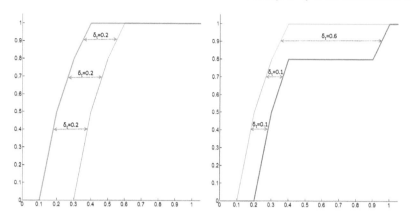

FIGURE 15.5
CDFs of histograms X and Y (left), and X and Z (right).

where $D(X_i, \hat{X}_i)$ is an appropriate distance such as the L_1 or L_2 Wasserstein distances.

The choice between the L_1 and L_2 Wasserstein distances depends on whether we want to assign higher weights to larger discrepancies between the quantile functions (or CDF) than to smaller ones. We will illustrate this point with an example.

Let be three histograms $X = \{([0.1,\ 0.2]\ 0.5),\ ([0.2,\ 0.3]\ 0.3),\ ([0.3,\ 0.4]\ 0.2)\}$, $Y = \{([0.3,\ 0.4]\ 0.5),\ ([0.4,\ 0.5]\ 0.3),\ ([0.5,\ 0.6]\ 0.2)\}$, and $Z = \{([0.2,\ 0.3]\ 0.5),\ ([0.3,\ 0.4]\ 0.3),([0.4,\ 0.9]\ 0),\ ([0.9,\ 1]\ 0.2)\}$. The $D_{W1}(X,Y) = 0.2$ represents the area between the respective quantile functions (or CDFs). In this example, the value for the L_2 distance is the same $D_{W2}(X,Y) = 0.2$. This happens because in the example, the discrepancy between both quantile functions remains constant at 0.2 for the whole $[0,1]$ range. On the other hand $D_{W1}(X,Z) = 0.2$ because the area of the discrepancy between quantile functions is again 0.2. However, in this case $D_{W2}(X,Z) > D_{W1}(X,Z)$ because the squared aggregation of the L_2 distance penalizes more the larger discrepancies. Figure 15.5 illustrates this example with δ_t representing the discrepancy between the quantile functions of the considered distributions at quantile t.

Since the measures to be considered are distances, they take positive values and can be aggregated without cancelation problems. However, the use of squared values can be considered if we want an aggregated error measure that penalizes large errors in the forecast of distributions at time i.

Definition 5 Let $\{X_i\}$ and $\{\hat{X}_i\}$ with $i = 1, ..., n$ be an observed and fore-casted time series of distributions, the Root Mean Square Distance Error can be defined as

$$RMSDE(\{X_i\}, \{\hat{X}_i\}) = \sqrt{\frac{\sum_{i=1}^{n} \left(D(X_i, \hat{X}_i)\right)^2}{n}}, \qquad (15.4)$$

where $D(X_i, \hat{X}_i)$ is the L_1 or the L_2 Wasserstein distance.

The error measures proposed for distributional time series are useful because they provide a single value that aggregates both the error for each distribution at time i and all the errors in the time series for $i = 1, ..., n$. A single-valued error measure can be used as an optimization criterion as well as for performance assessment.

Decomposition of the error based on the squared L_2 Wasserstein distance

While single-valued error measures are certainly manageable, they provide summaries that can mask interesting phenomena especially in complex cases where the error between an observed and a predicted distributions can be caused by different factors or by different distribution parts, for example, just by the location of the distributions or by differences in the tails. Thus, it would be useful to count with measures that help the analyst to assess where the difference between the observed and the predicted distributions mainly stem from.

Interestingly, as shown in [20, 23] the squared L_2 Wasserstein distance between distributions f and g can be decomposed in the following way

$$D_{W2}^2(f, g) = (\mu_f - \mu_g)^2 + (\sigma_f - \sigma_g)^2 + 2\sigma_f\sigma_g \left(1 - \rho_{QQ}(f, g)\right), \qquad (15.5)$$

where μ_f, μ_g are resp. the means of distributions f and g, σ_f, σ_g are resp. the standard deviations of distributions f and g, and $\rho_{QQ}(f, g)$ is the Pearson correlation of the quantile functions of f and g.

Thus, the squared L_2 Wasserstein distance between distributions can be decomposed in an addition, where the first element accounts for the dissimilarity between the means (i.e. locations) of the distribution, the second accounts for the dissimilarity between the spreads of the distributions, and the third one for the dissimilarity in the shape of the distribution.

As a result, the (squared) RMDSE using the L_2 Wasserstein distance can be decomposed into the average of these three dissimilarimities offering a clearer representation of where the discrepancy between f and g actually lies.

Furthermore, if we want to know how the error behaves for each time point i we can easily disaggregate the time dimension. In addition, we can also study the error for a set of quantiles of interest as it is shown below.

15.4.2 Quantile error measures for distributional time series

In a distributional time series there might be points of the distribution (quantiles) that we might want to pay special attention to, for example the tails or the central part of the distribution. In order to do so, a set of quantiles of interest can be established and we can measure the error for each of these quantiles at time i as follows.

Definition 6 Let X_i and $\hat{X}_i$ be the observed and the forecasted distribution at time i, and let $\psi_i(t)$ and $\hat{\psi}_i(t)$ be their respective quantile functions with $t \in [0, 1]$. The quantile error for the quantile t at time i can be defined as

$$q_{it} = \psi_i(t) - \hat{\psi}_i(t). \tag{15.6}$$

Since the quantiles are expressed in the same unit as the variable under study, the quantile error is a measure expressed in the same units. In addition, the q_{it} can be aggregated along the time dimension in order to provide error measures such as the Mean Error, the Mean Absolute Error or the Root Mean Square Error for quantile t.

These quantile error measures can be considered as a disaggregation of the L_1 and L_2 distances that measure the discrepancy between the quantile functions of the distributions. Instead of summarizing the discrepancy along the quantile range, they focus on specific quantiles in order to analyse their forecast.

15.5 Tools for the analysis of distributional time series

Time series of distributions can exhibit a huge diversity of behaviors. As a result, tools are needed to help the forecaster to analyse them and to categorize possible patterns. This section will review the tools proposed so far for that purpose. However, many more can be presented and are needed to shed light on the potential complexity that distributional time series can show.

15.5.1 Components in the distributional time series

Three components or patterns are usually identified in classic time series: trend, seasonal and cyclic component. The trend represents long-term increase or decrease in the series. The seasonal component represents a pattern that repeats over known and fixed periods of time within the data (e.g., month, day of the week, hour of the day, etc.). The cyclic component represents patterns where data rise and fall, but the duration of the fluctuations is not a fixed period and the length of the cyclic pattern is longer than the seasonal one (cyclic behavior is usually associated with business cycles). A classic time series $\{x_i\}$ can be viewed as the composition of these components:

$$x_i = u_i + t_i + e_t, \text{ or } x_i = u_i \times t_i \times e_t, \qquad (15.7)$$

where u_i is the seasonal component, t_i is the trend-cycle component and e_t is the error component (or the remainder that is not represented by the other components). Some forecasting models, like the classical decomposition method, the X-13 ARIMA developed by the US Census Bureau, the STL [10] or the exponential smoothing family [14] explicitly deal with these components.

It is not straightforward to adapt such decompositions to the case of distributional time series. First, the lack of an effective arithmetic makes it difficult to propose a composition equation. For example, it is not possible to add two components represented as a distribution, and the error cannot be represented in distribution form, as seen in Section 15.4. Second, the potential components may be distributions, but may also not be distributions. For example, it is possible to think of a trend or a seasonal component that only affects the location of the distribution, as we will see below; but it could be possible that such a component would only affect the spread of the distribution or other features such as its symmetry or kurtosis. As a result, there are no decomposition methods for distributional time series yet, but some efforts have been done so far to identify and forecast these components. They are reviewed next.

15.5.1.1 Trend

In [2], the additive trend is a real number that represents the long term movement in the location of the distribution and the location of a distribution is represented by its mean (or center of gravity of the distribution). The use of the minimum or maximum observed values in the distribution is ruled out because these values are sensitive to outliers. The mean or center of gravity of a histogram $X = \{(I_1 p_1), ..., (I_m p_m)\}$ is estimated as

$$c(X) = \sum_{\ell=1}^{m} \frac{l_\ell + u_\ell}{2} p_\ell. \qquad (15.8)$$

This definition requires to assume that within each interval the frequency (or probability) is uniformly distributed, as it is usually assumed in the Symbolic Data Analysis context.

In order to add the trend in the location to the component representing the distribution, it is needed to define an operation to deal with real values and distributions such as the *translation* for distributions in the real line.

Definition 7 Let $X = \{(I_1 p_1), ..., (I_m p_m)\}$ be distribution represented by a histogram and a real number d. The translation of d units of distribution X along the real line is given by

$$X \pm d = \{([u_\ell \pm d, l_\ell \pm d], p_\ell)\}, \qquad (15.9)$$

with $\ell = 1, ..., m$ and $\pm$ representing that the translation can be either positive or negative.

This kind of operation can be used to compose a distributional time series by means of two different (additive) components, one of them being a real value, and the other one a distribution.

Some forecasting methods, such as the k-NN or the autoregression, could require removing the trend in the location of the distributional time series before dealing with it. In order to do so, it is needed a location-differencing operation, such as the one proposed for classic time series.

Definition 8 Let $\{X_i\}$ with $i = 1, ..., n$ be a distributional time series, its location-differenced time series is obtained as

$$X'_i = X_i - c(X_{i-i}), \tag{15.10}$$

where $c(X_{i-i})$ is the center of gravity of distribution X_{i-i} as defined in Equation (15.8), and the negative translation is defined in Equation (15.9).

Interestingly, it would be possible to test for stationarity of the time series that results from the centers of gravity of the considered distributional time series. We could use, for example, the Dickey-Fuller test. In fact, the resulting (classic) time series can be analyzed and modelled using econometrics methods. Nonetheless, for an adaptation of the concept of stationarity to distributions, the curious reader is referred to [17] where the authors offer a first definition of stationarity for distributional time series represented in histogram form. While the definition is interesting from a theoretical perspective, it is complicated to put it in practice. Thus, it will not be presented here.

15.5.1.2   Seasonal component

In a distributional time series the seasonal component can affect only the location of the distribution or the distribution as a whole. Hence, it is possible to consider the distributional time series at time i as the composition of a real-valued trend-cycle t_i and a distribution-valued seasonal component U_i

$$X_i = t_i + U_i, \tag{15.11}$$

or alternatively, as the composition of a distribution-valued trend-cycle T_i and a real-valued seasonal component u_i

$$X_i = T_i + u_i, \tag{15.12}$$

where in both cases the composition uses the translation operation as defined in Equation (15.9).

The seasonal component is composed by the seasonal indexes of the seasonal period p. For example, if $p = 4$ and we have quarterly data, we need to estimate the seasonal indexes of each of the quarters, generically will call them seasons. They can be estimated as follows.

Definition 9 Let $\{X_i\}$ with $i = 1, ..., n$ be a (detrended) distributional time series and p the lenght of the seasonal period. The seasonal index of the time series can be real-valued or distribution-valued and is denoted as u_j or U_j, respectively, with $j = 1, ..., p$. The seasonal index of the season j is estimated as the average of all the observations of season j in the time series. If the seasonal component is real-valued, then it is estimated as the average of the centers of gravity, as in Equation (15.8), of the distributions considered. On the other hand, if the seasonal component is distribution-valued, then it is estimated as the L_2 Wasserstein barycenter that will be defined below in Equation (15.14).

If the seasonal variation is not important for our analysis or we want to study the remainder of the time series, we could want to remove the seasonal component. If we have a real-valued seasonal component, we can remove it from the distribution time series by means of a translation operation as defined in Equation (15.9); in this case, the remainder of the operation will be a distributional time series. However, if we have a distribution-valued seasonal component, we cannot subtract it from the distribution time series, as we do not have a suitable arithmetic for distributions (see Section 15.4). As it can be seen in this case, it is not straightforward to adapt classic time series concepts to distributional time series.

15.5.2 Autocorrelation in the distributional time series

In the context of classic time series, the autocorrelation function (ACF) measures the extent of a linear relationship between the time series and its lagged values. The ACF can be adapted for distributional time series with the help of the correlation definition for distributional data proposed in [26]. This definition is based on the variance and autocovariance defined as functions of the distances with respect to the L_1 barycentric distribution, which is a measure of the centrality for distributional data. In [17], an autocorrelation function based on these measures is used to analyse the time dependence of a distributional time series.

Let $\{X_i\}$ with $i = 1, ..., n$ be a distributional time series, the autocorrelation of order k is given by

$$\rho_k = \frac{\gamma_k}{\gamma_0} \qquad (15.13)$$

where γ_k is the autocovariance of order k that can be denoted as $Cov(X_i, X_{i-k})$ and γ_0 is the variance of time series $\{X_i\}$. The autocorrelation ρ_k is bounded by [-1, 1].

In particular, we will use the L_2 Wasserstein variance, covariance and correlation definitions as in [23] and Chapter 3 of the present book. In particular, the variance is defined as the mean of the squared L_2 Wasserstein distances between each observed distribution and the L_2 Wasserstein distance mean, which is the barycentric distribution of the observations as it will be defined in Equation (15.14) in the present chapter. Similarly, the covariance ot two distributional variables X and Y is the mean of the product of the L_2 Wasserstein distances between each observation and its respective L_2 Wasserstein distance mean of the two considered variables. The interested reader is referred to [23] or Chapter 3 of the present book for a detailed definition and an analysis of its properties.

The first p autocorrelation coefficients form the sample autocorrelation function (ACF) and the plot of the ACF is known as a correlogram.

Furthermore, the autocorrelation functions of a set of quantiles of interest can also be displayed. The quantile t with $t \in [0,1]$ can be considered as a classic time series, $\{\psi_i(t)\}$ with $i = 1, ..., n$, and its ACF can be represented. It is important to remark that the information of the ACF of a distributional time series and the one of the ACFs of a set of quantiles is not the same. The ACF for a given quantile only analyses the linear information in that quantile time series and does not take into account the information in the rest of the distribution. In this sense, the information of the ACF of the distributional time series is much more comprehensive.

Autocorrelation in the residuals

In classic time series, the ACF of the residuals is used to analyse whether there is (linear) information that has not been accounted for by the forecasting method. The residual is the difference between the observed values and the forecasted values. The adaptation of this idea to distributional time series is not straightforward since there is no suitable subtraction operation to represent the residual of a distribution (see Section 15.4). Two alternatives are suggested to analyse the residual linear dependence, but both of them have limitations.

In [17], it is used the ACF of the L_2 Wasserstein distance between the observed and the forecasted distribution. While this distance is the basis of the correlation measure for distributional data, what is measured by the ACF based on the distance is not exactly the same. Furthermore, the distance is always positive which makes a rough approximation to the concept of residual.

Alternatively, given the time series of the observed and forecasted quantile t, $\{\psi_i(t)\}$ and $\{\hat{\psi}_i(t)\}$, respectively, with $t \in [0,1]$, it can be displayed the ACF of the residuals for a set of quantiles of interest. This will reveal if there is any autocorrelation in the residuals of the considered quantile. However, this ACF does not tell whether there is other information in the distribution or in other quantiles that could be taken into account to produce better forecasts.

Given the inherent complexity of distributional data and the idea of correlation in distributional data, both of the aforementioned approaches are flawed. However, they provide information, which might be limited or coarse, but may be more useful than no information at all.

15.6 Forecasting methods for distributional time series

This section reviews the most significant contributions to forecast distributional time series proposed in the literature so far. The number of methods available is still scarce, but as it will be exemplified later, it is possible to obtain accurate forecasts on a wide variety of situations.

15.6.1 Exponential smoothing methods

In classic time series, smoothing methods attempt to capture important long-term patterns, such as trends and cyclic movements, in the time series, while removing noise and rapid phenomena. This kind of methods can also be used for forecasting. In that case, forecasts are a result of a weighted averaging process of the past observations in the time series. These methods include moving averages and exponential smoothing methods, which can be very sophisticated and deal with trend and different seasonal patterns at the same time. In exponential smoothing the weights assigned to each of the previous values decrease according to a geometric (exponential) progression. The origin of these methods dates back the '50s, see [9] and [19]. Despite their simplicity or perhaps because of it, smoothing methods are a benchmark in forecasting because they often provide good results. The interested reader is referred to the excellent review in [14].

In distributional time series, smoothing should also capture important long-term patterns. It means that it should average all the distributional features such as location, variability (or range) and shape. In [2] and [4], two approaches are proposed for that purpose. One of them, the use of histogram arithmetic, is considered inappropriate for similar reasons as those exposed in Section 15.4. The other is the use of a barycentric distribution as proposed in [21], which minimizes the sum of the distances between itself and the distributions in the set.

Definition 10 The barycentric distribution X_B is defined as the distribution that minimizes the distance between itself and a set of distributions X_i for $i = 1, ..., n$,

$$\min_{X_B} \sum_{i=1}^{n} \omega_i D^l(X_i, X_B), \tag{15.14}$$

where l is a positive integer, $D(X_i, X_B)$ is a distance measure for distributions, and ω_i is the weight associated with distribution X_i such that $\omega_i \geq 0$ and $\sum_i \omega_i = 1$.

For the estimation of the barycentric distribution, the work in [25] analyses the results that are obtained with different dissimilarity measures and considers that the ones obtained with the L_2 Wasserstein distance and $l = 2$ are suitable in a clustering context.

In [4], the use of the L_1 Wasserstein distance and $l = 1$ is considered inappropriate for smoothing distributional time series because the median-like behavior of the resulting barycentric distribution does not produce a satisfactory smoothing of the distribution. This is due to the fact that the L_1 Wasserstein barycenter is obtained as the barycenter of a new set where each distribution in the original set is repeated a number of times n_j proportional to its weight (e.g. $n_j = N\omega_j$ with $N = \sum_{j=1}^{n} n_j$), and the quantile function of the resulting distribution is the median of the quantile functions of the new set. Given that, if in the smoothing process the weight assigned to the most recent observed distribution is $\omega_j > 0.5$, the result of the smoothing process is a distribution that exactly is the most recent observed distribution. Such coarse smoothing behavior results inappropriate. The inappropriateness of using the L_1 Wasserstein distance and $l = 1$ for smoothing distributions is further discussed in [4]. The median-like behavior of the resulting prototype disqualifies the approach.

On the contrary, the properties of the L_2 Wasserstein barycenter with $l = 2$ are much more attractive for a smoothing process as the resulting distribution has a quantile function that is the result of averaging the quantile functions of the distributions involved. This means that the location will be the result of averaging the location of the distributions involved, the same will happen with the range (or spread) of the distribution and with the shape of the quantile functions of the distributions involved. This behavior is more adequate for smoothing methods and as a result, for the rest of the section, we will only consider the case of L_2 Wasserstein barycenter with $l = 2$ for that purpose.

As mentioned above, smoothing methods include moving averages and exponential smoothing. Moving averages are mainly used to filter out noise, but they can also be used for forecasting purposes. They assume that averages of past and recent observations provide a sensible prediction of the phenomenon under study assuming that it is somewhat stable. Moving averages can be adapted for distributional time series with the help of the barycentric distribution as shown in [2] and [5].

A moving average of order q is the barycentric distribution that results from Equation (15.14) considering the set of distributions $X_{i-(j-1)}$ with $j = 1, ..., q$. The higher the value of q the more observations are considered to obtain the forecast, the smoother will be the forecasted time series. The weights ω_j can be the same for each observed distribution $X_{i-(j-1)}$, i.e. $\omega_j = \frac{1}{q}$, or can decrease following an arithmetic progression $\omega_j = \frac{q-j+1}{\sum_{j=1}^{q} j}$, or an exponential one $\omega_j = \alpha(1-\alpha)^{j-1}$ where $\alpha = \frac{2}{q+1}$.

Furthermore, moving averages can be used to remove noise from the time series and estimate the trend-cycle component of the time series, following the ideas of time series decomposition. In this case, the order q of the moving average is an odd number such as $q = 2k + 1$ and the moving average is computed as the (unweighted) barycentric distribution of the the set of distributions X_{i+j} with $j = -k, ..., 0, ..., k$. Note that this estimation uses both observations from the the past and from the future.

In the next subsections, we will present the adaptation of exponential smoothing methods to forecast distributional time series.

15.6.1.1 Simple exponential smoothing

Let $\{X_i\}$ with $i = 1, ..., n$ be a distributional time series, the forecast $\hat{X}_{i+1}$ based on an exponential smoothing filter is obtained as

$$\hat{X}_{i+1} = \alpha X_i + (1 - \alpha)\hat{X}_i, \tag{15.15}$$

where $\alpha \in [0, 1]$. By backward substitution we obtain the equivalent moving average with exponentially decreasing weights

$$\hat{X}_{i+1} = \sum_{j=1}^{i} \alpha(1 - \alpha)^{j-1} X_{i-(j-1)}. \tag{15.16}$$

As it is shown in [4], these two equations can be adapted to distributional time series with the L_2 Wasserstein barycenter with $l = 2$ as shown in Equation (15.14). In the case of Equation (15.15), the set of distributions will be composed by distributions X_i and $\hat{X}_i$, while in Equation (15.16) the set includes all the observed distributions in the time series $X_{i-(j-1)}$ with $j = 1, ..., i$. The barycenter obtained in both cases is the same, so the equivalence between both equations in classic time series also holds in the distributional time series context.

However, another equivalent equation in the classical time series is the so-called error correction equation that could be written as

$$X_{i+1} = \hat{X}_i + \alpha(X_i - \hat{X}_i). \tag{15.17}$$

Since this equation is not written as a weighted or unweighted average, it is not possible to adapt it to the barycentric distribution.

In Equation (15.16), an initial value of $\hat{X}_1$ is required. The most simple initialization is $\hat{X}_1 = X_1$. However, we could also use the barycenter of the first three or four observations. A more sophisticated initialization consists of backcasting, which means, reverting the time series and obtain the forecast of the last observation, which is the first one in the original time series, and use that last forecast as initial value of $\hat{X}_1$.

15.6.1.2 Exponential smoothing with additive trend in the location

In classical time series, the work in [19] proposes an exponential smoothing method to forecast time series that exhibit a long term increase or decrease. The method consists of two smoothing equations one for the level of the time series and another for the trend. This idea was adapted to the context of distributional time series in [2], where the trend is considered as the long term movements in the location of the distribution and the level is expressed as a distribution. The location is the center of gravity of the distribution as in Equation (15.8). Furthermore, as the method models the distributional time series as two different components, one of them being a real value and the other one a distribution, it is required the use of the translation for distributions in the real line as in Definition 8.

Let $\{X_i\}$ with $i = 1, ..., n$ be a distributional time series, the forecast $\hat{X}_{i+1}$ based on an exponential smoothing with additive trend in the location is estimated as

$$S_i = \alpha X_i + (1 - \alpha)(S_{i-1} + t_{i-1}), \tag{15.18}$$

$$t_i = \delta(c(S_i) - c(S_{i-1})) + (1 - \delta)t_{i-1}, \tag{15.19}$$

$$\hat{X}_{i+h} = S_i + ht_i, \tag{15.20}$$

where $\alpha, \delta \in [0, 1]$; S_i is the smoothed level at time i and its value defined in Equation (15.18) is a distribution computed as the L_2 Wasserstein barycenter for distributions in Equation (15.14); t_i is the trend value at time i; $c(X)$ is the center of gravity of distribution X as defined in Equation (15.8); and h is the forecast horizon, which in this case is used as a multiplier to extrapolate the trend value and obtain the forecast at time $i + h$.

The values for $i = 1$ can be obtained by means of a backcasting process. The initialization of the backcasting can be done in a simple way $t_n = c(X_{n-1}) - c(X_n)$ and $S_n = X_n$.

Following the example of classical time series, it is straightforward to propose an exponential smoothing method that *dampens* the location trend. According to [15], Holt's method usually over-forecasts due to the extrapolation of the trend. This problem is more serious for longer forecasting horizons and in automatic forecasting. The authors propose the use of a parameter to nuance the trend effect and yield a more conservative forecast. The exponential smoothing method with damped trend in location can be defined as follows:

$$S_i = \alpha X_i + (1 - \alpha)(S_{i-1} + \phi t_{i-1}), \tag{15.21}$$

$$t_i = \delta(c(S_i) - c(S_{i-1})) + (1 - \delta)\phi t_{i-1}, \tag{15.22}$$

$$\hat{X}_{i+h} = S_t + \sum_{l=1}^{h} \phi^l t_t, \tag{15.23}$$

where α, $\delta \in [0, 1]$, $\phi \geq 0$, $c(X)$ is the center of gravity of distribution X as defined in Equation (15.8), and h is the forecasting horizon. If $\phi = 0$, the equations turn into the simple exponential smoothing method and if $\phi = 1$ we get the exponential smoothing method with trend in the location; if $\phi \in (0, 1)$, the trend *dampens*, the closer to 0 the sooner the trend eventually becomes a flat line; if $\phi > 1$, the trend will increase exponentially.

15.6.1.3 Exponential smoothing with seasonality

In distributional time series, seasonality can appear in different ways. For example, considering that seasonal changes only affect the distribution location, or assuming that seasonality affects the distribution as a whole. The work in [2] proposes an exponential smoothing method for forecasting distributional time series with these types of seasonality. The proposed methods use the concept of translation for distributions in the real line as in Definition 8 to model the seasonal and level components of the distributional time series. They are described below.

If we display the distributional time series and the seasonal effect appears as changes in the distribution location, it can be represented as a real value u_i. The distribution location can be given by its center of gravity, Equation (15.8), as in the case of the trend in exponential smoothing. In this case, we can decompose the distribution time series at time i into two components: the seasonal effect u_i represented as a real value, and the level of the time series represented as distribution S_i where the seasonal effect has been removed from. The equations of the exponential smoothing with additive seasonality in the location follow:

$$S_i = \alpha(X_i - u_{i-p}) + (1 - \alpha)S_{i-1}, \tag{15.24}$$

$$u_i = \delta(c(X_i) - c(S_i)) + (1 - \delta)u_{i-p}, \tag{15.25}$$

$$\hat{X}_{i+1} = S_i + u_{i-p+1}, \tag{15.26}$$

where $\alpha, \delta \in [0, 1]$, $c(X)$ is the center of gravity of distribution X as defined in Equation (15.8), and p is the number of observations in the seasonal period. The value of S_i in Equation (15.24) is obtained as the L_2 Wasserstein barycenter for distributions defined in Equation (15.14).

The first p observations of the time series are required for initialization. The initial values for the exponential smoothing method can be set by means of a backcasting process. We can set as follows the starting values of the backcasting for the level and for the p first values in the seasonal time series:

$$S_{i-(p-1)} = \frac{X_i + X_{i-1} + \ldots + X_{i-(p-1)}}{p},$$

$$u_i = c(X_i) - c(S_{i-(p-1)}),$$

$$u_{i-1} = c(X_{i-1}) - c(S_{i-(p-1)}), \ldots,$$

$$u_{i-(p-1)} = c(X_{i-(p-1)}) - c(S_{i-(p-1)}).$$

On the other hand, seasonality can affect the distribution as a whole. In this case, the seasonal component can be represented as a distribution U_i, and the level by means of a real value s_i. The forecast of the exponential smoothing with additive seasonality in distribution is computed as follows:

$$s_i = \alpha(c(X_i) - c(S_{i-p})) + (1 - \alpha)s_{i-1}, \tag{15.27}$$

$$U_i = \delta(X_i - s_i) + (1 - \delta)U_{i-p}, \tag{15.28}$$

$$\hat{X}_{i+1} = s_i + U_{i-p+1}, \tag{15.29}$$

where $\alpha, \delta \in [0, 1]$, and p is the number of observations in the seasonal period. Equation (15.28) is computed as the L_2 Wasserstein barycenter for distributions defined in Equation (15.14).

The first p values of the time series are needed for initialization. The backcasting process can estimate its starting values as follows:

$$s_{i-(p-1)} = \frac{c(X_i) + c(X_{i-1}) + \dots + c(X_{i-(p-1)})}{p},$$

$$U_i = X_i - s_{i-(p-1)},$$

$$U_{i-1} = X_{i-1} - s_{i-(p-1)}, \dots,$$

$$U_{i-(p-1)} = X_{i-(p-1)} - s_{i-(p-1)}.$$

15.6.2 The k Nearest Neighbors method

The k Nearest Neighbours (k-NN) algorithm is a machine learning algorithm that can be used for classification and regression.

The algorithm is based on the use of closer instances to locally approximate the output function. This family of learning methods is usually called instance-based learning or also lazy learning because computation is deferred until a new instance arrives.

In this case, we are interested in its use for regression tasks [1] and more precisely in time series forecasting [28]. The k-NN in time series forecasting must first identify the k past sequences most similar to the current ones, and, second, combine their future values to estimate the prediction for the current sequence. The adaptation of the k-NN algorithm to forecast distributional time series is proposed in [2] and [6].

Let $\{X_i\}$ with $i = 1, ..., n$ be a distributional time series, it can be transformed into a series of d-dimensional vectors as

$$X_i^d = (X_i, X_{i-1}, ..., X_{i-d+1}) \tag{15.30}$$

with $i = d, ..., n$. The parameter d defines the number of past distributions that are used to characterize the element at time i. The distance between the last vector of the series X_n^d and the rest of the vectors X_i^d with $i = d, ..., n-1$ is estimated as follows

$$D^q(X_n^d, X_t^d) = \left(\frac{\sum_{i=1}^{d}(D(X_{n-i+1}, X_{t-i+1}))^q}{d} \right)^{\frac{1}{q}}, \tag{15.31}$$

where $D(X_{n-i+1}, X_{t-i+1})$ is an appropriate distance and q is a parameter to measure the discrepancy between the vectors as in the Root Mean Square Error ($q = 2$) or as in the Mean Absolute Error ($q = 1$).

Once the $n - d$ distances are estimated, the k closest vectors to X_n^d are identified. These vectors will be denoted as $X_{T_p}^d$ with $p = 1, ..., k$.

The prediction $\hat{X}_{n+1}$ is computed as the baricentric distribution that minimizes the sum of the distances between itself and each of the subsequent distributions of the k distribution sequences $X_{T_p}^d$. In other words, $\hat{X}_{n+1}$ is the solution to the following equation

$$\min_{\hat{X}_{n+1}} \sum_{p=1}^{k} \left(\omega_p D^l (\hat{X}_{n+1}, X_{T_p+1}) \right)^l , \tag{15.32}$$

where $D(\hat{X}_{n+1}, X_{T_p+1})$ is a suitable distance for distributions, X_{T_p+1} is the next distribution in the sequence $X_{T_p}^d$, l is a positive integer, and ω_p is the weight assigned to neighbor p with $\omega_p \geq 0$ and $\sum_{p=1}^{k} \omega_p = 1$.

Different weighting schemes can be used. The simplest one is to assign the same weight to all the elements $\omega_p = 1/k, \ \forall p$.

Alternatively, we can assign to element p a weight that is inversely proportional to the distance between the neighboring sequence p and the current one, such as

$$\omega_p = \frac{\eta_p}{\sum_{l=1}^{k} \eta_l}, \tag{15.33}$$

where $\eta_p = (D(X_n^d, X_{T_p}^d) + \xi)^{-1}$ with $p = 1, ..., k$, $D(X_n^d, X_{T_p}^d)$ is given by Equation (15.31) and ξ is a constant with a *negligible* value that prevents η_p to take an infinite value in case of null distances, e.g. $\xi = 10^{-8}$.

The k-NN adaptation depends on the choice of the distance that is used to measure the similarity between distribution sequences and to estimate the forecast as is shown next. In addition, the distance can be used for error measurement and for estimating the optimal k in a cross-validation setting. Given that the three tasks are related, we recommend to use the same distance for all of them. The task of measuring similarity and error measurement is quite similar, so the conclusions in Section 15.4.1 also apply here. In [2] and [6] the L_1 and L_2 Wasserstein distances are considered suitable for the k-NN. In particular, with the L_1 Wasserstein distance and $l = 1$ a median-like barycenter is obtained, while with the L_2 Wasserstein distance and $l = 2$ we obtain a mean-like barycenter. The reader interested in the efficient estimation of the solution to the minimization problem for both weighting schemes and both distances can check [2] or [6].

15.6.3 Autoregressive approach

In recent years, regression methods for distributional data have been proposed in [13] and [23] (see Chapters 13 and 14, respectively). In this section, we will use them for forecasting purposes. These methods can be used for forecasting distributional time series using an autoregressive approach. In autoregression the variable of interest is forecasted using a linear combination of past values of the variable.

In our case, the prediction $\hat{X}_{n+1}$ of a distributional time series $\{X_i\}$ with $i = 1, ..., n$ can be estimated with an autoregressive model of order p such as

$$\hat{X}_{n+1} = lf(X_n, X_{n-1}, ..., X_{n-p+1}) \qquad (15.34)$$

where p is the number of lags and lf is a linear function for distributions such as the ones estimated by the regression methods for distributional data proposed in [13] and [23] and described in Chapters 13 and 14, respectively.

Additionally, the location trend of the distributional time series can be removed from the series, as explained in Section 15.5.1.1. Moreover, as it is shown in Section 15.5.2, it is also possible to describe the autocorrelations in distributional time series and there are approaches to determine the autocorrelation of the residuals.

The tools proposed so far do not add up to create a methodology such as the ARIMA methodology proposed in [8] for classic time series. However, they are a first approximation to the problem that is able to competently forecast a distributional time series.

15.7   Applications

In this section, two different applications are presented to illustrate how to forecast distributional time series. One example from meteorology and the other from finance. The features that exhibit the considered distributional time series are different, but the methods presented in this chapter manage to perform remarkably well in both cases.

15.7.1   Cairns rainfall

Rainfall daily data is suitable for a distributional approach since this data is usually presented as a montly total or a monthly mean. While these values provide information about the central tendency, they tell nothing about the variability. Does this value come from a month with almost daily (small) rain or from a month with some heavy-rain days? Or, using more precise words, is it a matter of precipitation frequency or intensity? The duality of rainfall frequency and variability has often been studied in meteorology, see for example [24]. In this example, we want to show that distributional time series are suitable to deal with both aspects and that the forecasting methods presented here are effective tools for that matter.

For our application we will study the rainfall in the city of Cairns in Queensland, Australia. Cairns experiences hot and humid summers and mild, dry winters. Most of the region's annual rainfall (around 2000mm) occurs during the wet season that begins building up around December and ends in April. We will use the public data provided by the Bureau of Meteorology of the Australian Goverment[1] and we will use the meteorology station of Cairns Aero due to the quality and availability of its data. More precisely, we will focus on the period from 1943 to 1993, because there is no missing data during that period and all data sucessfully completed the quality control according to the Bureau of Meteorology report. Bear in mind that very few stations have a complete unbroken record of climate information, because stations may have been closed, re-opened, or may have had damaged instrumentation during some periods.

We will aggregate daily rainfall (in mm) into monthly distributions. For each distribution we will only focus on the following list of quantiles: 0, 50, 60, 70, 80, 90, 95 and 100. It can be seen that the first half of the distribution has no interest at all for us. The reason behind this choice is because during the wet season it rains around 15 days (half month), but during the rest of the year we have between 5 and 8 of rainy days per month.[2]. Thus, we consider as uninteresting the first half of the distribution, while the second half is described by means of gaps of 10% quantiles, where each gap corresponds to 3 days, except the right tail of the distribution where we also want to focus on the 95% quantile to better account for outlying values. The resulting time series consists in 612 monthly distributions from January 1943 to December 1993, i.e. 51 years. Figure 15.6 shows only the median and the maximum values of the distribution to illustrate the phenomenom. It can be seen that the median is zero in some months, which means that in those months half of the days had no rain, but this is not the case for some months in the wet season. So the lower part of the distribution is illustrated by the median and it represents in a certain way the frequency of rainy days during the month, which we can say is reasonably stable. On the other hand, the rainfall intensity is described by the rest of the distribution and specially by its higher quantiles. It can be seen that the maximum value of the monthly distributions greatly varies through the years, even though the yearly maximum is always found during the wet season because the rainfall phenomenom is seasonal. So in this location intensity seems more important than frequency.

If we take a closer look into a small period, such as the period between 1980 and 1982 shown in Figure 15.7, we can see that while there is seasonality, the pattern is not perfectly clear as it slightly varies through the years. There is obviously a clear change between the dry and the wet seasons, but also the pattern changes in the observed years. To read the chart correctly, it is important to bear in mind that lines represent quantiles and that the distance between two consecutive quantiles represents the spread between both

---

[1] http://www.bom.gov.au/climate/data/
[2] See http://www.australia.com/en/facts/weather/cairns-weather.html for more details

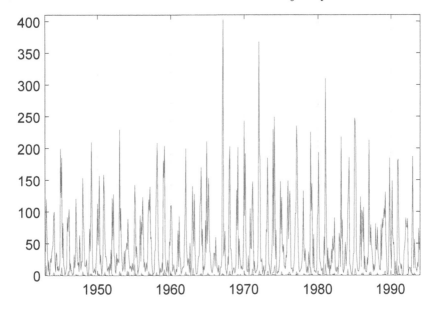

FIGURE 15.6
Median and maximum value of the monthly time series of aggregated daily rainfall (in mm).

quantiles. As we have seen already, the maximum (Q100) is subject to greater variability and it has no regular seasonal pattern. The seasonal pattern seems to be more stable and regular the lower the quantiles are, but this does not hold occasionally; see for example the first months of 1982 with an uncommon dry January and an uncommon wet April.

Given this interesting seasonal pattern, we are going to compare the performance of our methods in the distributional time series following a cross validation approach. Since k-NN looks for similar patterns in the past of the time series, it needs a substantial period of "past" to search for relevant neighbors. So the first part of the time series will be that initial period required by the k-NN. More precisely, we will consider the first 17 years (204 months) for the initialization of the k-NN. We will split the rest of the time series into two periods of 17 years, i.e. 204 months, for training and validation. More precisely, the training period will range from January 1960 to December 1976 and the validation from January 1977 to December 1993.

The training period will be used to find the optimal values of the considered forecasting methods, namely, the k-NN algorithm with the L_2 Wasserstein distance and the exponential smoothing method with seasonality in the distribution. As we can see in Figure 15.7, the seasonality affects the distribution shape and not only its location, hence we will not consider the exponential smoothing method with seasonality in the location.

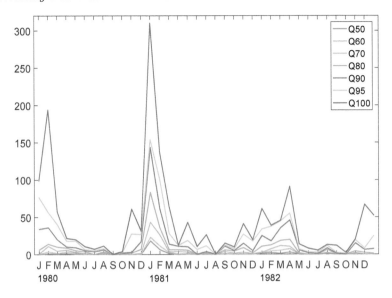

FIGURE 15.7
Quantiles of interest of the monthly time series of aggregated daily rainfall (in mm) from 1980 to 1982.

TABLE 15.2
MDE of the forecasting methods for the rainfall distributions (in mm).

Method	Train		Test	
	MDE	**Improv.**	**MDE**	**Improv.**
Seasonal naive	9.9	-	9.94	-
k-NN ($d = 21$, $k = 13$)	7.57	23.5%	7.83	21.2%
E. S. d ($\alpha = .01$, $\delta = .04$)	7.01	29.2%	7.15	28.1%

In order to determine the optimal values of the parameters we implement a grid search for minimizing the MDE using the L_2 Wasserstein distance. In the case of exponential smoothing, this is a two-dimensional search in the space defined by $\alpha \in [0, 1]$ and $\delta \in [0, 1]$, while in the case of k-NN this is again a two-dimensional search, but defined by $k \in [1, 15]$ and $d \in [1, 24]$. As benchmark we will use the naive method with seasonality, which means that we will consider as the forecast for a given month the distribution observed in the same month the previous year.

Table 15.2 shows the results for the training and test periods. Both methods improve the results of the naive method both in train and test, but the improvement of the exponential smoothing method is greater than the one obtained by the k-NN. If we analize the parameters of the exponential smoothing we see that both of them are very low, which means that the weights of the

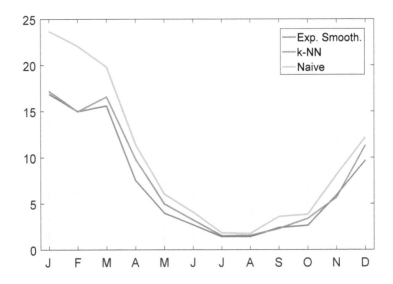

FIGURE 15.8
Monthly MDE (in mm) of the forecasting methods considered in the test set.

past observations decrease very slowly. This is especially relevant for δ that accounts for the seasonality, which means that distant observations of the same month are still relevant to yield accurate forecast. On the other hand, the k-NN only averages the next distribution of the 13 sequences more similar to the current one, where the sequences consist of the distributions of the last 21 months. It is interesting to remark that sequences are described by almost the last two years of monthly rainfall. Perhaps the length of the sequence informs about some kind of meteorological cycle in the region.

Now we will deepen into the results to shed some light into the phenomenon that we are studying. Figure 15.8 shows the MDE in each of the months for all the methods considered. First, it can be seen that the Seasonal naive method is outperformed by the k-NN and the Exponential Smoothing, however, the difference is very small in the drier months, i.e. July and August. The Exponential smoothing is the best method, but in six months its performance is similar to that of the k-NN. As it can be expected, the wet season is less stable and the error is higher. However, during these months the difference between the naive and the proposed methods is greater than in the dry season, so roughly we can say that the wetter the month, the more advantageous is the use of more sophisticated forecasting methods.

We can also analyse the error in the quantiles of interest, i.e. 50, 60, 70, 80, 90, 95 and 100. Figure 15.9 shows the RMSE in these quantiles for all the methods considered. As it can be seen the higher the quantile, the more

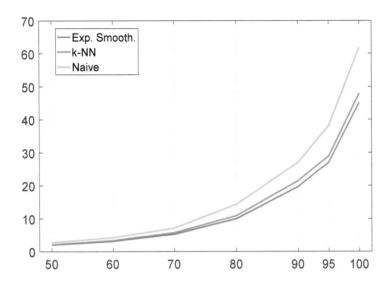

FIGURE 15.9
RMSE (in mm) of the forecasting methods considered for the quantiles of interest in the test set.

unpredictable it is, as often happens with extreme values in fields as meteorology. Similarly, the higher the quantile, the more advantageous is the use of the proposed forecasting methods for histograms, that is, the past of the time series is more useful to predict the future, even if that future is more uncertain, or precisely because of that.

15.7.2   IBM intra-daily returns

In this example we will aggregate intra-daily returns from the stock market by means of distributions. However, unlike other works, see [2], [4] and [17], that aggregate 5-min returns into daily distributions, we will follow a different approach here. In this work, we will aggregate 1-min returns into 6 intra-daily distributions that will represent different periods of the trading session. More precisely, we will work with the stock prices of IBM in the NYSE (New York Stock Exchange), whose trading session spans from 9:30 to 16:00, we will divide the trading session in histograms of 65 1-min returns. There is an exception to this general procedure, because there are special trading days that coincide with some American holidays (namely, Independence Day, Thanksgiving and Christmas), and if that happens, the trading session closes at 13:00. In these cases, we will only have 4 distributions, the first three of them will aggregate 65 1-min returns, while the last one will aggregate 80 returns. Another fact worth mentioning is that there are usually some trades shortly after

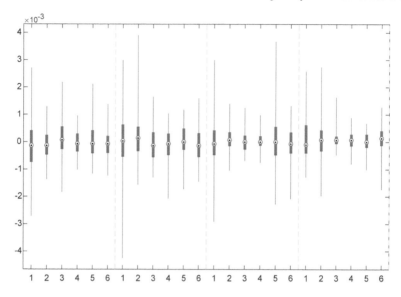

FIGURE 15.10
Intradaily distributions of the 1-min returns of IBM from 15th of December
to 18th of December, 2014.

the closing time, due to the high volume of trades at the end of the session.
The returns of these extra minutes have been taken into account, in the last
distribution in all the cases. In order to characterize the distributions we have
used boxplots, i.e. histograms of 4 bins where each bin accounts for the 25%
of the total frequency.

   We analysed a period of six years, from 2010 to 2015, for a total of 9021
intra-daily distributions. Figure 15.10 shows an excerpt of 4 days of the distributional time series represented as boxplots. As we said, there are six intradaily distributions per day. Interestingly, the first distribution of the day
shows more variance than the distributions of the rest of the day. This is a
hint of seasonality, even though it is not easy to see a clear seasonal pattern
in the rest of the six intradaily distributions.

   Figure 15.11 plots the mean (or center of gravity, see Equation (15.8))
and the standard deviation of the distributions according to the definition
in [22]. In this plot, we can appreciate that the distributional time series of
the returns is stationary in mean and that the mean is close to zero. In other
words, the trend that usually exhibit the time series of stock market prices is
removed when considering the (aggregated) time series of distributions. If we
look at the standard deviation time series, we can see that it is not stationary
as it ocassionaly exhibits the so-called volatility clusters typical in time series
returns, that is, volatile periods (i.e. periods with high standard deviation)
tend to be followed by volatile periods, while calm periods (i.e. periods with

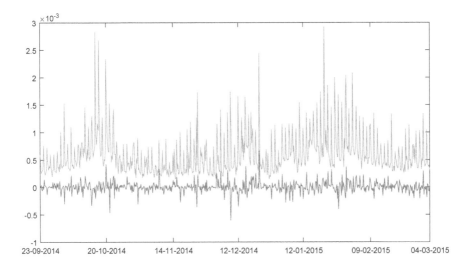

FIGURE 15.11
Time series of the mean and standard deviation of the distributions from 23rd of September, 2014 to 4th of March, 2015.

lower standard deviation) tend to be followed by calm periods. This behavior of mean and standard deviation is coherent with the well-known stylized facts of time series of stock market returns.

We will analyse the predictability of the resulting distributional intradaily time series with the proposed forecasting methods. We will try to see if considering a seasonal approach helps to forecast it. Hence, as benchmarks we will consider both the naive method and the naive method with seasonality.

The methods compared here are the k-NN with the L_2 Wasserstein distance, the simple exponential smoothing and the exponential smoothing with seasonality in the distribution, and the autoregression using the regression method of [23]. In this case, we will consider the first 2 years as initialization (needed for the k-NN), the next 3 years as training period and the last one for validation. In order to determine the optimal values of the parameters we implement a grid search for minimizing the MDE using the L_2 Wasserstein distance. In the case of the simple exponential smoothing, this is a one dimensional search in the space defined by $\alpha \in [0, 1]$, while the search space in the seasonal exponential smoothing is defined by $\alpha \in [0, 1]$ and $\delta \in [0, 1]$. In the case of the k-NN the space is two-dimensional and defined by $k \in [1, 15]$ and $d \in [1, 30]$. For the autoregression method we do not have an estimation theory as for classic time series, however we followed a procedure to avoid overfitting and find the lags relevant for the forecast. We estimated the error from 1 to 12 lags and we observed which lags produced a significant decrease in the error in

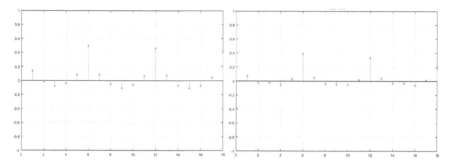

FIGURE 15.12
Autocorrelation function observed in the training set (left) and of the AR residuals (right).

the training period: these lags were the first two lags and lags 6 and 12, which are the first two seasonal lags. Finally, we used different combinations of AR with these four lags and determined the most suitable one taking into account the training error and the correlation function of the residuals, as explained in Section 15.5.2. Figure 15.12 shows that the linear information is mainly in the seasonal periods, that is heavily affected by the past values, as it decreases slowly. The AR residuals show less correlation in general and specially in the seasonal lags, but the seasonal lags still exhibit a slow decaying correlation, which means that there is linear information that the model ignores. Interestingly, similar peaks can be seen in the ACF of the residuals obtained with other forecasting methods, perhaps hinting some kind of non-stationary seasonal effect. This kind of phenomenom highlights the need of a more in-depth work in autoregression and histogram time series modeling.

In Table 15.3, we can see the results obtained by the methods applied. First, we can see that the seasonal naive method performs much better than the naive method, which confirms that the seasonal information matters. The same happens with the exponential smoothing with seasonality in the distribution that outperforms its simple counterpart. Due to these results and for the sake of clarity, we will omit the naive method and the simple exponential smoothing from the next figure. The best method is the k-NN which improves in more than 25% the results of the seasonal naive method. The seasonal exponential smoothing obtains an improvement around 18%, marginally better than the AR approach, which yields an improvement of 16.9%. According to the results of the seasonal naive, the test period is harder to forecast than the train period. Regardless this fact, the improvement of the forecasting methods proposed remains stable. Somehow this fact increases the merit of our methods, as the conditions of the time series to be forecasted seem to have changed considerably.

In Figure 15.13, we see the error in the quantiles of interest to try to ascertain whether it is easier to forecast the central tendency or its tails.

TABLE 15.3
MDE of the forecasting methods for the IBM intradaily distributions.

Method	Train MDE	Improv.	Test MDE	Improv.
Seasonal naive	$2.3 \cdot 10^{-4}$	-	$2.78 \cdot 10^{-4}$	-
Naive	$3.02 \cdot 10^{-4}$	-31.4%	$3.65 \cdot 10^{-4}$	-31.5%
S. E. S. ($\alpha = .05$)	$2.35 \cdot 10^{-4}$	-2.5%	$2.92 \cdot 10^{-4}$	-5.1%
E. S. d ($\alpha = .01$, $\delta = .19$)	$1.86 \cdot 10^{-4}$	19.1%	$2.27 \cdot 10^{-4}$	18.2%
AR (lags 1, 6, 12)	$1.88 \cdot 10^{-4}$	18.1%	$2.31 \cdot 10^{-4}$	16.9%
k-NN ($d = 13$, $k = 30$)	$1.68 \cdot 10^{-4}$	26.8%	$2.01 \cdot 10^{-4}$	27.5%

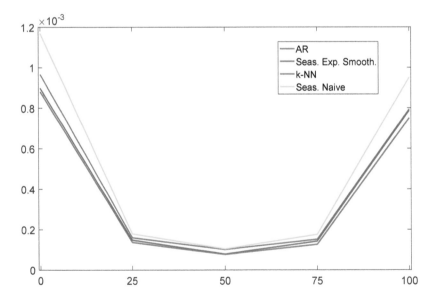

FIGURE 15.13
RMSE in the quantiles of interest of the forecasting methods considered in the test set.

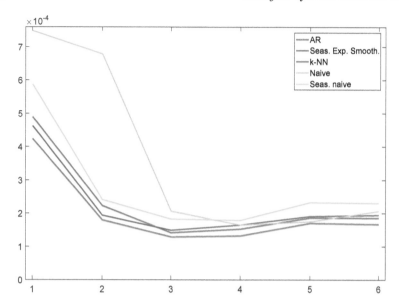

FIGURE 15.14
Seasonal MDE of the forecasting methods considered in the test set.

As we can see, the error in the central quantiles is less important than in the extreme ones, and the minimum value is harder to forecast than the maximum value, which makes perfect sense for the case of a time series of distributions. The k-NN is the best method for all the quantiles considered and the seasonal naive method is the worst. However, interestingly, the k-NN and the seasonal exponential smoothing perform similarly, but in the third quartile and in the maximum the difference between them increases. The AR seems to perform worst in the left tail of the distribution, but in the right tail it obtains similar results to that of the seasonal exponential smoothing.

Figure 15.14 show, the error for each of the 6 periods of the trading session. We can see a strong seasonal component in the error: the first period of the training session is much harder to predict than the rest of the trading session, while the easiest periods are the third and the fourth. In this figure we include both the naive and the seasonal naive, because we can draw interesting conclusions from their comparison: the first three periods of the trading session are better approximated by the first three periods of the previous trading session, i.e. by the seasonal naive method; however, the last three periods of the trading session are better approximated by the previous period in the session, i.e. by the naive method. In other words, the first half of the trading session is usually more similar to the previous session, while the second half of a session is more similar to the previous period of the current session. In fact, the naive method (the pink line in the figure) is almost as good as the best method (the k-NN represented by the red line) in the fifth period, which means that usually the intradaily returns of the IBM share remain stable from around 1PM to 3PM, and that it is difficult to yield better forecasts than that. Another interesting conclusion is that the k-NN is the best method for all the periods of the trading session. If we take a closer look to the parameters of the k-NN we see that the sequences are represented by the last 13 distributions, i.e. by the last two trading sessions and one extra period, which means that for a pattern-matching method as the k-NN the information of the two more recent trading sessions is relevant. Furthermore, the number of neighbors to be taken into account is 30, which means that a high number of neihgbors are relevant for the prediction. However, given the different behavior of the first two periods of the trading session and the rest of the session periods, perhaps better results could be obtained if we tried to adjust a different k-NN for each of the six periods of the trading session.

15.8 Concluding remarks

This chapter draws on the PhD thesis on forecasting distributional time series by Javier Arroyo [2] and the research on the topic already published in different articles [3, 4, 6, 17]. It sorted and summarized this previous research in a coherent document, where the methods and tools presented mostly rely on the L_2 Wasserstein distance. However, this chapter also introduced some new aspects, namely, a theoretical approximation of distributional time series, the use of quantile error measures and the decomposed error of the squared L_2 Wasserstein distance, and an early proposal of autoregressive methods based on regression methods for distributions [13, 23]. The applications on rainfall data and intra-daily stock returns are also new and serve to illustrate how distributional time series serve to represent real-life data in a meaningful way and to prove that the forecasting methods can produce accurate forecasts in contexts where it is difficult to beat naive approaches, such as precipitations and stock markets.

However, despite this significant research effort, the corpus of theory and tools to model and forecast distributional time series still requires further research. For example, some of the methods presented here, such as exponential smoothing and autoregressive methods, deserve further investigation and the development of time series decomposition and modeling tools needs to be completed and refined. We hope this chapter stimulate some curious readers to pursue this endeavor.

Bibliography

[1] N.S. Altman. An introduction to kernel and nearest-neighbor nonparametric regression. *The American Statistician*, 46(3):175–185, 1992.

[2] J. Arroyo. *Métodos de Predicción para Series Temporales de Intervalos e Histogramas*. PhD thesis, Universidad Pontificia Comillas (Spain), 2008.

[3] J. Arroyo, G. González-Rivera, and C. Maté. *Handbook of Empirical Economics and Finance*, chapter Forecasting with interval and histogram data. Some financial applications, pages 247–280. Chapman and Hall/CRC, 2010.

[4] J. Arroyo, G. González-Rivera, C. Maté, and A. Muñoz San Roque. Smoothing methods for histogram-valued time series: an application to value-at-risk. *Statistical Analysis and Data Mining*, 4(2):216–228, 2011.

[5] J. Arroyo and C. Maté. Forecasting time series of observed distributions with smoothing methods based on the barycentric histrogram. In *Computational Intelligence in Decision and Control. Proceedings of the 8th International FLINS Conference*, pages 61–66. World Scientific, 2008.

[6] J. Arroyo and C. Maté. Forecasting histogram time series with k-nearest neighbours methods. *International Journal of Forecasting*, 25(1):192–207, 2009.

[7] J. Berkowitz. Testing density forecasts, with applications to risk management. *Journal of Business & Economic Statistics*, 19(4):465–474, 2001.

[8] G.E.P. Box and G.M. Jenkins. *Time Series Analysis: Forecasting and Control*. Holden Day, San Francisco, 1970.

[9] R.G. Brown, editor. *Statistical Forecasting for Inventory Control*. McGraw-Hill, New York, 1959.

[10] R.B. Cleveland, W.S. Cleveland, J.E. McRae, and I. Terpenning. STL: a seasonal-trend decomposition procedure based on Loess. *Journal of Official Statistics*, 6(1):3 – 73, 1990.

[11] A.G. Colombo and R.J. Jaarsma. A powerful numerical method to combine random variables. *IEEE Transactions on Reliability*, 29(2):126–129, 1980.

[12] M.M. Deza and E. Deza. *Encyclopedia of Distances*, chapter Distances in Probability Theory, pages 233–245. Springer, 2013.

[13] S. Dias and P. Brito. Linear regression model with histogram-valued variables. *Statistical Analysis and Data Mining*, 8(2):75–113, 2015.

[14] E.S. Gardner. Exponential smoothing: The state of the art. Part 2. *International Journal of Forecasting*, 22(4):637–666, 2006.

[15] E.S. Gardner and E. McKenzie. Forecasting trends in time series. *Management Science*, 31:1237–1246, 1985.

[16] A.L. Gibbs and F.E. Su. On choosing and bounding probability metrics. *International Statistical Review*, 70(3):419–435, 2002.

[17] G. González-Rivera and J. Arroyo. Time series modeling of histogram-valued data: The daily histogram time series of S&P500 intradaily returns. *International Journal of Forecasting*, 28(1):20 – 33, 2012.

[18] S.G. Hall and J. Mitchell. Combining density forecasts. *International Journal of Forecasting*, 23(1):1–13, 2007.

[19] C.C. Holt. Forecasting trends and seasonals by exponentially weighted moving averages. Technical report, Carnegie Institute of Technology, 1957. O.N.R. Memorandum, vol. 52.

[20] A. Irpino and E. Romano. Optimal histogram representation of large data sets: Fisher vs piecewise linear approximation. In *Extraction et Gestion des Connaissances (EGC'2007), Actes des Cinquièmes Journées Extraction et Gestion des Connaissances*, pages 99–110, 2007.

[21] A. Irpino and R. Verde. A new Wasserstein based distance for the hierarchical clustering of histogram symbolic data. In V. Batagelj et al., editors, *Data Science and Classification, Proceedings of the IFCS 2006 Conference*, pages 185–192, Berlin, 2006. Springer.

[22] A. Irpino and R. Verde. Basic statistics for distributional symbolic variables: a new metric-based approach. *Advances in Data Analysis and Classification*, 9(2):143–175, 2015.

[23] A. Irpino and R. Verde. Linear regression for numeric symbolic variables: a least squares approach based on Wasserstein distance. *Advances in Data Analysis and Classification*, 9(1):81–106, 2015.

[24] E. Lu, Y. Ding, B. Zhou, X. Zou, X. Chen, W. Cai, Q. Zhang, and H. Chen. Is the interannual variability of summer rainfall in China dominated by precipitation frequency or intensity? An analysis of relative importance. *Climate Dynamics*, 47(1):67–77, 2016.

[25] R. Verde and A. Irpino. Dynamic clustering of histogram data: Using the right metric. In P. Brito, P. Bertrand, G. Cucumel, and F. de A. T. de Carvalho, editors, *Selected Contributions in Data Analysis and Classification*, pages 123–134. Springer, 2007.

[26] R. Verde and A. Irpino. Comparing histogram data using a Mahalanobis-Wasserstein distance. In *COMPSTAT 2008. Proceedings in Computational Statistics*, pages 77–89, Berlin, 2008. Springer-Verlag.

[27] R.C. Williamson. *Probabilistic Arithmetic*. PhD thesis, University of Queensland (Australia), 1989.

[28] S. Yakowitz. Nearest-neighbour methods for time series analysis. *Journal of Time Series Analysis*, 8(2):235–247, 1987.